Annals of Mathematics Studies
Number 154

Semiclassical Soliton Ensembles for the Focusing Nonlinear Schrödinger Equation

Spyridon Kamvissis
Kenneth D. T-R McLaughlin
Peter D. Miller

PRINCETON UNIVERSITY PRESS
PRINCETON AND OXFORD

Published by Princeton University Press, 41 William Street, Princeton, New Jersey 08540

In the United Kingdom: Princeton University Press, 3 Market Place, Woodstock,
Oxfordshire OX20 1SY

The Annals of Mathematics Studies are edited by John N. Mather and Elias M. Stein

Library of Congress Control Number: 2003108056

ISBN 0-691-11483-8 (cloth)
ISBN 0-691-11482-X (pbk.)

British Library Cataloging-in-Publication Data is available

Printed on acid-free paper. ∞

www.pupress.princeton.edu

Printed in the United States of America

10 9 8 7 6 5 4 3 2 1

Contents

Figures and Tables

Figures

Tables

Preface

We present a new generalization of the steepest-descent method introduced by Deift and Zhou [DZ93] for matrix Riemann-Hilbert problems and use it to study the semiclassical limit of the focusing nonlinear Schrödinger equation with real-analytic, even, bell-shaped initial data $\psi(x, 0) = A(x)$. We provide explicit, strong locally uniform asymptotics for a sequence of exact solutions $\psi(x, t)$ corresponding to initial data that has been modified in an asymptotically small sense. We call this sequence of exact solutions a semiclassical soliton ensemble. Our asymptotics are valid in those regions of the (x, t)-plane where a certain scalar complex phase function can be found. We characterize this complex phase function directly by a finite-gap ansatz and also via the critical point theory of a certain functional; the latter provides the correct generalization of the variational principle exploited by Lax and Levermore [LL83] in their study of the zero-dispersion limit of the Korteweg–de Vries equation.

For the special initial data $A(x) = A\operatorname{sech}(x)$, the scattering data was computed explicitly for all $\hbar$ by Satsuma and Yajima [SY74]. It turns out that for this case the modified initial data we use in general agrees with the true initial data. Thus our rigorous asymptotics for semiclassical soliton ensembles establish the semiclassical limit for this initial data.

Using a genus-zero ansatz for the complex phase function, we obtain strong asymptotics of general semiclassical soliton ensembles for small times independent of $\hbar$ in the form of a rapidly oscillatory and slowly modulated complex exponential plane wave. We show how, with the help of numerical methods, the ansatz can be verified for finite times up to a phase transition boundary curve in the (x, t)-plane called the primary caustic [MK98]. Using qualitative information obtained from the numerics concerning the mode of failure of the genus-zero ansatz at the primary caustic, we apply perturbation theory to show that a genus-two ansatz provides the correct asymptotic description of the soliton ensemble just beyond the caustic. Our analysis shows that the macrostructure in the genus-zero region is governed by the exact solution of the elliptic Whitham equations, and we obtain formulae solving this ill-posed initial-value problem in the category of analytic initial data. For the Satsuma-Yajima data, our solution of the Whitham equations reproduces that obtained many years ago by Akhmanov, Sukhorukov, and Khokhlov [ASK66], and our rigorous semiclassical analysis places their formal conclusions on sure footing.

The authors' joint work on this problem began when S. Kamvissis and P. D. Miller were members of the School of Mathematics at the Institute for Advanced Study during the 1997–1998 special year in geometric partial differential equations organized by Karen Uhlenbeck, and at the same time, K. T.-R. McLaughlin was visiting Princeton University. We are all very grateful to these institutions for their support of our research.

S. Kamvissis acknowledges the support of the Greek General Secretariat of Research and Technology on project number 97EL16, as well as the support of the French Ministry of Education and the Centre National de la Recherche Scientifique during his visit to the IAS. He also thanks the Max Planck Institute for Mathematics in the Sciences in Leipzig for hospitality and support in fall 1998. K. T.-R. McLaughlin acknowledges the support of the National Science Foundation under Postdoctoral Fellowship grant number DMS 9508946 and also under grant number DMS 9970328. P. D. Miller was supported by the NSF under grant number DMS 9304580 while at the IAS and grant number DMS 0103909 while at Michigan, and he was also supported by a Logan Fellowship while at Monash University.

The authors also wish to thank the organizers, Pavel Bleher and Alexander Its, of the special semester in random matrix models and their applications held at the Mathematical Sciences Research Institute in spring 1999, as well as the institute itself for hospitality during their stays there.

Finally, the authors want to acknowledge several of their colleagues for useful discussions: Percy Deift, Greg Forest, Arno Kuijlaars, Dave McLaughlin, John Nuttall, Evguenii Rakhmanov, Herbert Stahl, and Xin Zhou. Also we thank Dave Levermore and Nick Ercolani for giving us a copy of the preprint [EJLM93].

This work was submitted to the editors in November 2000 and was accepted in its original form. Since that time, there have been some relevant further developments. We have decided to include some additional material concerning these developments; these added notes are indicated clearly in the text.

Bonn, Chapel Hill, and Ann Arbor—July 2002

Semiclassical Soliton Ensembles for the Focusing Nonlinear Schrödinger Equation

Chapter One

Introduction and Overview

1.1 BACKGROUND

The initial-value problem for the focusing nonlinear Schrödinger equation is

$$i\hbar\partial_t\psi + \frac{\hbar^2}{2}\partial_x^2\psi + |\psi|^2\psi = 0, \qquad \psi(x,0) = \psi_0(x). \tag{1.1}$$

We are interested in studying the behavior of solutions of this initial-value problem in the so-called semiclassical limit. To make this precise, the initial data is given in the form

$$\psi_0(x) = A(x)\exp(iS(x)/\hbar), \tag{1.2}$$

where $A(x)$ is a positive real amplitude function that is rapidly decreasing for large $|x|$ and where $S(x)$ is a real phase function that decays rapidly to constant values for large $|x|$. Studying the semiclassical limit means: fix once and for all the functions $A(x)$ and $S(x)$, and then for each sufficiently small value of $\hbar > 0$, solve the initial-value problem (1.1) subject to the initial data (1.2), obtaining the solution $\psi(x,t;\hbar)$. Describe the collection of solutions $\psi(x,t;\hbar)$ in the limit of $\hbar \downarrow 0$.

The initial-value problem (1.1) is a key model in modern nonlinear optical physics and its increasingly important applications in the telecommunications industry. On the one hand, it describes the stationary profiles of high-intensity paraxial beams propagating in materials with a nonlinear response, the so-called Kerr effect. This is the realm of *spatial solitons*, which are envisioned as self-guided beams that can form the fundamental components of an all-optical switching system. In this context, the semiclassical scaling $\hbar \ll 1$ of (1.1) corresponds to the joint limit of paraxial rays and geometrical optics in the presence of nonlinear effects. On the other hand, (1.1) also describes the propagation of time-dependent envelope pulses in optical fibers operating at carrier wavelengths in the anomalous dispersion regime (usually infrared wavelengths near 1550 nm). These envelope pulses are known as *temporal solitons* and are envisioned as robust bits in a digital signal traveling through the fiber. In these fiber-optic applications, the semiclassical scaling $\hbar \ll 1$ is particularly appropriate for modeling propagation in certain dispersion-shifted fibers that are increasingly common. See [FM98] for a careful discussion of this point leading to a similarly scaled *defocusing* equation when the fiber parameters are indicative of weak normal dispersion; similar arguments with slightly adjusted parameters can lead to the focusing problem (1.1) just as easily. Of course in neither of these optical applications is the small parameter actually Planck's constant, but we write it as $\hbar$ in formal analogy with the quantum-mechanical interpretation of the linear terms in (1.1), which also gives rise to the description of the limit of interest as "semiclassical."

The independent variables x and t parametrize the semiclassical limit, and one certainly does not expect a pointwise asymptotic description of the solution to be uniform with respect to these parameters. The statement of the problem becomes more precise when one further constrains these parameters. For example, one might set $x = X/\hbar$ and $t = T/\hbar$ for X and T fixed as $\hbar \downarrow 0$. In this limit, several studies [B96, BK99] have suggested that for initial data with $|S(x)|$ sufficiently large, the field consists of trains of separated solitons, with the remarkable property that there is a well-defined relationship between the soliton amplitude and velocity (nonlinear dispersion relation) that is determined from the initial functions $A(x)$ and $S(x)$ via the asymptotic distribution of eigenvalues of the Zakharov-Shabat scattering problem. In general, solitons can have arbitrary amplitudes and velocities, so the observed correlation is a direct consequence of the semiclassical limit.

Here, we are concerned with a different asymptotic parametrization. Namely, we consider the sequence of functions $\psi(x, t; \hbar)$ in a fixed but arbitrary compact set of the (x, t)-plane in the limit $\hbar \downarrow 0$. In this scaling, the large number of individual solitons present in the initial data are strongly nonlinearly superposed, and interesting spatiotemporal patterns have been observed [MK98, BK99].

This choice of scaling has several features in common with similar limits studied in other integrable systems, for example, the zero-dispersion limit of the Korteweg–de Vries equation analyzed by Lax and Levermore [LL83], the continuum limit of the Toda lattice studied by Deift and K. T.-R. McLaughlin [DM98], and the semiclassical limit of the *defocusing* nonlinear Schrödinger equation studied by Jin, Levermore, and D. W. McLaughlin [JLM99]. In all of these cases, the challenge is to use the machinery of the inverse-scattering transform to prove convergence in some sense to a complicated asymptotic description that necessarily consists of two disparate space and time scales. One scale (the *macrostructure*) is encoded in the initial data, and the other scale (the *microstructure*) is introduced by the small parameter (the dispersion parameter in the Korteweg–de Vries equation, the lattice spacing in the Toda lattice, and Planck's constant, $\hbar$, in the nonlinear Schrödinger equation).

In Lax and Levermore's analysis of the zero-dispersion limit for the Korteweg–de Vries equation [LL83], a fundamental role was played by an explicit, albeit complicated, formula for the exact solution of the initial-value problem for initial data that has been modified in an asymptotically negligible sense. This formula directly represents the solution $u(x, t)$ of the problem in terms of the second logarithmic derivative of a determinant. When the determinant is expanded as a sum of principal minors, the minors are all positive, and the sum is shown to be asymptotically dominated by its largest term. This leads directly to a discrete maximization problem (the problem is discrete because the number of minors is finite but large when the dispersion parameter is small) in which the independent variables x and t appear as parameters. This maximization problem characterizes the determinant up to a controllable error. Leading-order asymptotics are obtained by letting the dispersion parameter go to zero and observing that the discrete maximization problem goes over into a variational problem in a space of admissible functions. It turns out that the weak limit of each member of the whole hierarchy of conserved local densities for the Korteweg–de Vries equation can be directly expressed in terms of the solution of the variational problem and its derivatives.

In all of the problems where the method of Lax and Levermore has been successful, the macrostructure parameters (or equivalently weak limits of various conserved local densities) have been shown to evolve locally in space and time as solutions of a hyperbolic system known as the *Whitham equations* or the *modulation equations*. The global picture consists of several regions of the (x, t)-plane in each of which the microstructure is qualitatively uniform (periodic or quasiperiodic) and the macrostructure obeys a system of modulation equations. The size of this system of equations (number of unknowns) is related to the complexity of the microstructure. The variational method of Lax and Levermore amounts to the global analysis showing how the solutions of the modulation equations are patched together at the boundaries of these various regions. By hyperbolicity and the corresponding local well-posedness of the modulation equations, it follows that, for example, the small-time behavior (sufficiently small, but independent of the size of the limit parameter) of the limit is connected with prescribed initial data in a stable fashion.

The modulation equations may be derived formally, without reference to initial data. For the focusing nonlinear Schrödinger equation, these quasilinear equations are *elliptic* [FL86], which makes the Cauchy initial-value problem for them ill-posed in common spaces. To illustrate this ill-posedness for the Whitham equations in their simplest version (genus zero), one might make the assumption that the microstructure in the solution of (1.1) resembles the modulated rapid oscillations present in the initial data. That is, one could suppose that for some order-one time the solution can be represented in the form

$$\psi(x,t) = A(x,t)\exp(iS(x,t)/\hbar), \tag{1.3}$$

where $A(x, 0) = A(x)$ and $S(x, 0) = S(x)$. Setting $\rho(x, t) = A(x, t)^2$ and $\mu(x, t) = A(x, t)^2 \partial_x S(x, t)$, one finds that the initial-value problem (1.1) implies

$$\partial_t \rho + \partial_x \mu = 0, \qquad \partial_t \mu + \partial_x \left(\frac{\mu^2}{\rho} - \frac{\rho^2}{2} \right) = \frac{\hbar^2}{4} \partial_x \left(\rho \partial_x^2 \log(\rho) \right), \tag{1.4}$$

with initial data $\rho(x, 0) = A(x)^2$ and $\mu(x, 0) = A(x)^2 S'(x)$. The modulation equations corresponding to our assumption about the microstructure are obtained by simply neglecting the terms that are formally of order $\hbar^2$ in these equations. That is, one supposes that for some finite time $\rho(x, t)$ and $\mu(x, t)$ are uniformly close, respectively, to functions $\rho_\circ(x, t)$ and $\mu_\circ(x, t)$ as $\hbar \downarrow 0$, where these latter two functions solve the system

$$\partial_t \rho_\circ + \partial_x \mu_\circ = 0, \qquad \partial_t \mu_\circ + \partial_x \left(\frac{\mu_\circ^2}{\rho_\circ} - \frac{\rho_\circ^2}{2} \right) = 0, \tag{1.5}$$

with initial data $\rho_\circ(x, 0) = A(x)^2$ and $\mu_\circ(x, 0) = A(x)^2 S'(x)$. This is a quasilinear nonlinear system, and it is easy to check that it is of elliptic type; that is, the characteristic velocities $\mu_\circ/\rho_\circ \pm i\sqrt{\rho_\circ}$ are complex at every point where $\rho_\circ$ is nonzero. This implies that the Cauchy problem posed here for the modulation equations is ill-posed.

This fact immediately makes the interpretation of the semiclassical limit of the initial-value problem (1.1) complicated; even if it turns out that one can prove convergence to the solutions of the modulation equations for some initial data, it is not clear that one can deduce anything at all about the asymptotics for "nearby" initial data. In this sense, the formal semiclassical limit of (1.1) is very unstable.

One feature that both the hyperbolic and elliptic modulation equations have in common is the possibility of singularities that develop in finite time from smooth initial data. This singularity formation seems physically correct in the context of spatial optical solitons, where the Kerr effect has been known for some time to lead to self-focusing of light beams and, in two transverse dimensions (the independent variable x), to the total collapse of the beam in finite propagation distance (the independent variable t). As long ago as 1966, this led Akhmanov, Sukhorukov, and Khokhlov [ASK66] to propose a certain exact solution of the modulation equations (1.5) as a model for the self-focusing phenomenon in one transverse dimension. They did not try to solve any initial-value problem for these equations; indeed they were clearly aware of the ellipticity of the system (1.5) and the coincident ill-posedness of its Cauchy problem. Rather, they introduced a clever change of variables (some insight into their possible reasoning was proposed by Whitham [W74]) and obtained a set of two real equations implicitly defining two real unknowns as functions of x and t. After the fact, they noted that their solution matched onto the initial data $A(x) = A\,\mathrm{sech}(x)$ and $S(x) \equiv 0$. The original paper of Akhmanov, Sukhorukov, and Khokhlov contains drawings of the solution at various times up to the formation of a finite-amplitude singularity (i.e., the singularity forms in the derivatives) at the time $t = t_{\mathrm{crit}} = 1/(2A)$. The authors even plotted their solution beyond the singularity, showing the onset of multivaluedness. They understood that the model solution cannot possibly be valid beyond the singularity and, in the physical context of interest in their study, ascribed this as much to the breakdown of the paraxial approximation leading to the nonlinear Schrödinger equation (1.1) as a beam propagation model in the first place as to the failure of the formal geometrical optics (semiclassical) limit for (1.1).

As is the case in all of the problems for which the method of Lax and Levermore has been successful, careful analysis of the semiclassical limit $\hbar \downarrow 0$ for (1.1) is possible in principle because the problem can be solved for each $\hbar$ by the inverse-scattering transform, as was first shown by Zakharov and Shabat [ZS72]. The small parameter necessarily enters the problem both in the forward-scattering step and in the inverse-scattering step. It is significant that the analysis of the semiclassical limit for (1.1) is frustrated in both steps. In the forward-scattering step, the difficulties are related to the nonselfadjointness of the scattering problem associated with (1.1). By contrast, in each of the cases mentioned previously, where calculations of this type were successfully carried out, the associated scattering problem is selfadjoint. In the inverse-scattering step, the difficulties are related to the limit being attained by a kind of furious cancellation in which no single term in the expansion of the solution is apparently dominant. In fact, in Zakharov and Shabat's paper [ZS72], there appears an explicit formula for the function $\rho(x, t)$ solving (1.4) that is qualitatively very similar to that solving the Korteweg–de Vries equation and taken as the starting point in Lax and Levermore's analysis. When $t = 0$, this formula has all of the properties required by the Lax-Levermore theory. Namely, the determinant can be expanded as a sum of positive terms, which is controlled by its largest term as $\hbar \downarrow 0$. This calculation is carried out in the paper of Ercolani, Jin, Levermore, and MacEvoy [EJLM93]. But when t is fixed at any nonzero value, the principal minors lose their positive definiteness, and it can no longer be proved that the sum is dominated by its largest term. If the weak limit exists, then all that can be said from this approach is that it arises out of subtle cancellation. In particular, from this point of view it appears that there is no obvious variational principle characterizing the limit.

1.2 APPROACH AND SUMMARY OF RESULTS

This book is primarily concerned with the semiclassical analysis of the inverse-scattering step. For simplicity, we restrict attention from the start to the case of initial data that satisfy $S(x) \equiv 0$. In this case it was observed already in Zakharov and Shabat's paper [ZS72] that while not strictly selfadjoint for any $\hbar > 0$ the scattering problem formally goes over into a semiclassically scaled selfadjoint linear Schrödinger operator in the limit $\hbar \downarrow 0$. In [EJLM93], this observation was exploited to propose WKB formulae that were subsequently used to study the zero-dispersion limit of the modified Korteweg–de Vries equation, an equation associated with the same scattering problem as (1.1), but whose inverse-scattering step is more straightforward because there is no cancellation of the type mentioned previously. (As already mentioned, this cancellation is also absent for the focusing nonlinear Schrödinger problem when $t = 0$, and the calculations in [EJLM93] hold in this case as well.) The WKB approximation amounts to the neglect of the reflection coefficient and the replacement of the true eigenvalues with a sequence of purely imaginary numbers that are obtained from an explicit Bohr-Sommerfeld-type quantization rule. These WKB formulae have not been rigorously established to date; their justification in [EJLM93] rests upon the fact that they reproduce the exact initial data when t is set to zero in the inverse-scattering step. There is, however, one function $A(x)$ for which all of the exact scattering data is known (assuming $S(x) \equiv 0$) exactly: $A(x) = A\,\mathrm{sech}(x)$. The spectrum corresponding to this potential in the nonselfadjoint Zakharov-Shabat scattering problem was computed exactly for all $\hbar$ by Satsuma and Yajima [SY74] and published in 1974. At face value this is a remarkable coincidence: the same initial data for which Akhmanov, Sukhorukov, and Khokhlov found (after the fact!) that they had an exact solution of the modulation equations turns out to be data for which the forward-scattering problem was later shown to be exactly solvable for all $\hbar$. Some additional special cases of potentials where the spectrum can be obtained exactly for all $\hbar$, including some cases with $S(x) \not\equiv 0$, have been found recently by Tovbis and Venakides [TV00].

It turns out that the exact scattering data for the special initial condition $\psi_0(x) = A\,\mathrm{sech}(x)$ coincides with the formal WKB approximation to the scattering data, as long as one restricts attention to a particular sequence of positive values of $\hbar \in \{\hbar_N\}$ converging to zero. For these special values of $\hbar$, the initial data is exactly reflectionless, there are exactly N eigenvalues all purely imaginary, and also the distance between the most excited state (the eigenvalue with the smallest magnitude) and the continuous spectrum is exactly half of the distance between each adjacent pair of eigenvalues. In particular, for $\hbar = \hbar_N$, there is no error incurred in reconstructing the corresponding solution of (1.1) using inverse-scattering theory *without reflection coefficient*; the true solution for these values of $\hbar$ is a pure ensemble of N solitons.

In this book, we develop a method that yields detailed strong asymptotics for the inverse-scattering problem corresponding to the scattering data just briefly described. Since this scattering data is the true scattering data corresponding to the Satsuma-Yajima potential, our results imply rigorous asymptotics for the corresponding initial-value problem (1.1). But since the scattering data for this case agrees with its WKB approximation, we prefer to approach the problem from the more general perspective of computing rigorous asymptotics for the inverse problem corresponding to a general family of WKB scattering data. Thus, our approach to the semiclassical limit for the initial-value problem (1.1) for quite general data satisfying $S(x) \equiv 0$ is essentially the familiar step of introducing modified reflectionless initial data whose scattering data is that predicted by the formal WKB approximation. This sort of modification was the first step in the pioneering work of Lax and Levermore [LL83]. Of course, for the Satsuma-Yajima initial data, no modification is necessary as long as $\hbar \in \{\hbar_N\}$.

The main idea that allows our analysis of the inverse-scattering problem to proceed for $t \neq 0$ where the Lax-Levermore method fails is to avoid the direct connection of the discrete scattering data with the solution of the problem via an explicit determinant formula and instead to introduce an intermediate object, namely, an appropriately normalized eigenfunction of the Zakharov-Shabat scattering problem. In general, this eigenfunction satisfies a certain matrix Riemann-Hilbert problem with poles encoding the discrete spectrum and a jump on the real axis of the eigenvalue corresponding to the reflection coefficient on the continuous spectrum. The solution of the nonlinear Schrödinger equation is obtained in turn from the solution of this Riemann-Hilbert problem. This is the essential content of inverse-scattering theory [FT87]. While it of course turns out that in the reflectionless case the Riemann-Hilbert problem may be explicitly solved in terms of meromorphic functions and ratios of determinants, leading to the formula that is the starting point for Lax-Levermore-type analysis, there is some advantage to ignoring this explicit solution and instead trying to obtain uniform asymptotics for the eigenfunction that is the solution of the Riemann-Hilbert problem. Only in studying this intermediate problem do we recover a variational principle that is a generalization of the one from Lax and Levermore's method.

The method we develop in this book to study the asymptotic behavior of the eigenfunction generalizes the steepest-descent method for matrix Riemann-Hilbert problems first proposed by Deift and Zhou in [DZ93] and subsequently developed and further applied in several papers [DVZ94, DZ95]. The generalization of the steepest-descent method that we present here has its basic features in common with the recent application of the method to the Korteweg–de Vries equation in [DVZ97], with recent applications in the theory of orthogonal polynomials and random matrices [DKMVZ97, DKMVZ99A, DKMVZ99B], and also with some applications to long-time asymptotics for soliton-free initial data in the focusing nonlinear Schrödinger equation [K95, K96]. These latter papers make use of an idea that was first introduced in [DVZ94]—using the special choice of a complex phase function to enable the asymptotic reduction of the Riemann-Hilbert problem to a simple form. Our work generalizes this approach because it turns out that an appropriate complex phase function typically does not exist at all relative to a given contour in the complex plane, unless this contour satisfies some additional conditions. In fact, we show that the existence of an appropriate complex phase function *selects portions of the contour on which the Riemann-Hilbert problem should be posed to begin with.* In this sense, the generalization of the method proposed in [DVZ94] that we present here further develops the analogy with the classical asymptotic method of steepest descent; the problem must be solved on a particular contour in the complex plane. In problems previously treated by the steepest-descent method of Deift, Zhou, and others, the problem of finding this special contour has simply not arisen because there is an obvious contour, often implied by the selfadjointness of a related scattering problem, for which the additional conditions that select the contour are *automatically* satisfied. The specification of this special contour can be given a variational interpretation that is the correct generalization of the Lax-Levermore variational principle.

Among our primary results are the following.

1. Strong leading-order semiclassical asymptotics for solutions of the focusing nonlinear Schrödinger equation corresponding to sequences of initial data whose spectral data are reflectionless and have a discrete spectrum obtained from a Bohr-Sommerfeld quantization rule. These asymptotics are valid even after wave breaking and come with a rigorous error bound. The explicit model we obtain—to which the semiclassical solutions are asymptotically close pointwise in x and t—displays qualitatively different behavior before and after wave breaking and in particular exhibits violent oscillations after breaking, confirming phenomena that have been observed in numerical experiments.
2. Formulae explicitly involving the initial data that solve the elliptic Whitham modulation equations. These formulae consequently provide the complete solution to the initial-value problem for the Whitham equations in the category of analytic initial data.
3. The characterization of the caustic curves in the (x, t)-plane where the nature of the microstructure changes suddenly. We also provide what amount to "connection formulae" describing the phase transition that occurs at the caustic. In particular our analysis shows that at first wave breaking there is a spontaneous transition from fields with smooth amplitude (genus zero) to oscillatory fields with intermittent concentrations in amplitude (genus two).
4. A significant extension of the steepest-descent method for asymptotic analysis of Riemann-Hilbert problems introduced by Deift and Zhou. For problems with analytic jump matrices, we show how the freedom of placement of the jump contour in the complex plane can be systematically exploited to asymptotically reduce the norms of the singular integral operators involved in the solution of the Riemann-Hilbert problem. Ultimately this expresses the solution as an explicit contribution modified by a Neumann series involving small bounded operators.

5. A new generalization of Riemann-Hilbert methods allowing the analysis of inverse-scattering problems in which there is an asymptotic accumulation of an unbounded number of solitons.
6. An interpretation of our asymptotic solution of the Riemann-Hilbert problem in terms of a new variational principle that generalizes the quadratic programming problem of Lax and Levermore and explicitly encodes the contour-selection mechanism. This interpretation also makes a strong connection with approximation theory, where variational problems of the same type appear when one tries to find sets of minimal weighted Green's capacity in the plane.
7. A proof that the systematic selection of an appropriate contour is guaranteed to succeed under certain generic conditions. Finding the correct contour amounts to solving a problem of geometric function theory, namely, the construction of "trajectories of quadratic differentials." We show that the existence of such trajectories is an open condition with respect to the independent variables x and t.

1.3 OUTLINE AND METHOD

We begin in chapter 2 by expressing the function $\psi(x, t; \hbar_N)$ in terms of the solution of a holomorphic matrix Riemann-Hilbert problem posed relative to a contour that surrounds the locus of accumulation of eigenvalues but is otherwise arbitrary a priori. The scattering data is introduced in chapter 3, where we present the formal WKB formulae for initial data satisfying $S(x) \equiv 0$ and appropriate functions $A(x)$. We carry out some detailed asymptotic calculations starting from the WKB approximations to the discrete eigenvalues that we require later in the book, and we compare these general calculations with the specific exact formulas of Satsuma and Yajima. With this WKB data in hand, we proceed in chapter 4 to study the asymptotics of the inverse-scattering problem for this (generally) approximate data. We introduce in §4.1 a certain complex scalar phase function, and in §4.2 we show how to choose it to capture the essentially wild asymptotic behavior of the solution of the Riemann-Hilbert problem. Factoring off a proper choice of the complex phase leads to a simpler Riemann-Hilbert problem whose leading-order asymptotics can be described explicitly. In §4.3 we solve this leading-order Riemann-Hilbert problem, the *outer model problem*, in terms of Riemann theta functions (and in fact for small time in terms of exponentials). Subject to proving the validity of this asymptotic reduction, the solution $\psi(x, t; \hbar_N)$ is then also given at leading order in terms of theta functions and exponentials.

Assuming the existence of the complex phase function on an appropriate contour, we continue with some detailed local analysis in §4.4, building local approximations near certain exceptional points in the complex plane. Patching these local approximations together with the outer approximation yields a uniform approximation of the solution of the Riemann-Hilbert problem that we prove is valid in §4.5.

This detailed error analysis is completely vacuous *unless* we can establish the existence of the complex phase function and its support contour. We carry out this construction in chapter 5 using a modification of the finite-gap ansatz familiar from the Lax-Levermore method. Temporarily tossing out the inequalities that the phase function must ultimately satisfy, we show how to write down equations for the endpoints of the bands and gaps along the contour and how the bands of the contour can be viewed as heteroclinic orbits of a particular explicit differential equation for contours in the complex plane (or trajectories of a quadratic differential). Some of the conditions we impose on the endpoints of the bands and gaps are precisely those that are necessary for the existence of the correct number of heteroclinic orbits. There is a finite-gap ansatz corresponding to any number of bands and gaps, and the idea is to choose this number so that the phase function satisfies certain inequalities as well. This choice then determines the local complexity (genus of the Riemann theta function) of the approximate solution of the initial-value problem (1.1). In §5.3 we show that in fixed neighborhoods of fixed x and t, the macrostructure parameters of the solutions (moduli of an associated hyperelliptic Riemann surface) satisfy a quasilinear system of partial differential equations that we believe to be the elliptic modulation (Whitham) equations for multiphase wavetrains [FL86].

In chapter 6, we investigate the simplest possible ansatz (i.e., genus zero), showing that for small time independent of $\hbar$ it does indeed satisfy all necessary inequalities. For the Satsuma-Yajima initial data, this completes the proof of convergence to the semiclassical limit for small time, ultimately justifying the geometrical optics approximation made by Akhmanov, Sukhorukov, and Khokhlov in 1966. For semiclassical soliton ensembles corresponding to more general data, we still obtain rigorous strong asymptotics, but the connection to initial data is more tenuous. The asymptotics formally recover the initial data and the successful ansatz persists for small time, but our scheme of essentially uniformly approximating the eigenfunction in the complex plane of the eigenvalue breaks down near $t = 0$, when the regions of the complex plane where the description of the eigenfunction requires detailed local analysis come into contact with the locus of accumulation of poles.[1] On the other hand, we know that asymptotics for $t = o(1)$ can be obtained by bypassing the Riemann-Hilbert problem and applying the Lax-Levermore method to the determinant solution formula [EJLM93]. Of course, even if the error is controlled uniformly near $t = 0$, the error present at $t = 0$ in cases where the WKB approximation is not exact can in principle be amplified by this unstable problem in ways that are not possible in "selfadjoint" integrable problems where the semiclassical limit is "hyperbolic." Using a computer program to construct the ansatz for finite times (as opposed to a perturbative calculation based at $t = 0$), we verify the ansatz in the special case of the Satsuma-Yajima data right up to the phase transition to more complicated local behavior termed the "primary caustic" in [MK98]. These computer simulations clearly demonstrate both the selection of the special contour and the breakdown of the ansatz when inequalities fail and/or integral curves of the differential equation determining the contour bands become disconnected. We use perturbation theory in chapter 7 to show that when the genus-zero ansatz fails at the primary caustic, the genus-two ansatz takes over. At such a transition, the smooth wave field "breaks" and gives way to a hexagonal spatiotemporal lattice of maxima.

[1] Note added: This difficulty for t near zero has been surmounted recently using a modification of the methods presented in this book. See [M02] for details.

The conditions that we use to specify the complex phase function are naturally obtained in chapter 8 as the Euler-Lagrange variational conditions describing a particular type of critical point for a certain functional related to potential theory in the upper half-plane. This makes the problem of computing the semiclassical limit equivalent to solving a certain problem of extreme Green's capacity and establishing regularity properties of the solution. Solving the variational problem can be given the physical interpretation of finding unstable electrostatic equilibria of a certain system of electric charges under the influence of an externally applied field which has an attractive component that is exactly the potential of the WKB eigenvalue distribution. The significance of variational problems in the characterization of singular limits of solutions of completely integrable partial differential equations was first established by Lax and Levermore [LL83] for the Korteweg–de Vries equation, and the method was subsequently extended to the Toda lattice [DM98] and the entire hierarchy of the *defocusing* nonlinear Schrödinger equation [JLM99].

The calculations presented in §4.4 and §4.5 rely on certain technical details of the Fredholm theory of Riemann-Hilbert problems posed in Hölder spaces and small-norm theory for Riemann-Hilbert problems in L^2 that we present in the appendices. In particular, the Hölder theory that we summarize unites some very classical results of the Georgian school of Muskhelishvili and others with the treatment of matrix Riemann-Hilbert problems posed on self-intersecting contours given by Zhou [Z89]. The Hölder theory appears to have fallen by the wayside in inverse-scattering applications, possibly because these problems are often posed from the start in L^p or Sobolev spaces. However, in local analysis one is always dealing with explicit piecewise-analytic jump relations on piecewise-smooth contours, and at the same time one requires uniform control on the solutions right up to the contour. In such cases, the compactness required for Fredholm theory comes almost for free (and significantly in a contour-independent way) in Hölder spaces at the cost of an arbitrarily small loss of smoothness. At the same time, once existence is established in a Hölder space, the required control up to the contour is built-in as a property of the solution. On the other hand, in the bigger L^p or Sobolev spaces, compactness depends on a rational approximation argument that can be a lot of work to establish (and in particular it seems that the argument must be tailored for each particular contour configuration). And then having established existence in these spaces, one must put in extra effort to obtain the required control up to the contour, with special care needing to be taken near self-intersection points.

In summary, our primary mathematical techniques include the following.

1. Techniques for the asymptotic analysis of matrix Riemann-Hilbert problems, including the steepest-descent method of Deift and Zhou.
2. The Fredholm theory of Riemann-Hilbert problems in the class of functions with Hölder-continuous boundary values on self-intersecting contours.
3. The use of Cauchy integrals (or Hilbert transforms) to solve certain scalar boundary value problems for sectionally analytic functions in the plane.
4. Careful perturbation theory to establish the semiclassical limit for small times and then to study the phase transition that occurs at a caustic curve in the (x,t)-plane.
5. Some theory of logarithmic potentials with external fields.

1.4 SPECIAL NOTATION

We use several different branches of the logarithm, distinguished one from another by notation. We only use the lowercase $\log(z)$ to refer to a generic branch (cut anywhere) when it makes no difference in an expression, that is, when it appears in an exponent or when its real part is considered. The uppercase $\text{Log}(z)$ always refers to the standard cut of the principal branch, defined for $z \in \mathbb{C}\backslash\mathbb{R}_-$ by the integral

$$\text{Log}(z) := \int_1^z \frac{dw}{w} . \tag{1.6}$$

All other branches of the function $\log(\lambda-\eta)$ considered as a function of λ for η fixed are written with notation like $L_\eta^s(\lambda)$. Each of these is also defined for $z = \lambda - \eta$ by (1.6), but with a particular well-defined branch cut in the λ-plane that is associated with the logarithmic pole η and the superscript s. Each of these branches is clearly defined when it first appears in the text. Exponential functions will *always* refer to the principal branch, $a^u = e^{u\,\text{Log}(a)}$.

We use the Pauli matrices throughout the book. They are defined as

$$\sigma_1 := \begin{bmatrix} 0 & 1 \\ 1 & 0 \end{bmatrix}, \quad \sigma_2 := \begin{bmatrix} 0 & -i \\ i & 0 \end{bmatrix}, \quad \sigma_3 := \begin{bmatrix} 1 & 0 \\ 0 & -1 \end{bmatrix}. \tag{1.7}$$

Chapter Two

Holomorphic Riemann-Hilbert Problems for Solitons

The initial-value problem (1.1) is solvable for arbitrary $\hbar$ because the focusing nonlinear Schrödinger equation can be represented as the compatibility condition for two systems of linear ordinary differential equations:

$$\hbar\partial_x\begin{bmatrix} u_1 \\ u_2 \end{bmatrix}=\begin{bmatrix} -i\lambda & \psi \\ -\psi^* & i\lambda \end{bmatrix}\begin{bmatrix} u_1 \\ u_2 \end{bmatrix}, \tag{2.1}$$

$$i\hbar\partial_t\begin{bmatrix} u_1 \\ u_2 \end{bmatrix}=\begin{bmatrix} \lambda^2-|\psi|^2/2 & i\lambda\psi-\hbar\partial_x\psi/2 \\ -i\lambda\psi^*-\hbar\partial_x\psi^*/2 & -\lambda^2+|\psi|^2/2 \end{bmatrix}\begin{bmatrix} u_1 \\ u_2 \end{bmatrix}, \tag{2.2}$$

where λ is an arbitrary complex parameter. The compatibility condition for (2.1) and (2.2) does not depend on the value of λ and is equivalent to the nonlinear Schrödinger equation.

The N-soliton solutions of the nonlinear Schrödinger equation are those complex functions $\psi(x,t)$ for which there exist simultaneous column vector solutions of (2.1) and (2.2) of the particularly simple form

$$\begin{aligned}\mathbf{u}^+(x,t,\lambda)&=\begin{bmatrix} \sum_{p=0}^{N-1}A_p(x,t)\lambda^p \\ \lambda^N+\sum_{p=0}^{N-1}B_p(x,t)\lambda^p \end{bmatrix}\exp(i(\lambda x+\lambda^2 t)/\hbar), \\ \mathbf{u}^-(x,t,\lambda)&=\begin{bmatrix} \lambda^N+\sum_{p=0}^{N-1}C_p(x,t)\lambda^p \\ \sum_{p=0}^{N-1}D_p(x,t)\lambda^p \end{bmatrix}\exp(-i(\lambda x+\lambda^2 t)/\hbar),\end{aligned} \tag{2.3}$$

satisfying the relations

$$\begin{aligned}\mathbf{u}^+(x,t,\lambda_k)&=\gamma_k\mathbf{u}^-(x,t,\lambda_k), \\ -\gamma_k^*\mathbf{u}^+(x,t,\lambda_k^*)&=\mathbf{u}^-(x,t,\lambda_k^*), \quad k=1,\ldots,N,\end{aligned} \tag{2.4}$$

for some distinct complex numbers $\lambda_0,\ldots,\lambda_{N-1}$ in the upper half-plane and nonzero complex numbers (not necessarily distinct) $\gamma_0,\ldots,\gamma_{N-1}$. It is easy to check that given the numbers $\{\lambda_k\}$ and $\{\gamma_k\}$, the relations (2.4) determine the coefficient functions $A_p(x,t)$, $B_p(x,t)$, $C_p(x,t)$, and $D_p(x,t)$ in terms of exponentials via the solution of a square inhomogeneous linear algebraic system. In Faddeev and Takhtajan [FT87], it is shown that this linear system is always nonsingular assuming the $\{\lambda_k\}$ are distinct and nonreal and the $\{\gamma_k\}$ are nonzero. The solution of the nonlinear Schrödinger equation for which the column vectors $\mathbf{u}^\pm(x,t,\lambda)$ are simultaneous solutions of (2.1) and (2.2) turns out to be

$$\psi(x,t) = 2iA_{N-1}(x,t). \tag{2.5}$$

Remark. This construction is equivalent to a classical problem of rational approximation, the construction of *multipoint Padé interpolants* for entire functions [B75]. Let $G(\lambda)$ be any polynomial satisfying $G(\lambda_k) = \mathrm{Log}(\gamma_k)$ and $G(\lambda_k^*) = \mathrm{Log}(-1/\gamma_k^*)$. Then, looking at the first row of (2.4), we see that we are seeking polynomials $P_{N-1}(\lambda)$ and $Q_{N-1}(\lambda)$ both of degree $N-1$ such that

$$\frac{P_{N-1}(\lambda)}{\lambda^N + Q_{N-1}(\lambda)} = \exp(G(\lambda))\exp(-2i(\lambda x + \lambda^2 t)/\hbar),$$
$$\text{for } \lambda = \lambda_0, \ldots, \lambda_{N-1}, \lambda_0^*, \ldots, \lambda_{N-1}^*. \tag{2.6}$$

The coefficients of $P_{N-1}(\lambda)$ are the $\{A_p(x,t)\}$, and the coefficients of $Q_{N-1}(\lambda)$ are the $\{C_p(x,t)\}$.

Whereas the usual Padé approximants are constructed by demanding sufficiently high-order agreement in the asymptotic expansion of (2.6) for large or small λ, the multipoint approximants are constructed by demanding simple agreement of the function values on the left- and right-hand sides of (2.6) at a sufficiently large number of distinct points. This latter version of the rational interpolation problem was first considered by Cauchy and Jacobi. The Cauchy-Jacobi problem in its most general form can fail to have a solution, that is, given a set of nodes of interpolation, there exist isolated "unreachable" function values. In the context of the N-soliton solution problem, however, this undesirable situation does not occur due to the complex conjugation symmetry of the interpolation points and corresponding symmetry properties of the assigned values at those points.

A typical initial condition $\psi_0(x)$ for (1.1) will not correspond exactly to a multisoliton solution. As is well known [ZS72, FT87], the procedure for solving (1.1) generally begins with the study the solutions of (2.1) for real λ and for $\psi = \psi_0(x)$. One obtains from this analysis a complex-valued *transmission coefficient* $T(\lambda) = 1/a(\lambda)$, $\lambda \in \mathbb{R}$. Now, after the fact it turns out that the function $a(\lambda)$ has an analytic continuation into the whole upper half-plane, and its zeros occur at values of λ for which (2.1) has an $L^2(\mathbb{R})$ eigenfunction. In this sense, the study of the scattering problem for real λ yields results for complex λ by unique analytic continuation. The function $a(\lambda)$ can be interpreted as a Wronskian between two particular solutions of (2.1) that have analytic continuations into the upper half-plane. Thus at each L^2 eigenvalue $\lambda = \lambda_k$, there is a complex number γ_k that is the ratio of these two analytic solutions. In addition to the transmission coefficient, one also finds a complex-valued function $b(\lambda)$ that gives rise to a *reflection coefficient* $r(\lambda) := b(\lambda)/a(\lambda)$, $\lambda \in \mathbb{R}$. The main results of Zakharov and Shabat [ZS72] are the following:

1. When $\psi(x,t)$ is the solution of (1.1) with initial data $\psi_0(x)$, then for each $t > 0$ one has different coefficients in the linear problem (2.1). Therefore, the eigenvalues $\{\lambda_k\}$, proportionality constants $\{\gamma_k\}$, and the function $b(\lambda)$ can be computed independently for each $t > 0$. However, it follows from (2.2) that

the eigenvalues $\{\lambda_k\}$ (more generally the function $a(\lambda)$) and also $|b(\lambda)|$, $\lambda \in \mathbb{R}$, are independent of t, and the proportionality constants $\{\gamma_k\}$ and $\arg(b(\lambda))$, $\lambda \in \mathbb{R}$, evolve simply in time. Thus, $r(\lambda, t) = r(\lambda, 0)\exp(-2i\lambda^2 t/\hbar)$ and $\gamma_k(t) = \gamma_k(0)\exp(-2i\lambda_k^2 t/\hbar)$.

2. The function $\psi(x,t)$ can be reconstructed at later times $t > 0$ in terms of the discrete spectrum $\{\lambda_k\}$, $\{\gamma_k\}$, and the reflection coefficient $r(\lambda)$.

If for the initial condition $\psi_0(x)$ we have $b(\lambda) \equiv 0$, then the step of reconstructing the solution of the initial-value problem (1.1) is essentially what we have already described. Namely, one solves the linear equations (2.4) for the coefficient $A_{N-1}(x,t)$ and then the solution of (1.1) is given by (2.5). Note that N is the number of L^2 eigenvalues for $\psi_0(x)$ in the upper half-plane.

In general, the reconstruction of ψ from the scattering data can be recast in terms of the solution of a matrix-valued meromorphic *Riemann-Hilbert problem.* One seeks (for each x and t, which play the role of parameters) a matrix-valued function $\mathbf{m}(\lambda)$ of λ that is jointly meromorphic in the upper and lower half-planes, and for which the following three conditions hold:

1. $\mathbf{m}(\lambda) \to \mathbb{I}$ in each half-plane as $\lambda \to \infty$.
2. The singularities of $\mathbf{m}(\lambda)$ are completely specified. There are simple poles at the eigenvalues $\{\lambda_k\}$ and the complex conjugates with residues of a certain specified type (see (2.11) and (2.12)).
3. On the real axis $\lambda \in \mathbb{R}$, there is the *jump relation*

$$\mathbf{m}_+(\lambda) = \mathbf{m}_-(\lambda)\mathbf{v}(\lambda), \quad \mathbf{m}_\pm(\lambda) := \lim_{\epsilon \downarrow 0} \mathbf{m}(\lambda \pm i\epsilon), \tag{2.7}$$

where $\mathbf{v}(\lambda)$ is a certain *jump matrix* built out of $r(\lambda)$ and depending *explicitly* on x and t. The jump matrix becomes the identity matrix for $b(\lambda) \equiv 0$.

However, if the boundary values $\mathbf{m}_\pm(\lambda)$ are continuous and if $b(\lambda) \equiv 0$, then it is easy to see that the solution $\mathbf{m}(\lambda)$ must be a rational function of λ. This is the case we now develop in more detail. Let $J = \pm 1$ be a free parameter. From the column vectors $\mathbf{u}^\pm(x,t,\lambda)$, we build a matrix solution of (2.1):

$$\Psi(\lambda) := [\mathbf{u}^-(x,t,\lambda), \mathbf{u}^+(x,t,\lambda)]\sigma_1^{\frac{1-J}{2}} \operatorname{diag}\left(\prod_{j=1}^N (\lambda - \lambda_j)^{-1}, \prod_{j=1}^N (\lambda - \lambda_j^*)^{-1}\right)$$
$$\times \sigma_1^{\frac{1-J}{2}} \exp(i\sigma_3\lambda^2 t/\hbar). \tag{2.8}$$

This special matrix solution of (2.1) is called a *Jost solution.* Note that $\Psi(\lambda)$ would also satisfy (2.2) if it were not for the exponential factor in this formula. The reason for this exponential factor is that the Jost solution matrix has simple large x asymptotics that are, to leading order, independent of t. Indeed, if we define a matrix $\mathbf{m}(\lambda)$ by

$$\mathbf{m}(\lambda) := \Psi(\lambda)\exp(i\sigma_3\lambda x/\hbar), \tag{2.9}$$

then we find using (2.4) that for all fixed complex λ different from the eigenvalues $\{\lambda_k\}$ and their complex conjugates, $\mathbf{m}(\lambda)$ is a uniformly bounded function of x that satisfies $\mathbf{m}(\lambda) \to \mathbb{I}$ as $x \to J\infty$. *Thus, the parameter J merely indicates whether the Jost solution matrix is normalized at $x = +\infty$ or $x = -\infty$.*

Remark. In the general case when $b(\lambda)$ does not necessarily vanish identically, the Jost solution matrix is defined for all $\lambda \in \mathbb{C}$ as the unique matrix solution $\Psi(\lambda)$ of (2.1) that satisfies the conditions

$$\begin{aligned} &\textbf{Normalization:} \quad \Psi(\lambda)\exp(i\sigma_3\lambda x/\hbar) \to \mathbb{I}, \quad \text{as } x \to J\infty, \\ &\textbf{Boundedness:} \quad \sup_{x\in\mathbb{R}} \|\Psi(\lambda)\| < \infty. \end{aligned} \tag{2.10}$$

The boundedness condition is superfluous when $\lambda \in \mathbb{R}$ but is absolutely necessary for uniqueness when $\Im(\lambda) \neq 0$. The definition (2.9) yields a matrix-valued function of λ that is meromorphic in the upper and lower half-planes. For $\lambda \in \mathbb{R}$, however, the three matrices $\mathbf{m}(\lambda)$, $\mathbf{m}_+(\lambda)$, and $\mathbf{m}_-(\lambda)$ (cf. (2.7)) are generally all different unless $b(\lambda) \equiv 0$.

Continuing with the pure soliton case of $b(\lambda) \equiv 0$, we can deduce from the explicit form (2.3) of the vectors $\mathbf{u}^\pm(x, t, \lambda)$ and from the relations (2.4) that $\mathbf{m}(\lambda)$ solves the following problem.

Riemann-Hilbert Problem 2.0.1 (Meromorphic problem). *Given the discrete data $\{\lambda_k\}$ and $\{\gamma_k\}$, find a matrix $\mathbf{m}(\lambda)$ with the following properties:*

1. **Rationality:** *$\mathbf{m}(\lambda)$ is a rational function of λ, with simple poles confined to the eigenvalues $\{\lambda_k\}$ and the complex conjugates. At the singularities,*

$$\begin{aligned} \operatorname*{Res}_{\lambda=\lambda_k} \mathbf{m}(\lambda) &= \lim_{\lambda\to\lambda_k} \mathbf{m}(\lambda)\sigma_1^{\frac{1-J}{2}} \begin{bmatrix} 0 & 0 \\ c_k(x,t) & 0 \end{bmatrix} \sigma_1^{\frac{1-J}{2}}, \\ \operatorname*{Res}_{\lambda=\lambda_k^*} \mathbf{m}(\lambda) &= \lim_{\lambda\to\lambda_k^*} \mathbf{m}(\lambda)\sigma_1^{\frac{1-J}{2}} \begin{bmatrix} 0 & -c_k(x,t)^* \\ 0 & 0 \end{bmatrix} \sigma_1^{\frac{1-J}{2}}, \end{aligned} \tag{2.11}$$

for $k = 0, \ldots, N-1$, with

$$c_k(x,t) := \left(\frac{1}{\gamma_k}\right)^J \frac{\prod_{n=0}^{N-1}(\lambda_k - \lambda_n^*)}{\prod_{\substack{n=0 \\ n\neq k}}^{N-1}(\lambda_k - \lambda_n)} \exp(2iJ(\lambda_k x + \lambda_k^2 t)/\hbar). \tag{2.12}$$

2. **Normalization:**

$$\mathbf{m}(\lambda) \to \mathbb{I}, \quad \textit{as } \lambda \to \infty. \tag{2.13}$$

Whereas we deduced the two properties characterizing Riemann-Hilbert Problem 2.0.1 from the explicit construction of the vector solutions $\mathbf{u}^\pm(x, t, \lambda)$, it is not difficult to see that these two properties actually characterize the matrix function $\mathbf{m}(\lambda)$ uniquely.

Proposition 2.0.1 *The meromorphic Riemann-Hilbert Problem 2.0.1 corresponding to the discrete data $\{\lambda_k\}$ and $\{\gamma_k\}$ has a unique solution whenever the λ_k*

are distinct in the upper half-plane and the γ_k *are nonzero. The function defined from the solution by*

$$\psi := 2i \lim_{\lambda\to\infty} \lambda m_{12}(\lambda) \tag{2.14}$$

(that this limit exists is part of the proposition) is a nontrivial N*-soliton solution of the focusing nonlinear Schrödinger equation.*

Proof. One obtains a solution of the meromorphic Riemann-Hilbert Problem 2.0.1 by making an ansatz of the form (2.3) and observing that the residue conditions (2.11) with (2.12) are equivalent to (2.4), that is, by reversing our steps. The solvability of the linear system for the coefficients implied by (2.4) is guaranteed by the distinctness of the λ_k and the conditions $\gamma_k \neq 0$ [FT87]. Under an ansatz of the form (2.3), the relation (2.14) is equivalent to (2.5). Note that the same solution of the nonlinear Schrödinger equation is obtained from the matrix $\mathbf{m}(\lambda)$ for *both cases* $J = \pm 1$ by the formula (2.14). Uniqueness follows from Liouville's theorem. □

Thus, we may drop the explicit algebraic relations (2.4) and instead view Riemann-Hilbert Problem 2.0.1 as the fundamental characterization of the N-soliton solutions of the focusing nonlinear Schrödinger equation.

Remark. In the general case when $b(\lambda) \not\equiv 0$, the meromorphic Riemann-Hilbert problem is altered. One only insists that $\mathbf{m}(\lambda)$ be piecewise-meromorphic in the upper and lower half-planes and that the boundary values taken from above and below on the real λ-axis satisfy a jump relation (cf. (2.7)) with a matrix $\mathbf{v}(\lambda)$ built out of $r(\lambda)$ and going over into the identity matrix when $b(\lambda) \equiv 0$ (and thus $r(\lambda) \equiv 0$). When $r(\lambda) \not\equiv 0$, the corresponding meromorphic Riemann-Hilbert problem cannot be solved by algebraic operations alone, and in general the solution can be obtained by solving a system of integral equations. But even in this more general case, the function $\psi(x,t)$ defined by (2.14) when $r(\lambda)$ decays sufficiently rapidly satisfies the nonlinear Schrödinger equation.

Another very important property of the matrix $\mathbf{m}(\lambda)$ is the following "reflection symmetry" in the real axis:

$$\mathbf{m}(\lambda^*) = \sigma_2 \mathbf{m}(\lambda)^* \sigma_2. \tag{2.15}$$

On the right-hand side and in other similar formulae, the star denotes componentwise complex conjugation; the matrix is not transposed.

We now show how to convert the meromorphic Riemann-Hilbert problem into a sectionally holomorphic Riemann-Hilbert problem, that is, how to remove the poles from the problem. The reader can find a similar construction in [DKKZ96] and a useful alternative construction in [M02]. Let C be a simple closed contour that is the boundary of a simply connected domain D in the upper half-plane that contains all of the eigenvalues $\{\lambda_k\}$. We assign to C an orientation ω ($\omega = +1$ means counterclockwise, and $\omega = -1$ means clockwise), and when this orientation is important (i.e., in contour integration and specifying Riemann-Hilbert jump relations), we write the contour as C_ω. By C^* and D^* we mean the corresponding complex-conjugate sets in the lower half-plane, and when we write $[C \cup C^*]_\omega$, we mean that both loops share the same orientation ω. See figure 2.1.

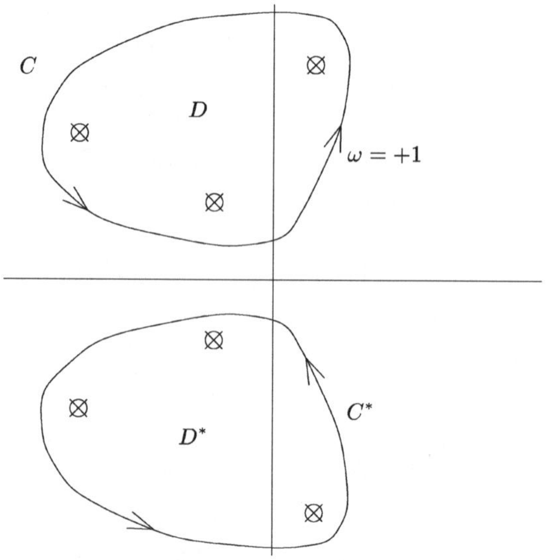

Figure 2.1 The complex λ-plane with three eigenvalues λ_k in the upper half-plane, their complex conjugates, and the contours C, C^* and domains D, D^*. The orientation in the figure is $\omega = +1$.

Next, we need to interpolate the proportionality constants at the eigenvalues. Choose a constant Q and a function $X(\lambda)$ analytic in D so that

$$\gamma_k = Q\exp(X(\lambda_k)/\hbar), \quad k = 0, \ldots, N-1. \tag{2.16}$$

In general, $X(\lambda)$ could be systematically constructed as an interpolating polynomial of degree $\sim N$. In other circumstances (see (3.11) and (3.12)), the phases γ_k can be highly correlated so that for very large N one can choose for $X(\lambda)$ a polynomial of low degree (or another simple expression). Note that the interpolant of the γ_k is not necessarily unique; for each K in some indexing set, there is a distinct pair $(Q_K, X_K(\lambda))$ such that for all j, $\gamma_j = Q_K \exp(X_K(\lambda_j)/\hbar)$. We make use of this freedom later; for now we just carry the subscript K.

With the help of the interpolant of the proportionality constants, we define a new matrix $\mathbf{M}(\lambda)$ for $\lambda \in \mathbb{C} \setminus (C \cup C^*)$ in the following way. First, for all $\lambda \in D$, set

$$\mathbf{M}(\lambda) := \mathbf{m}(\lambda)\sigma_1^{\frac{1-J}{2}} \times \begin{bmatrix} 1 & 0 \\ -\left(\frac{1}{Q_K}\right)^J \left(\prod_{n=0}^{N-1} \frac{\lambda-\lambda_n^*}{\lambda-\lambda_n}\right) \exp\left(\frac{J}{\hbar}(2i\lambda x + 2i\lambda^2 t - X_K(\lambda))\right) & 1 \end{bmatrix} \sigma_1^{\frac{1-J}{2}}. \tag{2.17}$$

Next, for all $\lambda \in D^*$, set

$$\mathbf{M}(\lambda) := \sigma_2 \mathbf{M}(\lambda^*)^* \sigma_2. \tag{2.18}$$

Finally, for all $\lambda \in \mathbb{C} \setminus (\overline{D} \cup \overline{D}^*)$ (i.e., in the rest of the complex plane minus $C \cup C^*$), simply set

$$\mathbf{M}(\lambda) := \mathbf{m}(\lambda). \tag{2.19}$$

It is straightforward to verify that by our choice of interpolants and the Blaschke factor appearing in (2.17) that $\mathbf{M}(\lambda)$ *has no poles in D or D^* and hence is sectionally holomorphic in the complex λ-plane*. By definition, we have preserved the reflection symmetry of $\mathbf{m}(\lambda)$ so that for all $\lambda \in \mathbb{C} \setminus (C \cup C^*)$ we have

$$\mathbf{M}(\lambda^*) = \sigma_2 \mathbf{M}(\lambda)^* \sigma_2. \tag{2.20}$$

The matrix $\mathbf{M}(\lambda)$ has continuous boundary values from either side on C and C^*. To describe these, let the left (respectively, right) side of the oriented contour $[C \cup C^*]_\omega$ be denoted by "+" (respectively, "−"). For $\lambda \in [C \cup C^*]_\omega$ define

$$\mathbf{M}_\pm(\lambda) := \lim_{\substack{\mu \to \lambda \\ \mu \in \pm \text{ side of } [C \cup C^*]_\omega}} \mathbf{M}(\mu), \tag{2.21}$$

that is, the nontangential limits from the left and right sides. Then using the fact that $\mathbf{m}(\lambda)$ is analytic on $C \cup C^*$ and the piecewise definition of $\mathbf{M}(\lambda)$ given by (2.17), (2.18), and (2.19), we find

$$\begin{aligned} \mathbf{M}_+(\lambda) &= \mathbf{M}_-(\lambda) \mathbf{v_M}(\lambda), \quad \lambda \in C_\omega, \\ \mathbf{M}_+(\lambda) &= \mathbf{M}_-(\lambda) \sigma_2 \mathbf{v_M}(\lambda^*)^* \sigma_2, \quad \lambda \in [C^*]_\omega, \end{aligned} \tag{2.22}$$

where, for $\lambda \in C$,

$$\mathbf{v_M}(\lambda) := \sigma_1^{\frac{1-J}{2}} \times \begin{bmatrix} 1 & 0 \\ -\omega\left(\frac{1}{Q_K}\right)^J \left(\prod_{n=0}^{N-1} \frac{\lambda - \lambda_n^*}{\lambda - \lambda_n}\right) \exp\left(\frac{J}{\hbar}(2i\lambda x + 2i\lambda^2 t - X_K(\lambda))\right) & 1 \end{bmatrix} \sigma_1^{\frac{1-J}{2}}. \tag{2.23}$$

Note that the orientation choice $\omega = \pm 1$ is arbitrary, leading to the *same* matrix $\mathbf{M}(\lambda)$. We allow for both possibilities of orientation for later convenience. Also, observe that if one introduces a discrete measure in the complex plane by

$$d\mu = \sum_{k=0}^{N-1} \left[\hbar \delta_{\lambda_k^*} - \hbar \delta_{\lambda_k}\right], \tag{2.24}$$

then for any branch of the logarithm,

$$\prod_{k=0}^{N-1} \frac{\lambda - \lambda_k^*}{\lambda - \lambda_k} = \exp\left(\frac{1}{\hbar} \int \log(\lambda - \eta)\, d\mu(\eta)\right). \tag{2.25}$$

The jump matrices are therefore conveniently written in terms of phases,

$$\alpha := \int \log(\lambda - \eta)\, d\mu(\eta) + J \cdot (2i\lambda x + 2i\lambda^2 t - X_K(\lambda)) \pmod{2\pi i \hbar}. \tag{2.26}$$

Suppose the eigenvalues $\{\lambda_k\}$ and proportionality constants $\{\gamma_k\}$ are given along with an appropriate interpolation $Q_K \exp(X_K(\lambda)/\hbar)$ of the γ_k and a smooth closed contour C enclosing the eigenvalues in the upper half-plane. We thus define a Riemann-Hilbert problem.

RIEMANN-HILBERT PROBLEM 2.0.2 (Holomorphic problem). *Given the eigenvalues* $\{\lambda_k\}$, *the interpolant* $Q_K \exp(X_K(\lambda)/\hbar)$, *and the oriented contour* C_ω, *find a matrix function* $\mathbf{M}(\lambda)$ *satisfying the following:*

1. **Analyticity:** $\mathbf{M}(\lambda)$ *is analytic in each component of* $\mathbb{C} \setminus (C \cup C^*)$;
2. **Boundary behavior:** $\mathbf{M}(\lambda)$ *assumes continuous boundary values on* $C \cup C^*$;
3. **Jump conditions:** *The boundary values taken on* $[C \cup C^*]_\omega$ *satisfy the relations* (2.22) *with* $\mathbf{v_M}(\lambda)$ *given explicitly by* (2.23);
4. **Normalization:** $\mathbf{M}(\lambda)$ *is normalized at infinity:*

$$\mathbf{M}(\lambda) \to \mathbb{I} \quad \text{as } \lambda \to \infty. \tag{2.27}$$

PROPOSITION 2.0.2 *The holomorphic Riemann-Hilbert Problem* 2.0.2 *has a unique solution* $\mathbf{M}(\lambda)$ *whenever the* λ_k *are distinct and nonreal and when the* γ_k *are nonzero. The function defined by*

$$\psi := 2i \lim_{\lambda \to \infty} \lambda M_{12}(\lambda) \tag{2.28}$$

is independent of the value of the index J, *as well as the particular choice of loop contour* C *and interpolant index* K, *and it is the N-soliton solution of the focusing nonlinear Schrödinger equation corresponding to the discrete data* $\{\lambda_k\}$ *and* $\{\gamma_k\}$.

Proof. The existence part of the proof of this proposition follows from the corresponding existence result for the meromorphic Riemann-Hilbert Problem 2.0.1, whose solution $\mathbf{m}(\lambda)$ yields a solution $\mathbf{M}(\lambda)$ of the holomorphic problem by the definitions (2.17), (2.18), and (2.19). The uniqueness part of the proof follows from the continuity of the boundary values and Liouville's theorem: the ratio $\mathbf{M}^{(1)}(\lambda)\mathbf{M}^{(2)}(\lambda)^{-1}$ of any two solutions is analytic in $\mathbb{C} \setminus (C \cup C^*)$ and continuous in $\mathbb{C}$. Therefore, this ratio is entire, and from the normalization condition we learn that $\mathbf{M}^{(1)}(\lambda) \equiv \mathbf{M}^{(2)}(\lambda)$. □

Note that it is possible to allow C to meet the real axis at one or more isolated points $u_k \in \mathbb{R}$, as long as at each u_k the incoming and outgoing parts of C make nonzero angles with the real axis and with each other. The contour C should thus meet the axis in "corners" (if at all).

Remark. The holomorphic Riemann-Hilbert Problem 2.0.2 with $J = +1$ is equivalent to that for $J = -1$ in the sense that they yield the same solution ψ of the nonlinear Schrödinger equation via (2.28). Note, however, that the solution matrix $\mathbf{M}(\lambda)$ for $J = +1$ is *not* the same as that for $J = -1$, even though they have similar leading-order asymptotics at $\lambda = \infty$. On the other hand, if J is given a fixed value, then the full matrix solution $\mathbf{M}(\lambda)$ of the Riemann-Hilbert problem corresponding to a contour C and interpolant K agrees identically with that corresponding to a contour C' and interpolant K' outside any circle containing both $C \cup C^*$ and $C' \cup C'^*$.

From the point of view of ease of analysis, the different formulations of the inverse problem corresponding to different choices of C, J, and K, although all equivalent, are not necessarily all equally valuable. For given values of x and t, it may turn out that one formulation of the Riemann-Hilbert problem results in a jump matrix that is very close to the identity in some parts of the complex plane, while in another equivalent formulation the jump matrix is large and oscillatory, with the solution being obtained by a kind of cancellation. The picture to have in mind here is that of evaluating a particularly complicated algebraic expression that one wants to evaluate and show is small. It may turn out that the expression can be written as a sum of residues of a contour integral in several different ways. In one integral it may turn out that the path of integration can be deformed in such a way that the integrand is uniformly very small, which is clearly useful in analysis. At the same time, it may be the case in another (equivalent!) integral that the integrand cannot be made uniformly small by any deformation, and the small result is always achieved by cancellation and consequently is more difficult to deduce. This is not an exact analogy, since matrix Riemann-Hilbert problems are not solved by direct contour integration, but it may help to illustrate the utility of having several possible formulations of the problem at hand.

Chapter Three

Semiclassical Soliton Ensembles

In this book, we present a technique for studying the behavior, near fixed x and t, of multisoliton solutions for which the number of solitons N is large but for which the solitons are highly *phase-correlated*. This means that for large N the discrete measure $d\mu$ converges and at the same time the interpolants $X_K(\lambda)$ converge and take on a simple limiting form.

3.1 FORMAL WKB FORMULAE FOR EVEN, BELL-SHAPED, REAL-VALUED INITIAL CONDITIONS

The highly correlated situation mentioned above arises naturally when one considers the semiclassical limit of the sequence of initial-value problems (1.1) for initial data of the form (1.2). Because $\hbar$ is present explicitly in the scattering problem (2.1) and generally in the initial data as well (if $S(x) \not\equiv 0$), the scattering data will depend on $\hbar$, and in particular the number of L^2 eigenvalues is a function of $\hbar$. Unfortunately, the question of how to rigorously extract all relevant asymptotic properties of the spectrum for (2.1) in the limit of $\hbar \downarrow 0$, given fixed functions $A(x)$ and $S(x)$, remains wide open (but see [M01] for some recent ideas in this direction).

On the other hand, some progress has been made in the *formal* analysis of the nonselfadjoint Zakharov-Shabat scattering problem (2.1) in the semiclassical limit. For example, the calculations in [EJLM93] suppose that $S(x) \equiv 0$ and then exploit the fact that, as $\hbar$ converges to zero, the eigenvalue problem (2.1) appears to go over into that of a semiclassically scaled selfadjoint Schrödinger operator with a nonselfadjoint and energy-dependent but bounded and small (formally $O(\hbar^2)$) correction (in fact, this was observed already in the paper of Zakharov and Shabat [ZS72]). This observation suggests that WKB formulae for the Schrödinger operator might be valid and, in particular, that the reflection coefficient $r(\lambda)$ is exponentially small for $\lambda \neq 0$ and that the discrete eigenvalues accumulate on the imaginary axis in the λ-plane with a certain asymptotic density. The true discrete eigenvalue measure $d\mu$ defined by (2.24) is presumed to converge in the sense of weak-$*$ convergence of measures to a measure $d\mu_0^{\mathrm{WKB}}$, which, in the case of functions $A(x)$ having a single local maximum (without loss of generality, we take it to occur for $x = 0$) for which there are always exactly two turning points, is given by the formula

$$d\mu_0^{\mathrm{WKB}}(\eta) := \rho^0(\eta)\chi_{[0,iA]}(\eta)\,d\eta + \rho^0(\eta^*)^*\chi_{[-iA,0]}(\eta)\,d\eta, \tag{3.1}$$

with

$$\rho^0(\eta) := \frac{\eta}{\pi}\int_{x_-(\eta)}^{x_+(\eta)} \frac{dx}{\sqrt{A(x)^2+\eta^2}} = \frac{1}{\pi}\frac{d}{d\eta}\int_{x_-(\eta)}^{x_+(\eta)} \sqrt{A(x)^2+\eta^2}\,dx, \tag{3.2}$$

for $\eta \in (0, iA)$, where $x_\pm(\eta)$ are the two real turning points and the square root is positive. In (3.1), $A = A(0)$ is the maximum amplitude of $A(x)$, and the imaginary segments $(-iA, 0)$ and $(0, iA)$ are both considered to be oriented from bottom to top to define the differential $d\eta$. In the two-turning-point case with $S(x) \equiv 0$, the confinement of the discrete spectrum to the imaginary axis has only recently been rigorously justified (exactly, for all $\hbar$). See [KS02]. The two-turning-point condition is essential for exact (as opposed to asymptotic) confinement.

Note that the function $\rho^0(\eta)$ defined by (3.2) is well behaved at $\eta = 0$ (in the sense of limit from positive imaginary values) if the function $A(x)$ decays sufficiently rapidly for large x (exponential is sufficient) and at $\eta = iA$ if $A(x)$ has nonvanishing curvature at its peak at $x = 0$. We assume both of these conditions on $A(x)$ in all that follows. Now, the formal WKB method not only provides a guess for the weak-$*$ limit of the discrete eigenvalue measures, but in fact it defines approximations to the discrete eigenvalues themselves for each value of $\hbar$. These are numbers $\lambda_{\hbar,n}^{\text{WKB}}$ lying on the positive imaginary axis in $(0, iA)$ that satisfy the Bohr-Sommerfeld quantization condition,

$$-\int_{\lambda_{\hbar,n}^{\text{WKB}}}^{iA} \rho^0(\eta)\,d\eta = \hbar(n+1/2), \tag{3.3}$$

for $n = 0, 1, 2, \ldots, N-1$, where N is the greatest integer such that

$$N \le -\frac{1}{\hbar}\int_0^{iA} \rho^0(\eta)\,d\eta + \frac{1}{2}. \tag{3.4}$$

Corresponding to these approximations, there is a discrete measure $d\mu_\hbar^{\text{WKB}}$ defined by the formula (cf. (2.24))

$$d\mu_\hbar^{\text{WKB}}(\eta) := \sum_{k=0}^{N-1}\left(\hbar\delta_{\lambda_{\hbar,k}^{\text{WKB}*}} - \hbar\delta_{\lambda_{\hbar,k}^{\text{WKB}}}\right). \tag{3.5}$$

The weak-$*$ convergence of these discrete measures to $d\mu_0^{\text{WKB}}$ is a direct matter to establish and analyze, in contrast with the convergence of the discrete measures of the true eigenvalues. We carry out a detailed convergence analysis of these approximate discrete measures in §3.2.

The "ground state" eigenvalue $\lambda_{\hbar,0}^{\text{WKB}}$ is characterized by

$$\int_{\lambda_{\hbar,0}^{\text{WKB}}}^{iA} \rho^0(\eta)\,d\eta = -\frac{\hbar}{2}, \tag{3.6}$$

and for symmetry it is useful to choose a sequence of values of $\hbar$ converging to zero so that the "most excited state" eigenvalue $\lambda_{\hbar,N-1}^{\text{WKB}}$ similarly satisfies

$$\int_0^{\lambda_{\hbar,N-1}^{\text{WKB}}} \rho^0(\eta)\,d\eta = -\frac{\hbar}{2}. \tag{3.7}$$

Thus, we can find a sequence of values $\hbar = \hbar_N$ so that for each $N = 1, 2, 3, \ldots$ there are exactly N WKB eigenvalues and the ground state and most excited state are equidistant from the endpoints of the imaginary interval $[0, iA]$ with respect to the measure $-\rho^0(\eta)\,d\eta$. This distance from the endpoints is exactly half of the distance between each of the eigenvalues (with respect to the same measure).

If, in addition to having a single local maximum and satisfying the decay and curvature conditions mentioned above, the function $A(x)$ is also even in x, then the proportionality constant γ_k associated with each eigenvalue λ_k, purely imaginary or not, is always equal to either ± 1. This follows from two facts. First, note that since the matrix in (2.1) is trace-free, the Wronskian determinant of any two solutions for the same value of λ is independent of x. Because the Wronskian of two bound states at the same value of λ necessarily vanishes as $x \to \pm\infty$, this implies that the $L^2(\mathbb{R})$ eigenspace for a given λ is at most 1-dimensional. This fact holds for completely arbitrary potentials $A(x)$ and $S(x)$. Second, when $S(x) \equiv 0$ and $A(x) = A(-x)$, then whenever $[u_1(x), u_2(x)]^T$ satisfies (2.1) for some λ, then so does $[v_1(x), v_2(x)]^T := [u_2(-x), u_1(-x)]^T$. Since bound states are nondegenerate, for $S(x) \equiv 0$ and $A(x) = A(-x)$ each bound state must be an eigenvector of this involution, whose eigenvalues are ± 1. Now, if the difference between the Zakharov-Shabat problem and the Schrödinger equation can be neglected for small $\hbar$, then from Sturm-Liouville oscillation theory, one expects that the proportionality constant simply alternates between the two values ± 1 from one eigenvalue to the next along the imaginary axis. Thus, one is led to propose an approximate interpolation formula $\gamma_j \approx \gamma_{\hbar,j}^{\rm WKB} := Q_K \exp(X_K(\lambda_{\hbar,j}^{\rm WKB})/\hbar)$, where

$$Q_K := -i(-1)^K, \quad X_K(\lambda) := i\pi(2K+1)\int_\lambda^{iA} \rho^0(\eta)\,d\eta, \tag{3.8}$$

and K is an arbitrary fixed integer. This formula gives values of the proportionality constant that vary from 1 to -1 from each WKB eigenvalue to the next, starting with $\gamma_{\hbar,0}^{\rm WKB} = -1$ for the WKB ground state $\lambda_{\hbar,0}^{\rm WKB}$.

We now collect these formal calculations into a definition for future reference.

Definition 3.1.1 *Let $A(x)$ be a positive, real-valued, even, bell-shaped function of x, and let $A = A(0)$ denote the maximum value. Let the function $\rho^0(\eta)$ be defined for $\eta \in (0, iA)$ by (3.1). Suppose further that $A(x)$ has nonvanishing curvature at its peak and decays sufficiently rapidly for large x so that $\rho^0(\eta)$ has a continuous extension to the closed imaginary interval $[0, iA]$. For each positive integer N, define $\hbar_N$ by*

$$\hbar_N := -\frac{1}{N}\int_0^{iA} \rho^0(\eta)\,d\eta. \tag{3.9}$$

Then, the WKB scattering data of the potential $\psi(x) = A(x)$ are defined as follows for $\hbar = \hbar_N$. The $L^2(\mathbb{R})$ eigenvalues are the N numbers $\lambda_{\hbar_N,n}^{\rm WKB}$ in the interval $(0, iA)$ that satisfy

$$-\int_{\lambda_{\hbar_N,n}^{\rm WKB}}^{iA} \rho^0(\eta)\,d\eta = \hbar_N(n + 1/2), \qquad \text{for } n = 0, 1, 2, \ldots, N-1. \tag{3.10}$$

The proportionality constant $\gamma_{\hbar_N,n}^{\rm WKB}$ corresponding to the eigenvalue $\lambda_{\hbar_N,n}^{\rm WKB}$ is given by

$$\gamma_{\hbar_N,n}^{\rm WKB} := Q_K \exp\big(X_K\big(\lambda_{\hbar_N,n}^{\rm WKB}\big)/\hbar_N\big), \tag{3.11}$$

where K is any integer and

$$Q_K := -i(-1)^K, \quad X_K(\lambda) := i\pi(2K+1)\int_\lambda^{iA} \rho^0(\eta)\,d\eta. \tag{3.12}$$

Finally, the reflection coefficient $r_{\hbar_N}^{\mathrm{WKB}}(\lambda) \equiv 0$. We call the exact solution of the focusing nonlinear Schrödinger equation corresponding to this set of scattering data for arbitrary N and $\hbar = \hbar_N$ the semiclassical soliton ensemble associated with the function $A(x)$.

We make no attempt here to discuss the validity of these formulae from the point of view of direct-scattering theory. Instead, we adopt the approach of beginning with the reflectionless approximate WKB spectrum and working out a completely rigorous inverse-scattering theory for this spectral data valid for sufficiently large N, which corresponds to sufficiently small $\hbar$. In the context of the semiclassical analysis of the initial-value problem for the focusing nonlinear Schrödinger equation, our procedure amounts to an a priori modification of the initial data, in which we replace the $\hbar$-independent initial data $\psi(x,0) \equiv A(x)$ by a sequence of $\hbar$-dependent initial data $\psi(x,0) \equiv A_N(x)$, which for each N is the unique potential whose *exact* scattering data is the reflectionless formal WKB approximation to the scattering data of $A(x)$ described in detail earlier. There are cases of particular functions $A(x)$, however, in which each element $A_N(x)$ of the sequence of functions turns out to be equal to $A(x)$, which means that the WKB approximation is exact. For these cases, the inverse theory that we develop here does indeed provide rigorous asymptotics for the semiclassical limit, without any further arguments.

Remark. Although rigorous statements about the errors of the WKB approximation are lacking for the case of $S(x) \equiv 0$, there are reasons for confidence that the formulae stated earlier are indeed valid. Moreover, the properties of these formulae are well understood since they are the same as in the classical Schrödinger case. By contrast, the asymptotic behavior of the spectrum of (2.1) when $S(x) \not\equiv 0$ is only beginning to be explored even at the qualitative level. For analytic potentials at least, the eigenvalues appear to accumulate on unions of curves in the complex plane that can be quite complicated. See [B96], [B01], and [M01] for more details on these spectra.

3.2 ASYMPTOTIC PROPERTIES OF THE DISCRETE WKB SPECTRUM

We begin by defining a particular branch of the logarithm.

Definition 3.2.1 *Let $L_\eta^0(\lambda)$ denote the particular branch of $\log(\lambda-\eta)$, considered as a function of λ, that agrees with the principal branch $\mathrm{Log}(\lambda-\eta)$ for $\lambda-\eta \in \mathbb{R}_+$ and that is cut from $\lambda = \eta \in i\mathbb{R}$ down along the imaginary axis to $-i\infty$. In terms of the standard cut of the principal branch,*

$$L_\eta^0(\lambda) := \mathrm{Log}(-i(\lambda-\eta)) + \frac{i\pi}{2}. \tag{3.13}$$

That is, we are taking $\arg(\lambda-\eta) \in (-\pi/2, 3\pi/2)$.

In our study of the inverse-scattering problem, we need to analyze carefully the integral

$$I^0(\lambda) := \frac{1}{\hbar}\int L_\eta^0(\lambda)d\mu_0^{\rm WKB}(\eta) = \frac{1}{\hbar}\int_0^{iA} L_\eta^0(\lambda)\rho^0(\eta)\,d\eta + \frac{1}{\hbar}\int_{-iA}^0 L_\eta^0(\lambda)\rho^0(\eta^*)^*\,d\eta \tag{3.14}$$

and its difference from

$$I^\hbar(\lambda) := \frac{1}{\hbar}\int L_\eta^0(\lambda)d\mu_\hbar^{\rm WKB}(\eta), \tag{3.15}$$

where λ lies in the complex plane away from the imaginary axis between $-iA$ and iA. Note that

$$\exp\big(I^\hbar(\lambda)\big) = \prod_{n=0}^{N-1}\frac{\lambda-\lambda_{\hbar,n}^{\rm WKB*}}{\lambda-\lambda_{\hbar,n}^{\rm WKB}}. \tag{3.16}$$

Since $S(x) \equiv 0$, the support of the WKB eigenvalue measure is confined to the imaginary axis, and we therefore have $\rho^0(\eta^*)^* \equiv -\rho^0(-\eta)$. Now, for $\eta \in [0, iA]$, define the *mass integral* by

$$m(\eta) := -\int_0^\eta \rho^0(\xi)\,d\xi. \tag{3.17}$$

Since $-i\rho^0(\eta) > 0$ for all $\eta \in [0, iA]$, this function is invertible, with an inverse that we denote by $\eta = e(m)$. Also, let $M := m(iA)$, so that $m([0, iA]) = [0, M]$. Then, with a change of variables, we find

$$\int_0^{iA} L_\eta^0(\lambda)\rho^0(\eta)\,d\eta = -\int_0^M L_{e(m)}^0(\lambda)\,dm. \tag{3.18}$$

Similarly, we find

$$\textstyle\int_{-iA}^0 L_\eta^0(\lambda)\rho^0(\eta^*)^*\,d\eta = \int_0^M L_{-e(m)}^0(\lambda)\,dm. \tag{3.19}$$

Therefore,

$$I^0(\lambda) = \frac{1}{\hbar}\int_0^M \big[L_{-e(m)}^0(\lambda) - L_{e(m)}^0(\lambda)\big]\,dm. \tag{3.20}$$

Note that by Definition 3.1.1 of the WKB spectrum, $e(m_n) = \lambda_{\hbar_N,n}^{\rm WKB}$, where $m_n := M - \hbar(n + 1/2)$. Because the sequence of values of $\hbar$ is such that the points m_n are symmetrically placed in the interval $[0, M]$, the integral can be easily represented in the form

$$I^0(\lambda) = \sum_{n=0}^{N-1} I_n^0(\lambda), \tag{3.21}$$

with

$$I_n^0(\lambda) := \frac{1}{\hbar}\int_{m_n-\hbar/2}^{m_n+\hbar/2}\big[L_{-e(m)}^0(\lambda) - L_{e(m)}^0(\lambda)\big]\,dm. \tag{3.22}$$

The midpoint rule approximation for $I^0(\lambda)$, as taught in every calculus course, comes from approximating each $I_n^0(\lambda)$ simply by the value of the integrand at $m = m_n$. That is, we write

$$I_n^0(\lambda) = L^0_{-e(m_n)}(\lambda) - L^0_{e(m_n)}(\lambda) + \text{ error terms,} \tag{3.23}$$

and we are reminded that $e(m_n) = \lambda^{\mathrm{WKB}}_{\hbar_N,n}$ and $-e(m_n) = \lambda^{\mathrm{WKB}*}_{\hbar_N,n}$. Let us introduce the notation

$$\widetilde{I}_n(\lambda) := I_n^0(\lambda) - \left[L^0_{-e(m_n)}(\lambda) - L^0_{e(m_n)}(\lambda)\right] \tag{3.24}$$

and

$$\widetilde{I}(\lambda) := \sum_{n=0}^{N-1} \widetilde{I}_n(\lambda) = I^0(\lambda) - I^{\hbar}(\lambda). \tag{3.25}$$

The following paragraphs describe the asymptotic behavior of $\widetilde{I}(\lambda)$ as $\hbar_N \downarrow 0$, for various regimes of λ.

3.2.1 Asymptotic Behavior for λFixed

For λ fixed, the midpoint rule is accurate to the second order in $\hbar = \hbar_N$. To see this, expand the integrand as

$$\begin{aligned} L^0_{-e(m)}(\lambda) &- L^0_{e(m)}(\lambda) \\ &= L^0_{-e(m_n)}(\lambda) - L^0_{e(m_n)}(\lambda) + \frac{2e'(m_n)\lambda}{\lambda^2 - e(m_n)^2}(m - m_n) \\ &\quad + \int_{m_n}^{m} d\zeta \int_{m_n}^{\zeta} d\xi \left[\frac{2e''(\xi)\lambda^3 - 2e''(\xi)e(\xi)^2\lambda + 4e'(\xi)^2 e(\xi)\lambda}{(\lambda^2 - e(\xi)^2)^2}\right]. \end{aligned} \tag{3.26}$$

Because the interval of integration for $I_n^0(\lambda)$ is symmetric about m_n, the linear term in the expansion integrates to zero. Thus,

$$\widetilde{I}_n(\lambda) = \frac{1}{\hbar}\int_{m_n-\hbar/2}^{m_n+\hbar/2} dm \int_{m_n}^{m} d\zeta \int_{m_n}^{\zeta} d\xi \left[\frac{2e''(\xi)\lambda^3 - 2e''(\xi)e(\xi)^2\lambda + 4e'(\xi)^2 e(\xi)\lambda}{(\lambda^2 - e(\xi)^2)^2}\right]. \tag{3.27}$$

For λ in the upper half-plane uniformly bounded away from the imaginary segment $[0, iA]$, the denominator of the quadratic term is bounded away from zero. From our assumptions on $A(x)$, $e'(\cdot)$ and $e''(\cdot)$ are uniformly bounded functions. Of course we automatically have $|e(\cdot)| \leq A$. Under these conditions, we easily get a bound on the quantity in square brackets that is uniform with respect to n:

$$\left|\frac{2e''(\xi)\lambda^3 - 2e''(\xi)e(\xi)^2\lambda + 4e'(\xi)^2 e(\xi)\lambda}{(\lambda^2 - e(\xi)^2)^2}\right| \leq K_1. \tag{3.28}$$

It follows that

$$\left|\widetilde{I}_n(\lambda)\right| \leq \frac{2K_1}{\hbar}\int_{m_n}^{m_n+\hbar/2} dm \int_{m_n}^{m} d\zeta \int_{m_n}^{\zeta} d\xi = \frac{2K_1}{\hbar}\frac{1}{6}\left(\frac{\hbar}{2}\right)^3 = \frac{K_1\hbar^2}{24}. \tag{3.29}$$

Since for sufficiently small $\hbar$ there exists a constant K_2 such that $N \leq K_2/\hbar$, summing over n gives

$$\left|\widetilde{I}(\lambda)\right| \leq \frac{K_1K_2\hbar}{24}. \tag{3.30}$$

It is the relative error that is of second order in $\hbar$ here, since the absolute error is order $\hbar$ and $I^0(\lambda)$ itself is of order $\hbar^{-1}$.

3.2.2 Letting λ Approach the Origin

This estimate fails if λ approaches the origin. We can improve upon the estimate by assuming first of all that we are dealing with values of $\lambda = |\lambda|e^{i\theta}$ in the upper half-plane such that $|\cos(\theta)| \geq \delta > 0$. That is, we prevent λ from coming within a symmetrical sector containing the imaginary axis.

The idea is to refine the estimate (3.28). We use the following estimates of the denominators. First, because λ lies in the upper half-plane and $e(\xi)$ is positive imaginary,

$$|\lambda + e(\xi)| \geq |\lambda - e(\xi)|. \tag{3.31}$$

Furthermore (see figure 3.1), we get both

$$|\lambda - e(\xi)| \geq \delta|\lambda| \quad \text{and} \quad |\lambda - e(\xi)| \geq \delta|e(\xi)|. \tag{3.32}$$

Also, we use the partial fraction expansion

$$\frac{2e''(\xi)\lambda^3 - 2e''(\xi)e(\xi)^2\lambda + 4e'(\xi)^2e(\xi)\lambda}{(\lambda^2 - e(\xi)^2)^2} = \frac{e''(\xi)}{\lambda + e(\xi)} + \frac{e''(\xi)}{\lambda - e(\xi)} - \frac{e'(\xi)^2}{(\lambda + e(\xi))^2} + \frac{e'(\xi)^2}{(\lambda - e(\xi))^2}. \tag{3.33}$$

From this we see that one estimate of the integrand in (3.26) is

$$\left|\frac{2e''(\xi)\lambda^3 - 2e''(\xi)e(\xi)^2\lambda + 4e'(\xi)^2e(\xi)\lambda}{(\lambda^2 - e(\xi)^2)^2}\right| \leq \frac{2E_2}{\delta|\lambda|} + \frac{2E_1^2}{\delta^2|\lambda|^2}, \tag{3.34}$$

where E_1 is the supremum of $|e'(\xi)|$ and E_2 is that of $|e''(\xi)|$, taken over the whole interval $\xi \in (0, M)$. As was the case with the estimate (3.28), this estimate does

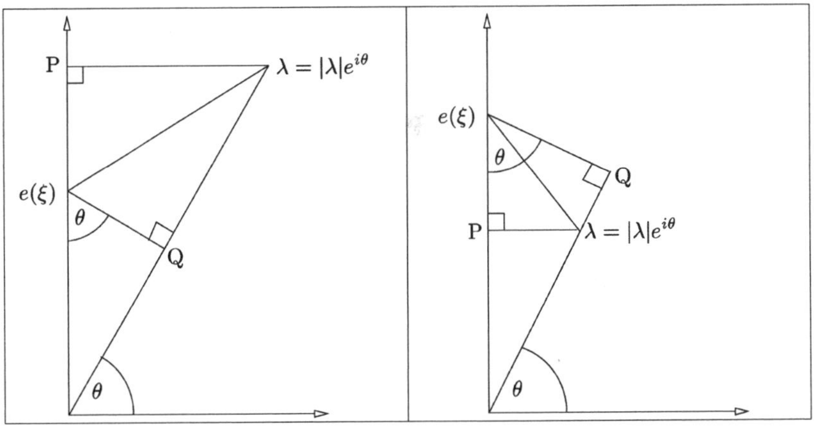

Figure 3.1 Justification of the geometrical estimates (3.32). The result is the same whether $\Im(\lambda) > |e(\xi)|$ (left) or $\Im(\lambda) < |e(\xi)|$ (right), although the trigonometry is slightly different. In both cases we see that $|\lambda - e(\xi)| \geq |\lambda - P| = |\lambda|\cos(\theta)$ and $|\lambda - e(\xi)| \geq |Q - e(\xi)| = |e(\xi)|\cos(\theta)$. Recall that $|\cos(\theta)| \geq \delta > 0$.

not depend on ξ nor on n, so by the same reasoning as that leading to (3.29), we get

$$\left|\widetilde{I}_n(\lambda)\right| \leq \frac{E_2\hbar^2}{12\delta|\lambda|} + \frac{E_1^2\hbar^2}{12\delta^2|\lambda|^2}. \tag{3.35}$$

If we further assume that $|\lambda|$ is uniformly bounded, then we can find a constant $C > 0$ depending on E_1, E_2, δ, and the bound on $|\lambda|$ so that the right-hand side of this estimate does not exceed $C\hbar^2/|\lambda|^2$, which gives

$$\left|\widetilde{I}_n(\lambda)\right| \leq \frac{C\hbar^2}{|\lambda|^2}. \tag{3.36}$$

On the other hand, another estimate of the integrand in (3.26) is

$$\left|\frac{2e''(\xi)\lambda^3 - 2e''(\xi)e(\xi)^2\lambda + 4e'(\xi)^2e(\xi)\lambda}{(\lambda^2 - e(\xi)^2)^2}\right| \leq \frac{2E_2}{\delta|e(\xi)|} + \frac{2E_1^2}{\delta^2|e(\xi)|^2}. \tag{3.37}$$

Since $|e(\xi)| \leq A < \infty$, we get $|e(\xi)| \geq |e(\xi)|^2/A$, and therefore with $D = 2AE_2/\delta + 2E_1^2/\delta^2$, we get

$$\left|\frac{2e''(\xi)\lambda^3 - 2e''(\xi)e(\xi)^2\lambda + 4e'(\xi)^2e(\xi)\lambda}{(\lambda^2 - e(\xi)^2)^2}\right| \leq \frac{D}{|e(\xi)|^2}. \tag{3.38}$$

Now, we use the fact that for any continuous function $g(\xi)$ we have

$$\int_{m_n-\hbar/2}^{m_n+\hbar/2} dm \int_{m_n}^{m} d\zeta \int_{m_n}^{\zeta} d\xi\, g(\xi) = \int_{m_n}^{m_n+\hbar/2} dm \int_{m_n}^{m} d\zeta \int_{m_n}^{\zeta} d\xi\, g(\xi) + \int_{m_n-\hbar/2}^{m_n} dm \int_{m}^{m_n} d\zeta \int_{\zeta}^{m_n} d\xi\, g(\xi), \tag{3.39}$$

where all volume elements are positively oriented in the integrals on the right-hand side, to get an estimate like (3.29). That is, we find

$$\left|\widetilde{I}_n(\lambda)\right| \leq \frac{D}{\hbar}\left(\int_{m_n}^{m_n+\hbar/2} dm \int_{m_n}^{m} d\zeta \int_{m_n}^{\zeta} \frac{d\xi}{|e(\xi)|^2} + \int_{m_n-\hbar/2}^{m_n} dm \int_{m}^{m_n} d\zeta \int_{\zeta}^{m_n} \frac{d\xi}{|e(\xi)|^2}\right). \tag{3.40}$$

Now, since $|e(\xi)|$ is by construction an increasing function of ξ, we have $|e(\xi)| \geq |e(m_n - \hbar/2)|$ in both integrals. Thus,

$$\left|\widetilde{I}_n(\lambda)\right| \leq \frac{D\hbar^2}{24|e(m_n - \hbar/2)|^2}. \tag{3.41}$$

To estimate the total error, we combine the two estimates (3.36) and (3.41). In particular, pick an integer L with $0 < L < N$, and use (3.41) to estimate the terms with $0 \leq n \leq L-1$ and (3.36) to estimate the terms with $n \geq L$. Thus,

$$\left|\widetilde{I}(\lambda)\right| \leq \frac{D\hbar^2}{24}\sum_{n=0}^{L-1}\frac{1}{|e(m_n - \hbar/2)|^2} + (N-L)\frac{C\hbar^2}{|\lambda|^2}. \tag{3.42}$$

Now, again since $|e(\xi)|$ is an increasing function, we easily obtain

$$\sum_{n=0}^{L-1}\frac{1}{|e(m_n - \hbar/2)|^2} \leq \frac{1}{\hbar}\int_{m_{L-1}-3\hbar/2}^{m_0-3\hbar/2}\frac{dm}{|e(m)|^2} = \frac{1}{\hbar}\int_{M-\hbar L-\hbar}^{M-2\hbar}\frac{dm}{|e(m)|^2}. \tag{3.43}$$

Also, since $|e(m)|$ is increasing and $e(0) = 0$ while $e'(0) \neq 0$, there is some constant $F > 0$ such that $|e(m)| \geq Fm$. Therefore,

$$\int_{M-\hbar L-\hbar}^{M-2\hbar} \frac{dm}{|e(m)|^2} \leq \frac{1}{F^2} \int_{M-\hbar L-\hbar}^{M-2\hbar} \frac{dm}{m^2} = \frac{1}{F^2}\left(\frac{1}{M-\hbar L-\hbar} - \frac{1}{M-2\hbar}\right). \quad (3.44)$$

Our estimate of the total error thus becomes

$$\left|\widetilde{I}(\lambda)\right| \leq \frac{D\hbar}{24F^2}\left(\frac{1}{M-\hbar L-\hbar} - \frac{1}{M-2\hbar}\right) + (N-L)\frac{C\hbar^2}{|\lambda|^2}. \quad (3.45)$$

We now consider how to optimally choose the cutoff L. Let $\beta > 0$, and take L to be the largest integer so that $M - \hbar L - \hbar \geq \hbar^\beta$. The quantity $N - L$ is then of order $\hbar^{\beta-1}$, and we thus have

$$\left|\widetilde{I}(\lambda)\right| = O(\hbar^{1-\beta}) + O(\hbar^{1+\beta}|\lambda|^{-2}). \quad (3.46)$$

If we assume that $|\lambda| \geq C\hbar^\gamma$ for some constant $C > 0$ and for some $\gamma \geq 0$, then the estimate becomes

$$\left|\widetilde{I}(\lambda)\right| = O(\hbar^{1-\beta}) + O(\hbar^{1+\beta-2\gamma}). \quad (3.47)$$

As long as $\gamma < 1$, the best error for $0 < \beta < 1$ is established via a dominant balance that occurs for $\beta = \gamma$. This gives an optimal estimate of

$$\left|\widetilde{I}(\lambda)\right| = O\left(\hbar^{1-\gamma}\right). \quad (3.48)$$

Thus, the error that we found to be order $\hbar$ for λ fixed is in fact small as long as $|\lambda| \gg \hbar$. This error estimate is uniform for λ big enough compared to $\hbar$ and lying in the upper half-plane outside of any given sector containing the imaginary axis.

3.2.3 Approximations Uniformly Valid for λ near the Origin

The error $\widetilde{I}(\lambda)$ is not uniformly small if $\lambda = O(\hbar)$. However, we may extract an additional contribution and then show that what remains is uniformly small for λ in some neighborhood of the origin.

Although $\widetilde{I}(\lambda)$ is not small, the total contribution of most of the terms $\widetilde{I}_n(\lambda)$ for which $|e(m_n)|$ is large enough compared to $\hbar$ will in fact be negligible. So again, we introduce a cutoff integer L, and then according to our previous results we get

$$\widetilde{I}(\lambda) = \sum_{n=0}^{L-1} \widetilde{I}_n(\lambda) + \sum_{n=L}^{N-1} \widetilde{I}_n(\lambda), \quad (3.49)$$

and we have the estimate

$$\left|\sum_{n=0}^{L-1} \widetilde{I}_n(\lambda)\right| \leq \frac{D\hbar}{24F^2}\left(\frac{1}{M-\hbar L-\hbar} - \frac{1}{M-2\hbar}\right). \quad (3.50)$$

As before, it is convenient to pick a number $\beta > 0$ and then choose L to be the largest integer so that $M - \hbar L - \hbar \geq \hbar^\beta$. The above estimate then becomes a λ-independent $O(\hbar^{1-\beta})$. Note that the constants L and β do not necessarily take the same values as in §3.2.2.

In each remaining term $\widetilde{I}_n(\lambda)$, both m_n and the integration variable m will be very close to zero, and it appears that it may be prudent to expand $\widetilde{I}_n(\lambda)$ to reflect this fact. Thus, we write

$$\widetilde{I}_n(\lambda) = J_n(\lambda) + \widetilde{J}_n(\lambda), \tag{3.51}$$

where

$$J_n(\lambda) := \frac{1}{\hbar}\int_{m_n-\hbar/2}^{m_n+\hbar/2} \left[L^0_{-e'(0)m}(\lambda) - L^0_{e'(0)m}(\lambda)\right] dm - \left[L^0_{-e'(0)m_n}(\lambda) - L^0_{e'(0)m_n}(\lambda)\right]. \tag{3.52}$$

We begin by estimating $\widetilde{J}_n(\lambda)$ for $L \le n \le N-1$. Explicitly,

$$\begin{aligned}\widetilde{J}_n(\lambda) = &\frac{1}{\hbar}\int_{m_n-\hbar/2}^{m_n+\hbar/2} \left[L^0_{-e(m)}(\lambda) - L^0_{-e'(0)m}(\lambda) - L^0_{e(m)}(\lambda) + L^0_{e'(0)m}(\lambda)\right] dm \\ &-\left[L^0_{-e(m_n)}(\lambda) - L^0_{-e'(0)m_n}(\lambda) - L^0_{e(m_n)}(\lambda) + L^0_{e'(0)m_n}(\lambda)\right].\end{aligned} \tag{3.53}$$

Recall Definition 3.2.1 of $L^0_\eta(\lambda)$. We can estimate $\widetilde{J}_n(\lambda)$ for $L \le n \le N-1$ by grouping the logarithms in pairs as follows. Imagining either $z = m$ or $z = m_n$ and using Taylor's theorem with remainder, we have

$$L^0_{\pm e(z)}(\lambda) - L^0_{\pm e'(0)z}(\lambda) = \text{Log}\left(1 \mp \frac{e''(\xi_\pm)z^2}{2(\lambda \mp e'(0)z)}\right), \tag{3.54}$$

where the $\xi_\pm$ lie between zero and z. Now, since λ is excluded from the symmetrical sector about the positive imaginary axis subtended by an angle of $2\sin^{-1}(\delta)$ and since $e'(0)$ is positive imaginary while $z > 0$ is real,

$$|\lambda \mp e'(0)z| \ge \delta|e'(0)|z. \tag{3.55}$$

Therefore, we find that for z sufficiently small there is a constant $G > 0$ such that

$$\left|L^0_{\pm e(z)}(\lambda) - L^0_{\pm e'(0)z}(\lambda)\right| \le \frac{G\sup|e''|}{2\delta|e'(0)|}z = Hz. \tag{3.56}$$

Consequently, we find

$$\left|\widetilde{J}_n(\lambda)\right| \le 2H(m_n + \hbar/2) + 2Hm_n, \tag{3.57}$$

because in the integral we have $m < m_n + \hbar/2$. Additionally, since $m_n \ge \hbar/2$, we have $m_n + \hbar/2 \le 2m_n$. Therefore, we also have

$$\left|\widetilde{J}_n(\lambda)\right| \le 6Hm_n. \tag{3.58}$$

So, summing over $L \le n \le N-1$, we find that

$$\left|\sum_{n=L}^{N-1}\widetilde{J}_n(\lambda)\right| \le 6H\sum_{n=L}^{N-1} m_n = 6H(N-L)\left[(M-\hbar L+\hbar/2)-\hbar\frac{N-L+1}{2}\right]. \tag{3.59}$$

Now, $N-L$ and $N-L+1$ are both order $\hbar^{\beta-1}$, while $M - \hbar L + \hbar/2$ is order $\hbar^\beta$. Therefore, we find that both terms are of the same order, and we have

$$\sum_{n=L}^{N-1}\widetilde{J}_n(\lambda) = O(\hbar^{2\beta-1}). \tag{3.60}$$

This estimate is uniform for bounded λ and will be small as long as $\beta > 1/2$.

Now we return to the quantities $J_n(\lambda)$ defined by (3.52). The integrals in $J_n(\lambda)$ can be evaluated exactly, and considerable simplification follows. In fact, if we set $\lambda = e'(0)\hbar w$, then we have, exactly,

$$\exp\left(\sum_{n=L}^{N-1} J_n(\lambda)\right) = w^{-w}(-w)^{-w}\frac{\Gamma(1/2+w)}{\Gamma(1/2-w)}\frac{(\overline{N}+w)^{\overline{N}+w}}{(\overline{N}-w)^{\overline{N}-w}}\frac{\Gamma(\overline{N}+1/2-w)}{\Gamma(\overline{N}+1/2+w)}, \tag{3.61}$$

where $\overline{N} := N - L$. Stirling's formula says

$$\Gamma(z) = e^{-z}z^{z-1/2}\sqrt{2\pi}\left(1 + O\left(|z|^{-1}\right)\right), \tag{3.62}$$

where the error is uniform with respect to direction for z in any sector $-\pi < -\phi \le \mathrm{Arg}(z) \le \phi < \pi$. If $\overline{N}$ is large, which is the same thing as saying that $\beta < 1$, we can apply this formula to (3.61), obtaining errors of the form $O((\overline{N}+1/2\pm w)^{-1})$. But because w is prevented from entering the symmetrical sector about the positive real axis with opening angle $2\sin^{-1}(\delta)$, it is not difficult to see that these terms are of order $O(\delta/\overline{N})$ or, for $\delta > 0$ fixed, simply $O(\overline{N}^{-1})$ uniformly for w in the right half-plane without the sector of angular width $2\sin^{-1}(\delta)$ about the positive real axis. The upshot of these considerations is that one finds

$$\exp\left(\sum_{n=L}^{N-1} J_n(\lambda)\right) = W(w)\left(1 + O\left(\overline{N}^{-1}\right)\right) \tag{3.63}$$

uniformly as $\overline{N}$ tends to infinity with w outside any fixed sector of the positive real axis of fixed angle, where

$$W(w) := e^{2w}w^{-w}(-w)^{-w}\frac{\Gamma(1/2+w)}{\Gamma(1/2-w)}. \tag{3.64}$$

The reader is reminded that all exponential functions are defined using the traditional cut of the principal branch of the logarithm; thus $W(w)$ is defined for $\Re(w) > 0$ and $\Im(w) \neq 0$. There is a cut on the positive real w-axis corresponding to the positive imaginary axis in the λ-plane. Note that $\overline{N} = O(\hbar^{\beta-1})$.

We thus see that while $\widetilde{I}(\lambda)$ itself is not small uniformly in λ, we can write

$$\begin{aligned}\exp\left(\widetilde{I}(\lambda)\right) &= W(w)\left(1 + O\left(\hbar^{1-\beta}\right)\right)\exp\left(O\left(\hbar^{2\beta-1}\right)\right)\exp\left(O\left(\hbar^{1-\beta}\right)\right)\\ &= W(w)\left(1 + O\left(\hbar^{1-\beta}\right) + O\left(\hbar^{2\beta-1}\right)\right).\end{aligned} \tag{3.65}$$

The dominant balance determining the optimal exponent occurs for $\beta = 2/3$, which gives the statement

$$\exp\left(\widetilde{I}(\lambda)\right) = W(w)\left(1 + O\left(\hbar^{1/3}\right)\right), \tag{3.66}$$

where the error is *uniformly* small for λ in any bounded region of the upper half-plane minus the sector about the imaginary axis of angular width $2\sin^{-1}(\delta)$.

3.2.4 Convergence Theorems for Discrete WKB Spectra

The work in the previous paragraphs establishes the following two theorems.

Theorem 3.2.1 (Near-field spectral asymptotics) *Let $\delta > 0$ be given, and consider $\lambda = |\lambda|e^{i\theta}$ to lie in a bounded set Λ in the upper half-plane such that $|\cos(\theta)| \geq \delta$. Then there is a constant $B_{\rm in} > 0$ such that for N sufficiently large, the WKB eigenvalues satisfy*

$$\left|\left[\prod_{n=0}^{N-1} \frac{\lambda - \lambda_{\hbar_N,n}^{\rm WKB*}}{\lambda - \lambda_{\hbar_N,n}^{\rm WKB}}\right] \exp(-I^0(\lambda))W(w) - 1\right| \leq B_{\rm in}\hbar_N^{1/3} \tag{3.67}$$

uniformly for $\lambda \in \Lambda$, where $w = \lambda/(e'(0)\hbar_N) = -\rho^0(0)\lambda/\hbar_N$ and where the canonical function $W(w)$ is defined by (3.64).

Theorem 3.2.2 (Far-field spectral asymptotics) *Suppose the conditions of Theorem 3.2.1, but also suppose that the set Λ contains only values λ satisfying $|\lambda|^{-1} = O(\hbar_N^{-\gamma})$ for some $\gamma > 0$, (i.e., λ is bounded away from the origin by an asymptotically small amount), then there is a constant $B_{\rm out} > 0$ such that for all N sufficiently large,*

$$\left|\left[\prod_{n=0}^{N-1} \frac{\lambda - \lambda_{\hbar_N,n}^{\rm WKB*}}{\lambda - \lambda_{\hbar_N,n}^{\rm WKB}}\right] \exp(-I^0(\lambda)) - 1\right| \leq B_{\rm out}\hbar_N^{1-\gamma}. \tag{3.68}$$

Furthermore, if λ is held fixed as N increases and does not lie on the imaginary segment $[-iA, iA]$, then the sectorial condition $|\cos(\theta)| \geq \delta$ can be dropped and for large enough N,

$$\left|\left[\prod_{n=0}^{N-1} \frac{\lambda - \lambda_{\hbar_N,n}^{\rm WKB*}}{\lambda - \lambda_{\hbar_N,n}^{\rm WKB}}\right] \exp(-I^0(\lambda)) - 1\right| \leq B_{\rm out}\hbar_N. \tag{3.69}$$

3.3 THE SATSUMA-YAJIMA SEMICLASSICAL SOLITON ENSEMBLE

There is at least one case of a real, even, single-maximum potential where the scattering data are known *exactly* for all $\hbar$. In 1974, Satsuma and Yajima [SY74] considered (essentially) the scattering problem (2.1) for arbitrary $\hbar > 0$ and for the special initial data

$$A(x) = A\,\mathrm{sech}(x), \qquad S(x) \equiv 0, \tag{3.70}$$

where $A > 0$ is an arbitrary constant. For this choice, they solved (2.1) explicitly in terms of hypergeometric functions and obtained

$$b(\lambda) = i\sin(\pi A/\hbar)\,\mathrm{sech}(\pi\lambda/\hbar), \qquad \lambda \in \mathbb{R}. \tag{3.71}$$

This implies that the reflection coefficient $r(\lambda)$ vanishes identically as long as

$$\hbar = \hbar_N = \frac{A}{N}, \qquad N = 1, 2, 3, \ldots. \tag{3.72}$$

Moreover, if $\hbar = \hbar_N$, then there are exactly N eigenvalues, and the relevant discrete data is given by

$$\lambda_k = i\hbar_N(N - k - 1/2), \quad \gamma_k = (-1)^{k+1}, \quad k = 0, \ldots, N-1. \tag{3.73}$$

When we consider the semiclassical limit in detail for this special initial data, we always assume in this book that $\hbar = \hbar_N$; that is, the parameter $\hbar$ always takes one of the "quantum" values where there is no contribution to the solution from the reflection coefficient. Satsuma and Yajima's calculations have recently been generalized to some other potentials by Tovbis and Venakides [TV00].

It is interesting to compare these exact results with their WKB approximations. Using $A(x) = A\,\mathrm{sech}(x)$ in (3.1), one finds

$$\rho^0(\eta) = \rho^0_{\mathrm{SY}}(\eta) \equiv i, \quad \eta \in (0, iA). \tag{3.74}$$

Therefore, it is easy to check directly that in this case the WKB approximations to the true eigenvalues for $\hbar = \hbar_N$ turn out to be exact:

$$\lambda^{\mathrm{WKB}}_{\hbar_N, n} = \lambda_n \quad \text{for } A(x) = A\,\mathrm{sech}(x). \tag{3.75}$$

Also, since $\hbar = \hbar_N$ implies that $b(\lambda) \equiv 0$, the true scattering data is reflectionless, as is the approximate WKB scattering data, that is, $r(\lambda) \equiv r^{\mathrm{WKB}}(\lambda) \equiv 0$.

For the Satsuma-Yajima initial data with $\hbar = \hbar_N$, we may interpolate the true proportionality constants $\{\gamma_j\}$ at the true eigenvalues $\{\lambda_j\}$ by very simple expressions. Thus, using (3.73) we have the exact expressions

$$\gamma_j = Q_{\mathrm{SY},K} \exp(X_{\mathrm{SY},K,K'}(\lambda_j)/\hbar), \tag{3.76}$$

where

$$Q_{\mathrm{SY},K} = -i(-1)^K, \quad X_{\mathrm{SY},K,K'}(\lambda) = (2K+1)\pi\lambda - (2K'+1)i\pi A, \tag{3.77}$$

with K and K' being arbitrary integers that index the interpolants. The key feature of these exact formulae in relation to the semiclassical limit is that the number $Q_{\mathrm{SY},K}$ and the analytic function $X_{\mathrm{SY},K,K'}(\lambda)$ are *independent of* $\hbar$. Now, when $\hbar = \hbar_N = A/N$, we see that the function $Q_{\mathrm{SY},K}\exp(X_{\mathrm{SY},K,K'}(\lambda)/\hbar)$ is independent of the parametrizing integer K'. However, the parametrizing integer K enters in a more useful way (this becomes clear in §5.1). For later convenience, we from now on set $K' = K$ and write $X_{\mathrm{SY},K}(\lambda)$ for $X_{\mathrm{SY},K,K}(\lambda)$. For $\hbar = \hbar_N$ and for the special initial data we are considering, the indexing set of interpolants is just $\mathbb{Z}$. Note that for the Satsuma-Yajima data, we have the exact relation (cf. (3.8))

$$X_{\mathrm{SY},K}(\lambda) = i\pi(2K+1)\int_\lambda^{iA} \rho^0_{\mathrm{SY}}(\eta)\,d\eta. \tag{3.78}$$

Thus, the WKB formula for the proportionality constants, although unjustified in general, is not only asymptotic but *exact* for all $\hbar = \hbar_N$ in this special case.

So, for the special case of the Satsuma-Yajima initial data $\psi(x,0) = A\,\mathrm{sech}(x)$ and for the sequence of "quantum" values of $\hbar = \hbar_N = A/N$, the true scattering data agrees exactly with its WKB approximation. This means that the solution $\psi(x,t)$ to the focusing nonlinear Schrödinger equation with initial data $\psi(x,0) = A\,\mathrm{sech}(x)$ is in this case exactly what we have called the semiclassical soliton ensemble corresponding to the function $A(x) = A\,\mathrm{sech}(x)$. Therefore, the rigorous asymptotics that we develop in the remainder of this book for semiclassical soliton ensembles corresponding to quite general analytic functions $A(x)$ provide without any further argument the rigorous semiclassical asymptotic description of the solution to the initial-value problem (1.1) with the special initial data $\psi_0(x) = A\,\mathrm{sech}(x)$.

Chapter Four

Asymptotic Analysis of the Inverse Problem

In this chapter, we study the asymptotic behavior, in the limit of N tending to infinity, of semiclassical soliton ensembles corresponding to analytic, even, bell-shaped functions $A(x)$ with nonzero curvature at the peak and sufficient decay for large x, so that the density $\rho^0(\eta)$ defined by (3.1) has an analytic extension from the imaginary interval $(0, iA)$. This means that we are going to study Riemann-Hilbert Problem 2.0.2 for the matrix $\mathbf{M}(\lambda)$ posed with discrete data $\{\lambda_{\hbar_N,0}^{\mathrm{WKB}}, \ldots, \lambda_{\hbar_N,N-1}^{\mathrm{WKB}}\}$ and $\{\gamma_{\hbar_N,0}^{\mathrm{WKB}}, \ldots, \gamma_{\hbar_N,N-1}^{\mathrm{WKB}}\}$ defined in terms of $A(x)$ by Definition 3.1.1, in the limit as N tends to infinity.

In our analysis, we would like to consider using a contour C that is held fixed as N grows and $\hbar_N$ becomes small. Because the set of complex numbers $\{\lambda_{\hbar_N,n}^{\mathrm{WKB}}\}$ has zero as an accumulation point, we need to ensure that $C \subset \mathbb{C}_+$ passes through zero, and in fact we need this to occur with some nonzero angle with respect to the imaginary axis (cusps are not allowed; see §4.4.3). See figure 4.1.

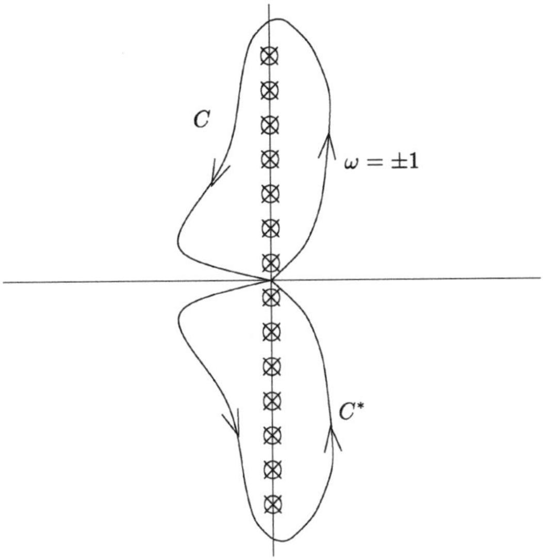

Figure 4.1 Examples of contours C and C^* appropriate for studying a semiclassical soliton ensemble in the limit $N \uparrow \infty$.

4.1 INTRODUCING THE COMPLEX PHASE

As pointed out in the first remark in chapter 2, the conditions (2.4) essentially describe a problem of rational interpolation of entire (exponential) functions; the numerator and denominator of the rational interpolant are the polynomials appearing in (2.3). These polynomials are then related to the matrix elements of the solution $\mathbf{M}(\lambda)$ of the Riemann-Hilbert problem. As $\hbar_N \downarrow 0$ and correspondingly as $N \uparrow \infty$, the degree of interpolation increases, and one expects the behavior of the interpolants to become increasingly wild (oscillatory and/or large in magnitude) away from the points of collocation.

In studying the Riemann-Hilbert problem for $\mathbf{M}(\lambda)$ in this limit, it is essential to capture this wild behavior. We therefore suppose that the matrix $\mathbf{M}(\lambda)$ can be written in the form

$$\mathbf{M}(\lambda) = \mathbf{N}(\lambda)\exp(g(\lambda)\sigma_3/\hbar), \tag{4.1}$$

where $g(\lambda)$ is some scalar function, yet to be determined, called the *complex phase*. At this point, the matrix $\mathbf{N}(\lambda)$ is just the quotient in (4.1). The complex phase function is intended to capture the wild behavior of $\mathbf{M}(\lambda)$ as $\hbar \downarrow 0$. *The guiding principle is that $g(\lambda)$ should be chosen so that the jump matrices for $\mathbf{N}(\lambda)$ implied by the change of variables (4.1) can be approximated by "simple" jump matrices and so that the error terms involved in the approximation can be controlled.*

There are several elementary constraints we place on the complex phase function $g(\lambda)$.

Definition 4.1.1 *The elementary properties of a complex phase function $g(\lambda)$ are*

1. **No $\hbar$-dependence:** *$g(\lambda)$ is independent of $\hbar$;*
2. **Analyticity:** *$g(\lambda)$ is analytic for $\lambda \in \mathbb{C}\setminus(C \cup C^*)$;*
3. **Decay:** *$g(\lambda) \to 0$ as $\lambda \to \infty$;*
4. **Boundary behavior:** *$g(\lambda)$ assumes continuous boundary values from both sides of $C \cup C^*$;*
5. **Symmetry:** *$g(\lambda^*) + g(\lambda)^* = 0$ for all $\lambda \in \mathbb{C}\setminus(C \cup C^*)$.*

Note that the analyticity in a deleted neighborhood of infinity and the decay at infinity together imply analyticity at infinity (i.e., g has a series representation in positive powers of λ^{-1} convergent for sufficiently large $|\lambda|$) and therefore uniformity of the decay with respect to direction. The symmetry condition on $g(\lambda)$ ensures that the matrix $\mathbf{N}(\lambda)$ inherits the reflection symmetry

$$\mathbf{N}(\lambda^*) = \sigma_2\mathbf{N}(\lambda)^*\sigma_2. \tag{4.2}$$

Let a function $g(\lambda)$ satisfying the five conditions of Definition 4.1.1 be given. For $\lambda \in C_\omega$, define the functions

$$\theta(\lambda) := iJ(g_+(\lambda) - g_-(\lambda)) \tag{4.3}$$

and

$$\phi(\lambda) := \int L_\eta^0(\lambda)\, d\mu_{\hbar_N}^{\mathrm{WKB}}(\eta) + J\cdot(2i\lambda x + 2i\lambda^2 t - X_K(\lambda) - g_+(\lambda) - g_-(\lambda)), \tag{4.4}$$

where $L^0_\eta(\lambda)$ is explained in Definition 3.2.1, $d\mu^{\rm WKB}_{\hbar_N}(\eta)$ is the discrete eigenvalue measure defined by (3.5) corresponding to the WKB eigenvalues for $\hbar = \hbar_N$, and $X_K(\lambda)$ is the interpolant of the WKB proportionality constants defined in Definition 3.1.1. From these functions, build a matrix for $\lambda \in C$:

$$\mathbf{v_N}(\lambda) := \sigma_1^{\frac{1-J}{2}} \begin{bmatrix} \exp(i\theta(\lambda)/\hbar) & 0 \\ ((-1)^K i)^{-J}\omega\exp(\phi(\lambda)/\hbar) & \exp(-i\theta(\lambda)/\hbar) \end{bmatrix} \sigma_1^{\frac{1-J}{2}}. \tag{4.5}$$

Then, in place of the holomorphic Riemann-Hilbert Problem 2.0.2 for $\mathbf{M}(\lambda)$, we may consider the following Riemann-Hilbert problem.

Riemann-Hilbert Problem 4.1.1 (Phase-conjugated problem). *Given a complex phase function $g(\lambda)$ satisfying the conditions of Definition 4.1.1, find a matrix function $\mathbf{N}(\lambda)$ satisfying following:*

1. **Analyticity:** *$\mathbf{N}(\lambda)$ is analytic for $\lambda \in \mathbb{C}\backslash(C \cup C^*)$;*
2. **Boundary behavior:** *$\mathbf{N}(\lambda)$ assumes continuous boundary values on $C \cup C^*$;*
3. **Jump conditions:** *The boundary values taken on $C \cup C^*$ satisfy*

$$\begin{aligned} \mathbf{N}_+(\lambda) &= \mathbf{N}_-(\lambda)\mathbf{v_N}(\lambda), & \lambda &\in C_\omega, \\ \mathbf{N}_+(\lambda) &= \mathbf{N}_-(\lambda)\sigma_2\mathbf{v_N}(\lambda^*)^*\sigma_2, & \lambda &\in [C^*]_\omega, \end{aligned} \tag{4.6}$$

   where $\mathbf{v_N}(\lambda)$ is defined by (4.5);
4. **Normalization:** *$\mathbf{N}(\lambda)$ is normalized at infinity:*

$$\mathbf{N}(\lambda) \to \mathbb{I} \quad \text{as } \lambda \to \infty. \tag{4.7}$$

Proposition 4.1.1 *For each given complex phase function $g(\lambda)$ satisfying Definition* 4.1.1, *the phase-conjugated Riemann-Hilbert Problem* 4.1.1 *has a unique solution and is equivalent to the holomorphic Riemann-Hilbert Problem* 2.0.2.

Proof. One finds a solution of the phase-conjugated problem by solving the holomorphic Riemann-Hilbert Problem 2.0.2 for $\mathbf{M}(\lambda)$ and then obtains $\mathbf{N}(\lambda)$ from (4.1). The analyticity and boundary behavior follow from the analogous properties of $\mathbf{M}(\lambda)$ and $g(\lambda)$. The jump conditions for $\mathbf{N}(\lambda)$ are verified using the formula (4.1) and the boundary conditions satisfied by $\mathbf{M}(\lambda)$, taking into account the discrepancy in boundary values of $g(\lambda)$ along the contour and the symmetry of $g(\lambda)$. Finally, the normalization condition follows from the corresponding property of $\mathbf{M}(\lambda)$ and the decay of $g(\lambda)$. Therefore, $\mathbf{N}(\lambda)$ so defined solves the phase-conjugated Riemann-Hilbert problem. Uniqueness of solutions for Riemann-Hilbert Problem 4.1.1 follows as before from Liouville's theorem using the continuity of the boundary values and the normalization at infinity. Clearly, the whole procedure can be reversed, and the unique solution of the holomorphic Riemann-Hilbert Problem 2.0.2 can be obtained from the solution $\mathbf{N}(\lambda)$ of the phase-conjugated Riemann-Hilbert Problem 4.1.1 by the same formula, (4.1). □

The next goal is to find a "good" contour C and then use the additional freedom afforded by the choice of the complex phase function $g(\lambda)$ so that, as $\hbar$ tends to zero, the phase-conjugated Riemann-Hilbert Problem 4.1.1 for $\mathbf{N}(\lambda)$ takes on a particularly simple form, namely, one that can be solved exactly.

4.2 REPRESENTATION AS A COMPLEX SINGLE-LAYER POTENTIAL. PASSING TO THE CONTINUUM LIMIT. CONDITIONS ON THE COMPLEX PHASE LEADING TO THE OUTER MODEL PROBLEM

We begin this section with an ad hoc assumption about $g(\lambda)$ that is justified only in chapter 6. Recall that the contour C lives in the upper half-plane and meets the origin in a corner point. Therefore, in a sufficiently small neighborhood of the origin, C is the union of two smooth arcs that join at $\lambda = 0$ at some angle. Near the origin there are two kinds of behavior that we consider.

Definition 4.2.1 *A complex phase function with parity σ is a function $g^\sigma(\lambda)$ satisfying the basic conditions of Definition* 4.1.1 *such that for $\sigma = +1$ there is a sufficiently small neighborhood U of the origin in which $g^\sigma(\lambda)$ is analytic on the part of $C \cap U$ in the left half-plane and has a nonconstant difference in its boundary values on the part of $C \cap U$ in the right half-plane. For $\sigma = -1$, the roles of the left and right half-planes are reversed.*

In particular, this means that the domain of analyticity of $g^\sigma(\lambda)$ is simply connected due to a (possibly small) gap in its contour of discontinuity on one side of the origin or the other. The two cases, $\sigma = +1$ and $\sigma = -1$, yield, for each value of x and t (as well as for each value of J and K), different functions $g^\sigma(\lambda)$; hence, most important quantities such as $\phi(\lambda)$, $\theta(\lambda)$, $\mathbf{N}(\lambda)$, and $\mathbf{v_N}(\lambda)$ inherit this dependence on σ. Thus, from now on we write $\phi(\lambda) = \phi^\sigma(\lambda)$, $\theta(\lambda) = \theta^\sigma(\lambda)$, $\mathbf{N}(\lambda) = \mathbf{N}^\sigma(\lambda)$, and $\mathbf{v_N}(\lambda) = \mathbf{v}_\mathbf{N}^\sigma(\lambda)$.

We now introduce a representation of $g^\sigma(\lambda)$ as the complex single-layer potential of a measure supported on $C \cup C^*$. First, we define a new branch of the logarithm.

Definition 4.2.2 *Suppose $\eta \in C \cup C^*$. Let $L_\eta^{C,\sigma}(\lambda)$ be the branch of* $\log(\lambda - \eta)$ *that is given by the principal branch integral* (1.6) *but when considered as a function of λ is cut from the point $\lambda = \eta$ backwards, using the orientation σ along C (if $\eta \in C$) and C^* to $\lambda = 0$ and then along the negative real axis to $-\infty$ for $\sigma = +1$ or along the positive real axis to $+\infty$ for $\sigma = -1$. See figure* 4.2.

Without loss of generality, we represent $g^\sigma(\lambda)$ in the form of a contour integral along $[C \cup C^*]_\sigma$ (both components have orientation σ):

$$g^\sigma(\lambda) = \frac{J}{2} \int_{[C\cup C^*]_\sigma} L_\eta^{C,\sigma}(\lambda) \rho^\sigma(\eta)\, d\eta \tag{4.8}$$

for some complex-valued *density function* $\rho^\sigma(\eta)$. The basic symmetry and decay conditions from Definition 4.1.1 then require that we assume

$$\int_{[C\cup C^*]_\sigma} \rho^\sigma(\eta)\, d\eta = 0 \qquad \text{and} \qquad \rho^\sigma(\eta^*) = \rho^\sigma(\eta)^*, \quad \eta \in C \cup C^*. \tag{4.9}$$

Thus, in particular, it is sufficient to determine $\rho^\sigma(\eta)$ for $\eta \in C$. If we define for $\lambda \in [C \cup C^*]_\sigma$

$$g_\pm^\sigma(\lambda) = \lim_{\substack{\mu \to \lambda \\ \mu \in \pm \text{ side of } [C\cup C^*]_\sigma}} g^\sigma(\mu), \tag{4.10}$$

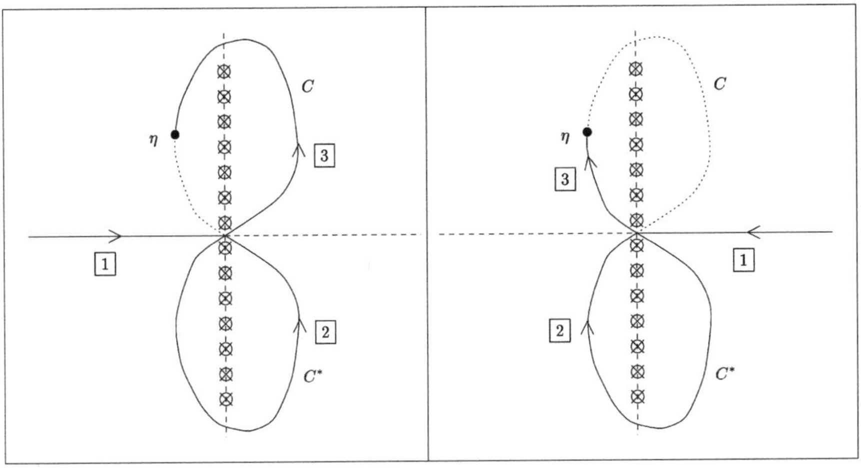

Figure 4.2 The branch cut of the functions $L_\eta^{C,\sigma}(\lambda)$ in the λ-plane is determined by the value of η, the shape of the contour C, and the orientation σ. For $\sigma = +1$ (left diagram), it connects to infinity along the negative real axis, while for $\sigma = -1$ (right diagram), it connects to infinity along the positive real axis. In each case, the arrows indicate the orientation σ of C and C^*, and the cut can be viewed as starting from infinity and ending at $\lambda = \eta$, proceeding as indicated along the real axis $\boxed{1}$, followed by C^* $\boxed{2}$, followed by C $\boxed{3}$.

then we have for $\lambda \in C \cup C^*$

$$\theta^\sigma(\lambda) = iJ(g_+^\sigma(\lambda) - g_-^\sigma(\lambda)) = -\pi \int_\lambda^0 \rho^\sigma(\eta)\, d\eta, \tag{4.11}$$

with the path of integration lying on $[C \cup C^*]_\sigma$. More precisely, if $\lambda \in C$, then the path of integration in (4.11) continues from $\eta = \lambda$ along C_σ to $\eta = 0$. On the other hand, if $\lambda \in C^*$, then the path of integration in (4.11) begins at $\eta = \lambda$, continues along $[C^*]_\sigma$ to $\eta = 0$, and then includes *all* of C_σ.

The fact that this representation (4.8) of the complex phase function is general can be seen by solving (4.11) for the density $\rho^\sigma(\eta)$ in terms of a given absolutely continuous jump $g_+^\sigma(\lambda) - g_-^\sigma(\lambda)$ on $[C \cup C^*]_\sigma$ that satisfies the symmetry condition. This condition guarantees that the density $\rho^\sigma(\eta)$ we compute will satisfy (4.9). The difference between $g^\sigma(\lambda)$ and the integral on the right-hand side of (4.8) is then an entire function of λ that vanishes at infinity and is therefore zero.

Using the representation of $g^\sigma(\lambda)$ as a complex potential, we now describe how to choose the density function $\rho^\sigma(\eta)$ to obtain a simple Riemann-Hilbert problem in the limit $\hbar \downarrow 0$. Having broken symmetry in the two distinct alternatives $\sigma = \pm 1$ in the phase function, we now do the same in the Riemann-Hilbert problem by fixing the arbitrary orientation of the contours supporting the jumps as a matter of convenience: from now on, we choose to set

$$\omega := \sigma. \tag{4.12}$$

This choice allows us to ultimately describe the asymptotics in §4.3 using formulae that do not depend on σ.

We first propose a "nearby" Riemann-Hilbert problem. Given a complex phase function $g^\sigma(\lambda)$, define for $\lambda \in C_\sigma$ the scalar function

$$\begin{aligned}\tilde{\phi}^\sigma(\lambda) := &\int_0^{iA} L_\eta^0(\lambda)\rho^0(\eta)\,d\eta + \int_{-iA}^0 L_\eta^0(\lambda)\rho^0(\eta^*)^*\,d\eta \\ &+ J\cdot\left(2i\lambda x + 2i\lambda^2 t - (2K+1)i\pi\int_\lambda^{iA}\rho^0(\eta)\,d\eta\right. \\ &\left. - g_+^\sigma(\lambda) - g_-^\sigma(\lambda)\right),\end{aligned} \tag{4.13}$$

and for $\lambda \in C$ then define a matrix function by

$$\mathbf{v}_{\widetilde{\mathbf{N}}}^\sigma(\lambda) := \sigma_1^{\frac{1-J}{2}}\begin{bmatrix}\exp(i\theta^\sigma(\lambda)/\hbar) & 0 \\ ((-1)^K i)^{-J}\sigma\exp(\tilde{\phi}^\sigma(\lambda)/\hbar) & \exp(-i\theta^\sigma(\lambda)/\hbar)\end{bmatrix}\sigma_1^{\frac{1-J}{2}}. \tag{4.14}$$

Note that $\tilde{\phi}^\sigma(\lambda)$ is what one would get by replacing the discrete measure $d\mu_{\hbar_N}^{\mathrm{WKB}}$ in the formula for $\phi^\sigma(\lambda)$ by its weak-$*$ limit $d\mu_0^{\mathrm{WKB}}$.

Riemann-Hilbert Problem 4.2.1 (Formal continuum limit). *Given a complex phase function $g^\sigma(\lambda)$, find a matrix function $\widetilde{\mathbf{N}}^\sigma(\lambda)$ satisfying the following:*

1. **Analyticity:** $\widetilde{\mathbf{N}}^\sigma(\lambda)$ *is analytic for* $\lambda \in \mathbb{C}\setminus(C \cup C^*)$;
2. **Boundary behavior:** $\widetilde{\mathbf{N}}^\sigma(\lambda)$ *assumes continuous boundary values on* $C \cup C^*$;
3. **Jump conditions:** *The boundary values taken on* $C \cup C^*$ *satisfy*

$$\begin{aligned}\widetilde{\mathbf{N}}_+^\sigma(\lambda) &= \widetilde{\mathbf{N}}_-^\sigma(\lambda)\mathbf{v}_{\widetilde{\mathbf{N}}}^\sigma(\lambda), && \lambda \in C_\sigma, \\ \widetilde{\mathbf{N}}_+^\sigma(\lambda) &= \widetilde{\mathbf{N}}_-^\sigma(\lambda)\sigma_2\mathbf{v}_{\widetilde{\mathbf{N}}}^\sigma(\lambda^*)^*\sigma_2, && \lambda \in [C^*]_\sigma;\end{aligned} \tag{4.15}$$

4. **Normalization:** $\widetilde{\mathbf{N}}^\sigma(\lambda)$ *is normalized at infinity:*

$$\widetilde{\mathbf{N}}^\sigma(\lambda) \to \mathbb{I} \quad \text{as } \lambda \to \infty. \tag{4.16}$$

Thus, we have introduced a *new* Riemann-Hilbert problem where we have taken the "continuum limit" of the discrete eigenvalue measure. According to Theorem 3.2.2, the error in replacing the discrete sums by integrals amounts to an $O(\hbar_N)$ error for fixed λ. Thus, the matrix $\mathbf{v}_{\widetilde{\mathbf{N}}}^\sigma(\lambda)$ is an $O(\hbar_N)$-accurate approximation to the original jump matrix $\mathbf{v}_{\mathbf{N}}^\sigma(\lambda)$ for all fixed λ on the contours. But, as we have seen in §3.2, the approximation necessarily breaks down in neighborhoods of radius $O(\hbar_N)$ near $\lambda = 0$, and in this region other corrections need to be taken into account (see §4.4.3).

Proposition 4.2.1 *If the continuum limit Riemann-Hilbert Problem* 4.2.1 *has a solution* $\widetilde{\mathbf{N}}^\sigma(\lambda)$, *then the solution is unique and satisfies the symmetry relation*

$$\widetilde{\mathbf{N}}^\sigma(\lambda^*) = \sigma_2\widetilde{\mathbf{N}}^\sigma(\lambda)^*\sigma_2. \tag{4.17}$$

Proof. Uniqueness follows from the continuity of boundary values and the normalization condition via Liouville's theorem. The symmetry of the solution then follows from the corresponding symmetry of the jump relations and uniqueness. □

Remark. Note that it is by no means clear that the Riemann-Hilbert Problem 4.2.1 for $\tilde{\mathbf{N}}^{\sigma}(\lambda)$ has any solution at all. Unlike the problem for $\mathbf{N}^{\sigma}(\lambda)$, this Riemann-Hilbert problem has not been obtained from an explicit rational matrix by a well-defined sequence of transformations. Rather, it is simply introduced as a reasonable asymptotic model for the Riemann-Hilbert problem of interest. We bypass questions of existence of solutions to the continuum limit Riemann-Hilbert Problem 4.2.1 here because its role is only to lead us, at a formal level, to an approximation of $\mathbf{N}^{\sigma}(\lambda)$ that we *prove* is valid in §4.5.

The following two conditions are fundamental for the viability of our asymptotic analysis, as well as for the actual determination of the complex phase function $g^{\sigma}(\lambda)$ and the contour C:

$$\textbf{Measure reality condition: } \rho^{\sigma}(\eta)\,d\eta \in \mathbb{R}, \quad \eta \in [C \cup C^*]_{\sigma}; \tag{4.18}$$

$$\textbf{Variational inequality condition: } \Re(\tilde{\phi}^{\sigma}(\lambda)) \leq 0, \quad \lambda \in C. \tag{4.19}$$

The measure reality condition (4.18) can be understood from the following heuristic argument. Note that $\tilde{\mathbf{N}}^{\sigma}(\lambda)$ and the corresponding jump matrix $\mathbf{v}^{\sigma}_{\tilde{\mathbf{N}}}(\lambda)$ have determinant one. This implies that if the complex phase $g^{\sigma}(\lambda)$ is to be chosen so that the matrix $\tilde{\mathbf{N}}^{\sigma}(\lambda)$ has uniformly bounded elements in the limit of $\hbar \downarrow 0$, then it is necessary for the jump matrix to also be bounded as $\hbar \downarrow 0$. This is only possible if the function $\theta^{\sigma}(\lambda)$ is *real-valued* for $\lambda \in C$, which further constrains the density function $\rho^{\sigma}(\eta)$ by requiring that (4.18) holds true. The variational inequality condition (4.19) is so named for reasons that are explained in detail in chapter 8. Heuristically, this condition can be understood in a similar fashion: If (4.19) were to fail somewhere, then the jump matrix $\mathbf{v}^{\sigma}_{\tilde{\mathbf{N}}}(\lambda)$ would have terms that are exponentially large for $\hbar$ small—a situation to be avoided.

Remark. Note that the measure reality condition (4.18) depends as much on the choice of the oriented contour C_{σ} via the complex-valued differential $d\eta$ as on the complex-valued density function $\rho^{\sigma}(\eta)$. Also notice that reality of $\rho^{\sigma}(\eta)\,d\eta$, when considered along with the second of the conditions (4.9), actually implies a stronger version of the first of the conditions (4.9): The integral over $[\gamma \cup \gamma^*]_{\sigma}$, where γ is any subarc of C, vanishes.

For reasons that become clear later in the book, we want to admit the possibility that the variational inequality (4.19) is not strict everywhere in C. We do, however, assume for simplicity that the subset of C where the inequality fails to be strict forms a system of closed subintervals of the regular curve C, whose topology is defined, say, by the arc-length parametrization. If $\lambda \in C$ is in this system of closed subintervals, then we say that λ lies in a *band*. Any other value of $\lambda \in C \cup C^*$ (that is, where the variational inequality (4.19) is strict if $\lambda \in C$) is said to lie in a *gap*.

DEFINITION 4.2.3 (Bands and gaps). *A band is a maximal connected component of the system of closed subintervals of C, where $\Re(\tilde{\phi}^{\sigma}(\lambda)) \equiv 0$. By symmetry, we say that λ^* lies in a band if λ does (although not in the same band). A gap is a maximal connected component of C minus the union of the bands, an open interval of C in which the strict inequality $\Re(\tilde{\phi}^{\sigma}(\lambda)) < 0$ holds. By symmetry, λ^* lies in a gap if λ does. We always assume the number of bands and gaps on C to be finite.*

If $\lambda \in C$ lies in a gap, then the off-diagonal entry $((-1)^K i)^{-J} \sigma \exp(\tilde{\phi}^\sigma(\lambda)/\hbar)$ in the jump matrix $\mathbf{v}^\sigma_{\tilde{\mathbf{N}}}(\lambda)$ is exponentially small as $\hbar \downarrow 0$ with λ held fixed. The diagonal entries are bounded but become increasingly oscillatory with respect to any fixed parametrization of C as $\hbar$ tends to zero. These wild oscillations in the jump matrix lead to growth of $\tilde{\mathbf{N}}^\sigma(\lambda)$ as one moves away from C, and consequently it is impossible to obtain uniform control of the solution in the complex λ-plane unless it is further assumed that $\theta^\sigma(\lambda)$ is *constant* (independent of $\lambda \in C$) throughout each gap. The constant value may be different in each gap. By the definition (4.11) of the function $\theta^\sigma(\lambda)$, we are therefore assuming that

$$\Re(\tilde{\phi}^\sigma(\eta)) < 0 \quad \text{and} \quad \rho^\sigma(\eta) \equiv 0, \quad \text{for } \eta \text{ in a gap of } C. \tag{4.20}$$

On the other hand, if $\lambda \in C$ lies in a band, then $\Re(\tilde{\phi}^\sigma(\lambda)) \equiv 0$, but the matrix does not appear to simplify more as $\hbar$ tends to zero. At this point, we want to resist the temptation to take $\theta^\sigma(\lambda)$ to be constant in the bands as well as in the gaps, since we would then have $\rho^\sigma(\eta) \equiv 0$ for all $\eta \in C$, which would imply $g^\sigma(\lambda) \equiv 0$, thus defeating the purpose of introducing the complex phase. Instead, we proceed by factoring the jump matrices as follows:

$$\mathbf{v}^\sigma_{\tilde{\mathbf{N}}}(\lambda) = \mathbf{a}^{\sigma,+}(\lambda)\mathbf{t}^\sigma(\lambda)\mathbf{a}^{\sigma,-}(\lambda), \tag{4.21}$$

where

$$\mathbf{a}^{\sigma,\pm}(\lambda) := \sigma_1^{\frac{1-J}{2}} \begin{bmatrix} 1 & i^J(-1)^K \sigma \exp(-\tilde{\phi}^\sigma(\lambda)/\hbar) \exp(\pm i\theta^\sigma(\lambda)/\hbar) \\ 0 & 1 \end{bmatrix} \sigma_1^{\frac{1-J}{2}} \tag{4.22}$$

and

$$\mathbf{t}^\sigma(\lambda) = \sigma_1^{\frac{1-J}{2}} \begin{bmatrix} 0 & -i^J(-1)^K \sigma \exp(-\tilde{\phi}^\sigma(\lambda)/\hbar) \\ -i^J(-1)^K \sigma \exp(\tilde{\phi}^\sigma(\lambda)/\hbar) & 0 \end{bmatrix} \sigma_1^{\frac{1-J}{2}}. \tag{4.23}$$

Let I_k^+ denote one of the bands on C. Suppose that in the band I_k^+ the functions $\tilde{\phi}^\sigma(\lambda)$ and $\theta^\sigma(\lambda)$ are the restrictions to the band of two functions $q_k(\lambda)$ and $r_k(\lambda)$, respectively, each of which is analytic in λ in a neighborhood of the interior of the band.

Definition 4.2.4 *We denote by $q_k(\lambda)$ the analytic extension of $\tilde{\phi}^\sigma(\lambda)$ from the interior of the band I_k^+ when such an extension exists. Likewise we denote by $r_k(\lambda)$ the analytic extension of $\theta^\sigma(\lambda)$ from the band I_k^+.*

Let C_{k+}^+ (respectively, C_{k-}^+) denote a contour connecting the two endpoints of the band, sharing the same orientation as C_σ, and lying within the domain of analyticity of q_k and r_k to the left (respectively, right) of the band (see figure 4.3). We think of C_{k+}^+ and C_{k-}^+ as being independent of $\hbar$ but lying sufficiently close to I_k^+ to allow us to draw conclusions concerning the behavior of the analytic functions q_k and r_k on these contours from the Cauchy-Riemann equations on I_k^+ (see the discussion prior to (4.31)).

With these definitions, we may now make a change of variables from the matrix $\tilde{\mathbf{N}}^\sigma(\lambda)$ to a new matrix $\mathbf{O}^\sigma(\lambda)$. Let $C_{k\pm}^-$ denote the complex-conjugate contours

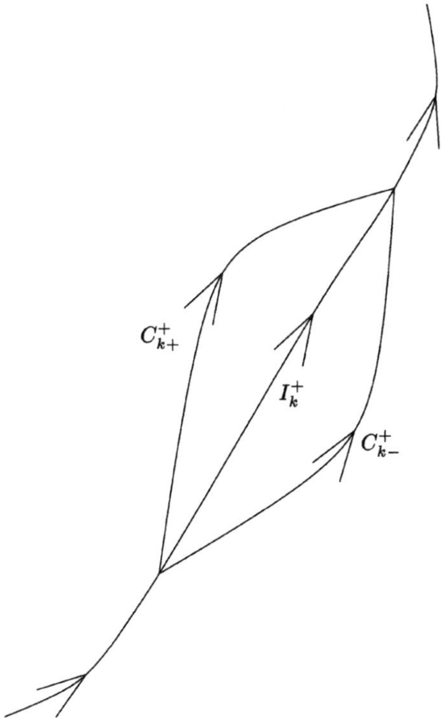

Figure 4.3 The lens-shaped region enclosed by the oriented contours C_{k+}^+ and C_{k-}^+ about the band $I_k^+ \subset C$.

$(C_{k\pm}^+)^*$, which get their orientation not from the conjugation operation but from $[C^*]_\sigma$. Define the matrix $\mathbf{O}^\sigma(\lambda)$ by

$$\mathbf{O}^\sigma(\lambda) := \widetilde{\mathbf{N}}^\sigma(\lambda)\sigma_1^{\frac{1-J}{2}} \begin{bmatrix} 1 & -i^J(-1)^K\sigma\exp(-q_k(\lambda)/\hbar)\exp(-ir_k(\lambda)/\hbar) \\ 0 & 1 \end{bmatrix} \sigma_1^{\frac{1-J}{2}},$$

for λ in between C_{k+}^+ and I_k^+,

$$\mathbf{O}^\sigma(\lambda) := \widetilde{\mathbf{N}}^\sigma(\lambda)\sigma_1^{\frac{1-J}{2}} \begin{bmatrix} 1 & i^J(-1)^K\sigma\exp(-q_k(\lambda)/\hbar)\exp(ir_k(\lambda)/\hbar) \\ 0 & 1 \end{bmatrix} \sigma_1^{\frac{1-J}{2}}, \tag{4.24}$$

for λ in between C_{k-}^+ and I_k^+,

$$\mathbf{O}^\sigma(\lambda) := \sigma_2\mathbf{O}^\sigma(\lambda^*)^*\sigma_2,$$

for λ^* in between C_{k-}^+ and C_{k+}^+, and $\mathbf{O}^\sigma(\lambda) := \widetilde{\mathbf{N}}^\sigma(\lambda)$, otherwise. Here, k ranges over all bands of C. This change of variables, in conjunction with Riemann-Hilbert Problem 4.2.1, suggests a new problem posed on the contour $C \cup C^*$ with "lenses."

Riemann-Hilbert Problem 4.2.2 (Open lenses). *Given a complex phase function $g^\sigma(\lambda)$, find a matrix function $\mathbf{O}^\sigma(\lambda)$ satisfying the following:*

1. **Analyticity:** $\mathbf{O}^\sigma(\lambda)$ *is analytic for* $\lambda \in \mathbb{C}$ *minus* $C \cup C^* \cup \{$*lens boundaries*$\}$.
2. **Boundary behavior:** $\mathbf{O}^\sigma(\lambda)$ *assumes boundary values from each connected component of the domain of analyticity that are uniformly continuous, including those at corner points corresponding to self-intersections of the contour* $C \cup C^* \cup \{$*lens boundaries*$\}$.
3. **Jump conditions:** *For* λ *in any gap of* C_σ,
$$\mathbf{O}_+^\sigma(\lambda) = \mathbf{O}_-^\sigma(\lambda)\mathbf{v}_{\tilde{\mathbf{N}}}^\sigma(\lambda)\,. \tag{4.25}$$
For λ *in the band* I_k^+ *of* C_σ,
$$\mathbf{O}_+^\sigma(\lambda) = \mathbf{O}_-^\sigma(\lambda)\mathbf{t}^\sigma(\lambda). \tag{4.26}$$
For λ *in either of the lens boundaries* $C_{k\pm}^+$ *of any band of* C_σ,
$$\mathbf{O}_+^\sigma(\lambda) = \mathbf{O}_-^\sigma(\lambda) \times \sigma_1^{\frac{1-J}{2}} \begin{bmatrix} 1 & i^J(-1)^K\sigma\exp(-q_k(\lambda)/\hbar)\exp(\mp i r_k(\lambda)/\hbar) \\ 0 & 1 \end{bmatrix} \sigma_1^{\frac{1-J}{2}}\,. \tag{4.27}$$
Finally, for λ^* *on any of the above contours, we have*
$$\mathbf{O}_+^\sigma(\lambda) = \mathbf{O}_-^\sigma(\lambda)\sigma_2\mathbf{v}_{\mathbf{O}}^\sigma(\lambda^*)^*\sigma_2\,, \tag{4.28}$$
where the orientation used to define the boundary values is induced by $[C^*]_\sigma$ *and where* $\mathbf{v}_{\mathbf{O}}^\sigma(\lambda)$ *is the jump matrix defined on the contours in the upper half-plane by* $\mathbf{v}_{\mathbf{O}}^\sigma(\lambda) := \mathbf{O}_-^\sigma(\lambda)^{-1}\mathbf{O}_+^\sigma(\lambda)$.
4. **Normalization:** $\mathbf{O}^\sigma(\lambda)$ *is normalized at infinity:*
$$\mathbf{O}^\sigma(\lambda) \to \mathbb{I} \quad \text{as } \lambda \to \infty. \tag{4.29}$$

Proposition 4.2.2 *The Riemann-Hilbert Problem* 4.2.2 *has at most one solution* $\mathbf{O}^\sigma(\lambda)$, *which necessarily satisfies the symmetry*
$$\mathbf{O}^\sigma(\lambda^*) = \sigma_2\mathbf{O}^\sigma(\lambda)^*\sigma_2. \tag{4.30}$$
The Riemann-Hilbert Problem 4.2.2 *is equivalent to the Riemann-Hilbert Problem* 4.2.1 *for the matrix* $\tilde{\mathbf{N}}^\sigma(\lambda)$.

Proof. Uniqueness and symmetry are proved as for $\tilde{\mathbf{N}}^\sigma(\lambda)$. The equivalence is established by the explicit triangular change of variables via (4.24); the transformation is clearly invertible, and it is a direct calculation to show that any solution of the Riemann-Hilbert Problem 4.2.1 leads via (4.24) to a solution of the Riemann-Hilbert Problem 4.2.2. □

Remark. On notation: Throughout this section, we are introducing a sequence of Riemann-Hilbert problems. Whenever the unknown in a new problem is related to the unknown in a previous problem by an explicit transformation, we denote the new unknown with a new letter. Whenever the unknown in a new problem is not directly related to that of the previous problem but the jump matrices are an ad hoc approximation of the previous jump matrices, the new unknown is written with the same letter as the old but with a tilde. Thus, each appearance of a new tilde denotes a new formal approximation that will need to be justified later.

Now we will argue that with two additional constraints on $\rho^\sigma(\lambda)$ in the bands, the jump relations in the Riemann-Hilbert Problem 4.2.2 simplify dramatically as $\hbar$ tends to zero. We know that the analytic functions $q_k(\lambda)$ are purely imaginary while $r_k(\lambda)$ are purely real for $\lambda \in I_k^+$. In particular, it follows from the Cauchy-Riemann equations that for $C_{k\pm}^+$ sufficiently close to I_k^+, the real part of $-ir_k(\lambda)$ will be strictly negative on C_{k+}^+ except at the endpoints, and at the same time the real part of $ir_k(\lambda)$ will be strictly negative on C_{k-}^+ except at the endpoints, *if the real-valued function $\theta^\sigma(\lambda)$ is assumed to be strictly decreasing along I_k^+ with its orientation σ.* If we apply similar arguments to the analytic functions $q_k(\lambda)$, we see that the only sure way to prevent the real part of $-q_k(\lambda)$ from being positive either on C_{k+}^+ or on C_{k-}^+ is to insist that *the imaginary part of the function $\tilde{\phi}^\sigma(\lambda)$ is constant along I_k^+.* Therefore, we are assuming that

$$\tilde{\phi}^\sigma(\lambda) = \text{an imaginary constant} \quad \text{and} \quad \rho^\sigma(\lambda)\,d\lambda < 0, \quad \text{for } \lambda \text{ in a band of } C. \tag{4.31}$$

The latter conditions are equivalent to the required monotonicity of $\theta^\sigma(\lambda)$, and they give us another interpretation of the bands. Their union is the support of the real measure $\rho^\sigma(\eta)\,d\eta$ in the complex plane. Under the condition (4.31), it is clear that as $\hbar \downarrow 0$ the jump matrices on the outsides $C_{k\pm}^+$ and $C_{k\pm}^-$ of the lenses converge pointwise to the identity matrix, uniformly in any neighborhood that does not contain the endpoints. At the same time, the jump for the matrix $\mathbf{O}^\sigma(\lambda)$ in each band is a constant (with respect to λ) matrix whose elements remain bounded as $\hbar$ tends to zero. Note that the jump matrix in each band, the restriction of $\mathbf{t}^\sigma(\lambda)$, may be a different constant in each band.

We collect these observations into a definition for future reference.

Definition 4.2.5 *An admissible density function for $g^\sigma(\lambda)$ is a complex-valued function $\rho^\sigma(\eta)$ defined on a loop contour C in the upper half-plane and its complex-conjugate C^* such that the following five conditions are satisfied:*

1. *$\rho^\sigma(\eta)\,d\eta$ is a real differential for $\eta \in C_\sigma$, and $\rho^\sigma(\eta^*) = \rho^\sigma(\eta)^*$;*
2. *The support of $\rho^\sigma(\eta)\,d\eta$ consists of a finite system of intervals, the bands, whose complement in $C \cup C^*$ is the system of gaps, and for $\sigma = +1$ (respectively, $\sigma = -1$) the origin is contained in a band that emerges only in the right half-plane (respectively, left half-plane);*
3. *In the interior of each band of C_σ, the function $\tilde{\phi}^\sigma(\lambda)$ evaluates to an imaginary constant, possibly different in each band, and the differential $\rho^\sigma(\eta)\,d\eta$ is strictly negative;*
4. *In the interior of each gap of C, where $\rho^\sigma(\eta)\,d\eta$ vanishes, we have the strict inequality $\Re(\tilde{\phi}^\sigma(\lambda)) < 0$;*

5. *The restriction of the complex-valued function $\rho^\sigma(\lambda)$ to the interior of each band of $C \cup C^*$ has an analytic continuation in some lens-shaped neighborhood on either side of the band.*

If for some indices J, K, and σ a contour C and an admissible density function $\rho^\sigma(\eta)$ for $\eta \in C \cup C^*$ can be found, then the jump relations in the Riemann-Hilbert problem for $\mathbf{O}^\sigma(\lambda)$ become very simple asymptotically as $\hbar \downarrow 0$, at least away from the interval endpoints.

Remark. The reader will observe that an elementary consequence of the definitions (4.3) of $\theta^\sigma(\lambda)$ and (4.13) of $\tilde{\phi}^\sigma(\lambda)$ is that the function $\theta^\sigma(\lambda) + i\tilde{\phi}^\sigma(\lambda)$, while defined on the contour $C \cup C^*$, is the boundary value of a function analytic on the "$-$" side of the whole contour $[C \cup C^*]_\sigma$ and similarly that $\theta^\sigma(\lambda) - i\tilde{\phi}^\sigma(\lambda)$ is the boundary value of a function analytic on the "$+$" side of $[C \cup C^*]_\sigma$. On the outside of the contour loops, the region of analyticity is in fact the entire exterior of $C \cup C^*$, while on the inside of the loops the region of analyticity excludes only the support of the asymptotic eigenvalue measure $\rho^0(\eta)\,d\eta$ on the imaginary axis.

This fact has interesting implications if a complex phase function can be found corresponding to an admissible density function $\rho^\sigma(\eta)$ as described in Definition 4.2.5. In each band $\tilde{\phi}^\sigma(\lambda)$ is an imaginary constant, while in each gap $\theta^\sigma(\lambda)$ is a real constant. It follows that, for example, the restrictions of $\tilde{\phi}^\sigma(\lambda)$ to two different gaps extend to a given side of the contour as two analytic functions of λ *whose difference is a constant.* Similarly, the restrictions of $\theta^\sigma(\lambda)$ to two different bands have analytic extensions that differ only by a constant. Differentiating with respect to λ, one then sees from (4.3) that the density function $\rho^\sigma(\lambda)$ is the same analytic function in each band and that this function has an extension to the whole complex λ-plane except the gaps of $C \cup C^*$ and the support of the measure $\rho^0(\eta)\,d\eta$. Thus, in particular, the third condition in Definition 4.2.5 implies the fifth.

A very important consequence of the fact that the function $\tilde{\phi}^\sigma(\lambda)$ has an analytic continuation to either side of each gap is that *the gap segments of the contour C may be deformed slightly with the endpoints held fixed without violating the strict inequality $\Re(\tilde{\phi}^\sigma) < 0$ on the interior of the gap.* On the other hand, the band segments of C cannot be freely deformed at all without violating the condition that the differential $\rho^\sigma(\eta)\,d\eta$ should be real.

These continuation arguments also give relations between the analytic extension of $\theta^\sigma(\lambda)$ from a band I_k^+ and the analytic extension of $\tilde{\phi}^\sigma(\lambda)$ from a gap Γ_j^+. If θ_j^σ denotes the constant value of $\theta^\sigma(\lambda)$ in the gap Γ_j^+ and $\tilde{\phi}_k^\sigma$ denotes the constant value of $\tilde{\phi}^\sigma(\lambda)$ in the band I_k^+, then the continuations of these two functions to the left ("$+$" side) of C_σ are simply related:

$$\theta^\sigma(\lambda) - i\tilde{\phi}_k^\sigma \equiv \theta_j^\sigma - i\tilde{\phi}^\sigma(\lambda). \tag{4.32}$$

Likewise, continuing these two functions to the right ("$-$" side) of C_σ, one finds

$$\theta^\sigma(\lambda) + i\tilde{\phi}_k^\sigma \equiv \theta_j^\sigma + i\tilde{\phi}^\sigma(\lambda). \tag{4.33}$$

These relations are particularly useful in the local analysis that must be undertaken near a point λ_k separating a band from a gap.

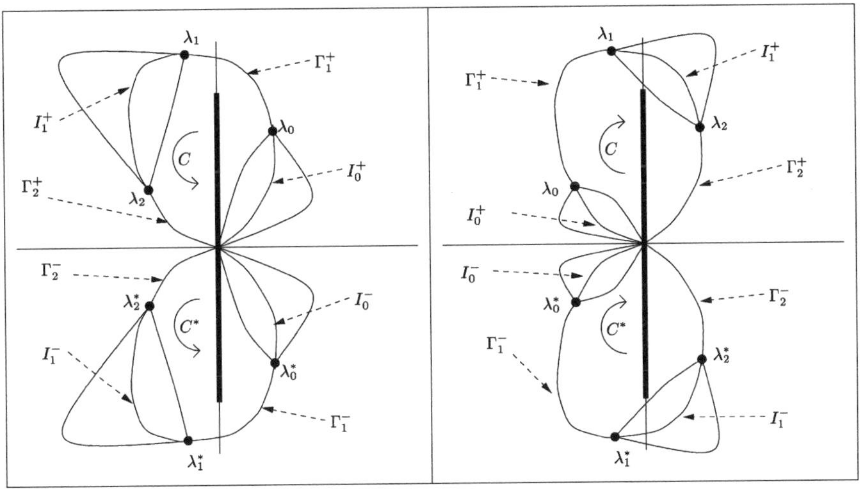

Figure 4.4 Notational conventions for the bands and gaps illustrated for $\sigma = +1$ (left) and $\sigma = -1$ (right).

Note that, by the very meaning of the index σ on the complex phase function $g^\sigma(\lambda)$ as introduced at the beginning of this chapter, the origin $\lambda = 0$ is a boundary point between a band and a gap of C. If we consider C_σ as an oriented loop beginning (and ending) at the origin, then by definition the band occupies the initial part of C_σ, while the final part of C_σ is a gap. This leads us to introduce some notation for the bands and gaps. For some even nonnegative integer G, the bands on C are labeled, in order of the orientation σ starting from $\lambda = 0$, $I_0^+, I_1^+, I_2^+, \ldots, I_{G/2}^+$. The gaps interlacing these bands are labeled in order along C_σ as $\Gamma_k^+, k = 1, \ldots, G/2+1$. The endpoints of the bands, enumerated along C_σ are denoted $0, \lambda_0, \ldots, \lambda_G$. On $[C^*]_\sigma$, we have by symmetry bands $I_k^- = I_k^{+*}$ for $k = 0, \ldots, G/2$ and gaps $\Gamma_k^- = \Gamma_k^{+*}$. By convention, we set $I_0 := I_0^+ \cup I_0^-$. With this notation, the orientation of each band and gap is expressed in terms of the endpoints $\{\lambda_k\}$ in a way that avoids direct reference to σ. Namely, each band I_k^+ is an oriented contour segment from λ_{2k-1} to λ_{2k}, except for I_0^+, which is an oriented contour from the origin to λ_0. Similarly, I_k^- is an oriented contour from λ_{2k}^* to λ_{2k-1}^*, except for I_0^-, which goes from λ_0^* to the origin. The gap Γ_k^+ begins at λ_{2k-2} and ends at λ_{2k-1}, except for $\Gamma_{G/2+1}^+$, which ends at the origin. Similarly, Γ_k^- begins at λ_{2k-1}^* and ends at λ_{2k-2}^*, except for $\Gamma_{G/2+1}^-$, which begins at the origin. These conventions regarding the contour where $\mathbf{O}^\sigma(\lambda)$ has jumps are illustrated in figure 4.4.

Remark. The complex numbers $\lambda_0, \ldots, \lambda_G$ should not be confused with $L^2(\mathbb{R})$ eigenvalues of the Zakharov-Shabat eigenvalue problem or their WKB approximations, which we denote by $\lambda_{\hbar_N,n}^{\mathrm{WKB}}$.

As $\hbar \downarrow 0$, the jump matrices for $\mathbf{O}^\sigma(\lambda)$ simplify as just described, subject to the availability of an appropriate complex phase function $g^\sigma(\lambda)$. Assuming the existence of such a $g^\sigma(\lambda)$, we now yield to the temptation of the pointwise asymptotics of the jump matrices to propose on an ad hoc basis a new model Riemann-Hilbert problem for an approximation $\widetilde{\mathbf{O}}(\lambda)$ to $\mathbf{O}^{\sigma}(\lambda)$. In the gap $\Gamma_k^+ \subset C$, let θ_k denote the real constant value of the function $J \cdot \theta^{\sigma}(\lambda)$. Note that it follows from (4.11) and the conditions imposed on $\rho^{\sigma}(\lambda)$ in Definition 4.2.5 that $\theta_{G/2+1} \equiv 0$. In the band $I_k^+ \subset C$ for $k \geq 0$, let $i\alpha_k$ denote the imaginary constant value of $J \cdot \tilde{\phi}^{\sigma}(\lambda)$ in that band.

Riemann-Hilbert Problem 4.2.3 (Outer model problem). *Find a matrix function $\widetilde{\mathbf{O}}(\lambda)$ satisfying the following:*

1. **Analyticity:** *$\widetilde{\mathbf{O}}(\lambda)$ is analytic for $\lambda \in \mathbb{C}\backslash((C\backslash\Gamma_{G/2+1}^+) \cup (C^*\backslash\Gamma_{G/2+1}^-))$.*
2. **Boundary behavior:** *$\widetilde{\mathbf{O}}(\lambda)$ assumes boundary values that are continuous except at the endpoints of the bands and gaps, where at worst inverse fourth-root singularities are admitted.*
3. **Jump conditions:** *For $\lambda \in I_k^+ \cup I_k^-$ and $k = 0, \ldots, G/2$,*
$$\widetilde{\mathbf{O}}_+(\lambda) = \widetilde{\mathbf{O}}_-(\lambda)\begin{bmatrix} 0 & i\exp(-i\alpha_k/\hbar) \\ i\exp(i\alpha_k/\hbar) & 0 \end{bmatrix}. \tag{4.34}$$
For $\lambda \in \Gamma_k^+ \cup \Gamma_k^-$ and $k = 1, \ldots, G/2$,
$$\widetilde{\mathbf{O}}_+(\lambda) = \widetilde{\mathbf{O}}_-(\lambda)\begin{bmatrix} \exp(i\theta_k/\hbar) & 0 \\ 0 & \exp(-i\theta_k/\hbar) \end{bmatrix}. \tag{4.35}$$
4. **Normalization:** *$\widetilde{\mathbf{O}}(\lambda)$ is normalized at infinity:*
$$\widetilde{\mathbf{O}}(\lambda) \to \mathbb{I} \quad \text{as } \lambda \to \infty. \tag{4.36}$$

Remark. The necessity of dropping the condition of uniform continuity of the boundary values is explained in §4.3.

Remark. Note that under the assumption
$$i^J(-1)^K\sigma = -i, \tag{4.37}$$
which we see in chapter 5 is sufficient for our purposes (one can think of this relation as defining, say, the interpolant index K in terms of J and σ), the jump matrices for $\widetilde{\mathbf{O}}(\lambda)$ are obtained from those for $\mathbf{O}^{\sigma}(\lambda)$ simply by explicitly computing pointwise leading-order asymptotics as $\hbar \downarrow 0$. Moreover, with this choice of parameters, the pointwise convergence yields the Riemann-Hilbert Problem 4.2.3 in which the parameter σ no longer explicitly appears.

Remark. The formal continuum limit Riemann-Hilbert Problem 4.2.1 and the open lenses Riemann-Hilbert Problem 4.2.2 are not solved directly in this book. In particular, details of behavior of boundary values and so on are included for completeness of presentation but do not immediately constrain our analysis. Both of these problems are posed primarily as intermediate steps in passing from the phase-conjugated Riemann-Hilbert Problem 4.1.1, for which we have existence and uniqueness, and the outer model Riemann-Hilbert Problem 4.2.3, for which we prove existence and uniqueness. But then, we use the explicit solution for $\tilde{\mathbf{O}}(\lambda)$ together with some local models we obtain in §4.4 to build an approximation to $\mathbf{N}^\sigma(\lambda)$ that we *prove* is uniformly valid in §4.5.

Given the pointwise convergence of the jump matrices, one might expect that a solution of the Riemann-Hilbert Problem 4.2 might yield a good approximation of $\mathbf{O}^\sigma(\lambda)$, which by an explicit change of variables approximates $\tilde{\mathbf{N}}^\sigma(\lambda)$ and thus $\mathbf{N}^\sigma(\lambda)$. The same caveats hold for the relationship of the Riemann-Hilbert Problem 4.2 to the Riemann-Hilbert Problem 4.2 for $\mathbf{O}^\sigma(\lambda)$ as mentioned when we introduced $\tilde{\mathbf{N}}^\sigma(\lambda)$ in place of $\mathbf{N}^\sigma(\lambda)$. Namely, it is not clear at the moment whether there exists a solution at all or whether it should be a good approximation to $\mathbf{O}^\sigma(\lambda)$ anywhere in the complex plane. We put aside the justification of these formal approximations, as that comes in §4.5. We now turn to the issue of existence of solutions for the model Riemann-Hilbert Problem 4.2.

4.3 EXACT SOLUTION OF THE OUTER MODEL PROBLEM

The model Riemann-Hilbert Problem 4.2 for the matrix $\tilde{\mathbf{O}}(\lambda)$ is the result of finding an appropriate contour C and density function $\rho^\sigma(\lambda)$ for some choice of indices J, K, and σ satisfying (4.37), and then neglecting small elements of the resulting jump matrices. This is a good approximation of the jump matrices everywhere except near the endpoints λ_k and λ_k^* and near $\lambda = 0$. The same sort of thing can be said about the "continuum limit" approximation made when substituting the matrix $\tilde{\mathbf{N}}^\sigma(\lambda)$ for $\mathbf{N}^\sigma(\lambda)$; the approximation of the jump matrices is only good outside of a small neighborhood near $\lambda = 0$. Reversing the approximation steps, we might optimistically expect the matrix $\tilde{\mathbf{O}}(\lambda)$ to provide a good approximation to $\mathbf{N}^\sigma(\lambda)$ everywhere in the complex λ-plane except near $\lambda = 0$ and the endpoints. The choice of terminology in calling the Riemann-Hilbert Problem 4.2 an "outer" model problem borrows from the theory of matched asymptotic expansions. In order to obtain a uniformly valid leading-order asymptotic description of $\mathbf{N}^\sigma(\lambda)$, we need to develop "inner" model problems as well to describe the behavior in the neighborhoods where the outer approximation fails. We carry out this program in §4.4.

In this section, however, we solve the outer model problem. The reader will observe that similar Riemann-Hilbert problems are solved in [DIZ97, DKMVZ99A, DKMVZ99B], but here we emphasize a slightly different, more algebro-geometric point of view. We do this for two concrete reasons. First, we want to clearly motivate the use of various tools and techniques from Riemann surface theory that we need (and that are used in [DIZ97, DKMVZ99A, DKMVZ99B]) by explicitly introducing a Riemann surface and building functions on it, rather than working on a complex plane with cuts. Second, we want to strengthen the connection between the Riemann-Hilbert approach to semiclassical theory for integrable systems on the one hand and the self-consistent Whitham modulation theory developed by Dubrovin, Novikov, Krichever, and others on the other hand. The latter theory has a strong algebro-geometric flavor, and a central role is played by the Baker-Akhiezer function, a unique and canonically defined function on a given Riemann surface that can be constructed using the Riemann theta function of the surface and certain abelian integrals. One of the results of this section is an explicit construction of the slowly modulating Baker-Akhiezer function via the Riemann-Hilbert problem for $\widetilde{\mathbf{O}}(\lambda)$. We continue to assume throughout this section that G is an even integer satisfying $0 \leq G < \infty$.

The main advantage here is that the jump matrix is now piecewise-constant as a function of λ. On the other hand, the Riemann-Hilbert Problem 4.2 for $\widetilde{\mathbf{O}}(\lambda)$ has one important flaw: the jump matrix is not continuous at the endpoints λ_k. This implies that all solutions blow up as $\lambda \to \lambda_k$. However, we are able to find a solution such that near each endpoint λ_k, the elements of $\widetilde{\mathbf{O}}(\lambda)$ grow like $(\lambda-\lambda_k)^{-1/4}$ and for which the boundary values are smooth on any open subset of the contour not containing any endpoints. All other matrix functions with the same domain of analyticity and satisfying the same jump relations almost everywhere are proportional via a meromorphic matrix-valued function with all singularities confined to the contour. Thus, the condition that a solution of the Riemann-Hilbert Problem 4.2 should have at worst inverse fourth-root singularities at all endpoints picks out the only solution that has finite boundary values between the endpoints and at the same time treats all endpoints on a symmetrical basis.

Remark. The simple fact that the jump matrices for $\widetilde{\mathbf{O}}(\lambda)$ are piecewise constant also suggests that special functions play a role in the solution. To see this, consider the derivative $\partial_\lambda \widetilde{\mathbf{O}}(\lambda)$ of the solution. This matrix shares the same domain of analyticity as its primitive and also satisfies the same jump relations on each segment of the contour where the jump matrix is constant. This means that the quotient $\partial_\lambda \widetilde{\mathbf{O}}(\lambda) \cdot \widetilde{\mathbf{O}}(\lambda)^{-1}$ is a meromorphic function on the complex plane, with poles at the endpoints λ_k and their conjugates and vanishing as $\lambda \to \infty$ (this follows from the normalization condition). It follows that the elements of the matrix $\widetilde{\mathbf{O}}(\lambda)$ satisfy a 2×2 linear system of ordinary differential equations in λ with rational coefficients.

4.3.1 Reduction to a Problem in Function Theory on Hyperelliptic Curves

Suppose that $h(\lambda)$ is an analytic function in the finite λ-plane wherever $\widetilde{\mathbf{O}}(\lambda)$ is supposed to be analytic, taking on continuous boundary values *that are uniformly bounded*, and that satisfies

$$\begin{aligned} h_+(\lambda) - h_-(\lambda) &= -\theta_k, \ \lambda \in \Gamma_k^+ \cup \Gamma_k^-, \quad k = 1, \ldots, G/2, \\ h_+(\lambda) + h_-(\lambda) &= -\alpha_k, \ \lambda \in I_k^+ \cup I_k^-, \quad k = 0, \ldots, G/2. \end{aligned} \tag{4.38}$$

Consider the matrix defined by

$$\mathbf{P}(\lambda) := \widetilde{\mathbf{O}}(\lambda) \exp(ih(\lambda)\sigma_3/\hbar)\,. \tag{4.39}$$

It is straightforward to verify that the matrix $\mathbf{P}(\lambda)$ has the identity matrix as the jump matrix in all gaps Γ_k^+ and Γ_k^-. Since the boundary values of $\widetilde{\mathbf{O}}(\lambda)$ and $h(\lambda)$ are assumed to be continuous, it follows that $\mathbf{P}(\lambda)$ is in fact analytic in the gaps. In the bands, the jump relation becomes simply

$$\mathbf{P}_+(\lambda) = i\mathbf{P}_-(\lambda)\sigma_1, \tag{4.40}$$

so the jump relation is the same in all bands. Next, suppose that $\beta(\lambda)$ is a scalar function analytic in the λ-plane except at the bands, where it satisfies

$$\beta_+(\lambda) = -i\beta_-(\lambda). \tag{4.41}$$

Suppose further for the sake of concreteness that $\beta(\lambda) \to 1$ as $\lambda \to \infty$. Then, setting

$$\mathbf{Q}(\lambda) := \beta(\lambda)\mathbf{P}(\lambda), \tag{4.42}$$

we see that the jump relations for $\mathbf{Q}(\lambda)$ take on the elementary form

$$\mathbf{Q}_+(\lambda) = \mathbf{Q}_-(\lambda)\sigma_1\,, \qquad \lambda \in \bigcup_k (I_k^+ \cup I_k^-). \tag{4.43}$$

Our purpose in reducing the jump relations to this universal constant form is to be able to move from the cut plane onto a compact Riemann surface on which the function theory is trivial by comparison.

Before continuing to study $\mathbf{Q}(\lambda)$, let us describe the scalar functions $h(\lambda)$ and $\beta(\lambda)$. For $\beta(\lambda)$ we propose the formula

$$\beta(\lambda)^4 = \frac{\lambda - \lambda_0^*}{\lambda - \lambda_0} \prod_{k=1}^{G/2} \frac{\lambda - \lambda_{2k-1}}{\lambda - \lambda_{2k-1}^*} \cdot \frac{\lambda - \lambda_{2k}^*}{\lambda - \lambda_{2k}}\,, \tag{4.44}$$

selecting the branch that tends to unity for large λ and that is cut along the bands I_k^+ and I_k^-. It is easily checked that $\beta(\lambda)$ as defined here is the only function satisfying the required jump condition and normalization at infinity that has continuous boundary values (except at half of the endpoints). To find $h(\lambda)$, we introduce the function $R(\lambda)$ defined by

$$R(\lambda)^2 := \prod_{k=0}^{G} (\lambda - \lambda_k)(\lambda - \lambda_k^*)\,, \tag{4.45}$$

choosing the particular branch that is cut along the bands I_k^+ and I_k^- and satisfies

$$\lim_{\lambda \to \infty} \frac{R(\lambda)}{\lambda^{G+1}} = -1 \quad \text{or, equivalently,} \quad R(0_+) = \prod_{k=1}^{G} |\lambda_k|. \tag{4.46}$$

This defines a real function, that is, one that satisfies $R(\lambda^*) = R(\lambda)^*$. At the bands, we have $R_+(\lambda) = -R_-(\lambda)$, while $R(\lambda)$ is analytic in the gaps. Setting

$$h(\lambda) = k(\lambda)R(\lambda), \tag{4.47}$$

we see that $k(\lambda)$ satisfies the jump relations

$$\begin{aligned} k_+(\lambda) - k_-(\lambda) &= -\frac{\theta_n}{R(\lambda)}, \qquad \lambda \in \Gamma_n^+ \cup \Gamma_n^-, \\ k_+(\lambda) - k_-(\lambda) &= -\frac{\alpha_n}{R_+(\lambda)}, \qquad \lambda \in I_n^+ \cup I_n^-, \end{aligned} \tag{4.48}$$

and is otherwise analytic. Such a function is given by the Cauchy integral

$$k(\lambda) = \frac{1}{2\pi i} \sum_{n=1}^{G/2} \theta_n \int_{\Gamma_n^+ \cup \Gamma_n^-} \frac{d\eta}{(\lambda - \eta)R(\eta)} + \frac{1}{2\pi i} \sum_{n=0}^{G/2} \alpha_n \int_{I_n^+ \cup I_n^-} \frac{d\eta}{(\lambda - \eta)R_+(\eta)}. \tag{4.49}$$

This function blows up like $(\lambda-\lambda_n)^{-1/2}$ near each endpoint, has continuous boundary values in between the endpoints, and vanishes like $1/\lambda$ for large λ. It is the only such solution of the jump relations (4.48). For concreteness, we accept exactly this solution of (4.48) and construct $h(\lambda)$ by using (4.47). The factor of $R(\lambda)$ renormalizes the singularities at the endpoints, so that, as desired, the boundary values of $h(\lambda)$ are bounded continuous functions. Near infinity, there is the asymptotic expansion

$$\begin{aligned} h(\lambda) &= h_G\lambda^G + h_{G-1}\lambda^{G-1} + \cdots + h_1\lambda + h_0 + O(\lambda^{-1}) \\ &= p(\lambda) + O(\lambda^{-1}), \end{aligned} \tag{4.50}$$

where all coefficients h_j of the polynomial $p(\lambda)$ can be found explicitly by expanding $R(\lambda)$ and the Cauchy integral (4.49) for large λ. It is easy to see from the reality of θ_j and α_j that $p(\lambda)$ is a polynomial with real coefficients.

Now, let us return to the matrix $\mathbf{Q}(\lambda)$ and determine what properties it must have in order for $\widetilde{\mathbf{O}}(\lambda)$ to have the appropriate boundary behavior and asymptotic behavior at infinity. Since $\widetilde{\mathbf{O}}(\lambda)$ should be $O((\lambda-\lambda_n)^{-1/4})$ at each endpoint, it follows from the behavior of $\beta(\lambda)$ that at λ_{2k}^* for $k=0,\ldots,G/2$ and λ_{2k-1} for $k=1,\ldots,G/2$, we need to ask that $\mathbf{Q}(\lambda)$ be bounded. Similarly, near λ_{2k} for $k=0,\ldots,G/2$ and λ_{2k-1}^* for $k=1,\ldots,G/2$, we need to require that $\mathbf{Q}(\lambda)$ blow up no worse than an inverse square root. Near $\lambda=\infty$, the simple asymptotic behavior required of $\widetilde{\mathbf{O}}(\lambda)$ implies that

$$\mathbf{Q}(\lambda)\exp(-ip(\lambda)\sigma_3/\hbar) = \mathbb{I} + O(\lambda^{-1}), \quad \lambda\to\infty, \tag{4.51}$$

where we recall that $p(\lambda)$ is a polynomial of degree G in λ with coefficients expressed explicitly in terms of the θ_k and α_k.

In fact, the jump relation (4.43), the asymptotic relation (4.51), and the condition that $\mathbf{Q}(\lambda)$ be holomorphic outside of the bands with boundary values for which the only allowable singularities are of inverse square root type near $\lambda_0, \lambda_1^*, \lambda_2, \lambda_3^*, \ldots, \lambda_G$ determine the matrix $\mathbf{Q}(\lambda)$ uniquely. This leads us to pose a new problem.

Riemann-Hilbert Problem 4.3.1 (Hyperelliptic problem). *Let $p(\lambda)$ be a given polynomial of degree G. Find a matrix function $\mathbf{Q}(\lambda)$ satisfying the following:*

1. **Analyticity:** *$\mathbf{Q}(\lambda)$ is analytic for $\lambda\in\mathbb{C}\setminus\cup_k I_k^\pm$;*
2. **Boundary behavior:** *$\mathbf{Q}(\lambda)$ takes continuous boundary values on $\cup_k I_k^\pm$ except at the alternating sequence of endpoints $\lambda_0, \lambda_1^*, \lambda_2, \lambda_3^*, \ldots, \lambda_G$, where inverse square root singularities in the matrix elements are admitted;*
3. **Jump conditions:** *On the interior of each oriented band $I_k^\pm$, the boundary values of $\mathbf{Q}(\lambda)$ satisfy the canonical jump conditions* (4.43)*;*
4. **Normalization:** *$\mathbf{Q}(\lambda)$ has an essential singularity at infinity, where it is normalized so that* (4.51) *holds.*

From the above explicit transformations relating $\widetilde{\mathbf{O}}$ and $\mathbf{Q}$, we have proved the following.

PROPOSITION 4.3.1 *When the polynomial $p(\lambda)$ is the principal part of the Laurent expansion of $h(\lambda)$ at infinity (cf. (4.50)) and where $h(\lambda)$ is given in terms of the constant parameters α_k and θ_k by the formula $h(\lambda) = R(\lambda)k(\lambda)$ with $k(\lambda)$ given by* (4.49)*, the Riemann-Hilbert Problem* 4.3.1 *is equivalent to the outer model Riemann-Hilbert Problem* 4.2.3.

In view of the jump relation (4.43) and the continuity of the boundary values within the bands, we may solve Riemann-Hilbert Problem 4.3.1 by considering the two columns of the matrix $\mathbf{Q}(\lambda)$ as two projections of a *single-valued* vector function defined on a hyperelliptic Riemann surface X, which is a double covering of the complex λ-plane. To achieve this, introduce X as two copies of the complex plane, individually cut and then mutually identified along the bands. From $\mathbf{Q}(\lambda)$, define a vector-valued function $\mathbf{v}(P)$ for $P \in X$ by arbitrarily labeling one copy of the cut complex plane in X as the "first sheet" and the other copy as the "second sheet," and then setting

$$\mathbf{v}(P) := \begin{cases} \text{the first column of } \mathbf{Q}(\lambda) \text{ if } P \in \text{the first sheet of } X, \\ \text{the second column of } \mathbf{Q}(\lambda) \text{ if } P \in \text{the second sheet of } X. \end{cases} \tag{4.52}$$

With suitable interpretations at the cuts, each point in the λ-plane has two preimages on X, except for the $2G+2$ branch points $\{\lambda_k\}$ and $\{\lambda_k^*\}$. Denote the preimage of $\lambda = \infty$ on the first (respectively, second) sheet of X by $P = \infty_1$ (respectively, $P = \infty_2$). With the inclusion of these two points, X is a compact Riemann surface of genus G.

The function $\mathbf{v}(P)$ so defined on X is holomorphic on X away from $P = \infty_1$, $P = \infty_2$, and the points $\lambda_0, \lambda_1^*, \lambda_2, \lambda_3^*, \ldots, \lambda_G$. At the two infinite points of X, $\mathbf{v}(P)$ has essential singularities, whereas at the other singular points the elements of $\mathbf{v}(P)$ grow, in terms of the hyperelliptic projection $\lambda(P)$, at worst like an inverse square root. In view of the double ramification of X at these isolated points, we see that in terms of *holomorphic charts* i.e., as a function on the *complex manifold* X, $\mathbf{v}(P)$ *has at worst simple poles at exactly half of the branching points of* X. Thus, $\mathbf{v}(P)$ is a meromorphic function on $X \setminus \{\infty_1, \infty_2\}$. Its poles in this finite part of X are at worst simple and are confined to the branch points $\lambda_0, \lambda_1^*, \lambda_2, \lambda_3^*, \ldots, \lambda_G$.

The two scalar components of $\mathbf{v}(P)$ have the elementary properties that they are meromorphic functions on $X \setminus \{\infty_1, \infty_2\}$ with the formal sum

$$\mathcal{D}^0 = \lambda_0 + \lambda_1^* + \lambda_2 + \lambda_3^* + \cdots + \lambda_G \tag{4.53}$$

as the *divisor* of the poles. The asymptotic behavior near the two infinite points of X is given by expansions of the form

$$\begin{aligned} v_1(P) &\sim \exp(ip(\lambda)/\hbar)(1 + O(\lambda^{-1})), && P \to \infty_1, \\ v_1(P) &\sim \exp(-ip(\lambda)/\hbar)O(\lambda^{-1}), && P \to \infty_2, \\ v_2(P) &\sim \exp(ip(\lambda)/\hbar)O(\lambda^{-1}), && P \to \infty_1, \\ v_2(P) &\sim \exp(-ip(\lambda)/\hbar)(1 + O(\lambda^{-1})), && P \to \infty_2. \end{aligned} \tag{4.54}$$

Now, from the Riemann-Roch theorem and the nonspeciality of the divisor $\mathcal{D}^0 - \infty_2$ (see the remark that follows), it follows that there exists a one-dimensional linear space of meromorphic functions on X having $G+1$ simple poles at the points P_k of the divisor $\mathcal{D}^0$ and a simple zero at $P = \infty_2$. This implies that there exists a unique meromorphic function $f_1(P)$ with these properties and normalized so that $f_1(\infty_1) = 1$. Similarly, there exists a unique function $f_2(P)$ on X with simple poles at the same points as $f_1(P)$, with a simple zero at $P = \infty_1$, and normalized so that $f_2(\infty_2) = 1$. Each of these functions has exactly G zeros on X (counted with multiplicities) in addition to the specified zero. Let $\mathcal{D}_1 = P_1^1 + \cdots + P_G^1$ and $\mathcal{D}_2 = P_1^2 + \cdots + P_G^2$ denote the divisors of these zeros. These divisors are necessarily nonspecial (again, see the remark that follows for details). Define $z_1(P)$ and $z_2(P)$ by

$$z_1(P) := \frac{v_1(P)}{f_1(P)}, \qquad z_2(P) := \frac{v_2(P)}{f_2(P)}. \tag{4.55}$$

These two functions are called *Baker-Akhiezer functions.* They are meromorphic functions on the finite part of X with poles confined to the divisors $\mathcal{D}_1$ and $\mathcal{D}_2$, respectively. Near the two infinite points of X,

$$\begin{aligned}
z_1(P) &\sim \exp(ip(\lambda)/\hbar)(1 + O(\lambda^{-1})), && P \to \infty_1,\\
z_1(P) &\sim \exp(-ip(\lambda)/\hbar)O(1), && P \to \infty_2,\\
z_2(P) &\sim \exp(ip(\lambda)/\hbar)O(1), && P \to \infty_1,\\
z_2(P) &\sim \exp(-ip(\lambda)/\hbar)(1 + O(\lambda^{-1})), && P \to \infty_2.
\end{aligned} \tag{4.56}$$

Remark. The integral divisors $\mathcal{D}_1$ and $\mathcal{D}_2$ are nonspecial if $\mathcal{D}^0 - \infty_2$ and $\mathcal{D}^0 - \infty_1$ are, since the divisors are then linearly equivalent by the meromorphic functions $f_1(P)$ and $f_2(P)$, respectively. So it remains to show the nonspeciality of $\mathcal{D}^0 - \infty_k$ for $k = 1$ and $k = 2$. By definition, this divisor is nonspecial if every abelian differential $\omega(P)$ on X whose worst singularity is a simple pole at $P = \infty_k$ and that vanishes at all the points P of the divisor $\mathcal{D}^0$, is necessarily equal to the zero differential. But since the residues of any abelian differential on X sum to zero, as a consequence of Cauchy's theorem on the boundary contour of the canonical dissection of X into a $4G$-gon in the complex plane, any such differential $\omega(P)$ is in fact holomorphic. Now, because X is a two-sheeted cover of the complex plane with square root branch points at $\lambda_0, \ldots, \lambda_G$ and $\lambda_0^*, \ldots, \lambda_G^*$, the holomorphic differential $\omega(P)$ can be written on either sheet in the form $\omega = p_{G-1}(\lambda)\, d\lambda / R(\lambda)$, where $p_{G-1}(\lambda)$ is a polynomial of degree $G-1$ in λ. For $\omega(P)$ to vanish at the points of $\mathcal{D}^0$, it is then necessary for the polynomial $p_{G-1}(\lambda)$ to vanish at the $G+1$ distinct values of λ corresponding to the points of $\mathcal{D}^0$. Of course this makes $p_{G-1}(\lambda)$ and hence $\omega(P)$ identically zero. This completes the argument of the nonspeciality of $\mathcal{D}^0 - \infty_k$ and hence of the integral divisors $\mathcal{D}_1$ and $\mathcal{D}_2$.

Given the polynomial $p(\lambda)$, each of the Baker-Akhiezer functions $z_1(P)$ and $z_2(P)$ is uniquely determined by these elementary properties. The algebro-geometric argument for uniqueness goes as follows. From the existence of any such function with minimal degree at the points of the divisor characterizing its admissible poles (that is, a function having poles of the largest admissible degree at these points), the uniqueness follows again from the Riemann-Roch theorem. For example, if one presumes the existence of *two* functions satisfying the conditions of, say, $z_1(P)$, one of which has minimal degree at the points of $\mathcal{D}_1$, and constructs their ratio with the minimal-degree function in the denominator, then this ratio is a meromorphic function on *all* of X (the essential singularities cancel) with degree G, and all poles of the ratio come from the zeros of the denominator. These zeros of the denominator move around on X as the parameters x and t vary, and it is reasonable to assume that the motion of the degree-G divisor of these zeros avoids the codimension-1 locus of divisors for which the Abel mapping (see below) fails to be invertible, the *special divisors*. In this sense, the most abstract form of the argument holds only for generic complex values of the parameters x and t. However, with some additional information about the Riemann surface X and the divisor $\mathcal{D}_1$ (so-called reality conditions), it is possible to prove that as long as x and t are real, the divisor of the zeros of the denominator will always be nonspecial. In any case, once it is known or assumed that the divisor of zeros (which is the pole divisor of the ratio) is nonspecial, then it follows from the Riemann-Roch theorem that the ratio is a constant function on X. By the normalization at $P = \infty_1$ (for $z_1(P)$), this ratio is exactly unity. Of course, from another point of view uniqueness is not really an issue at all here, because the Riemann-Hilbert problem for $\tilde{\mathbf{O}}(\lambda)$ has itself been used to prove uniqueness.

We have just specified two functions $z_1(P)$ and $z_2(P)$ on a Riemann surface X, and if we can prove that such functions exist, then we have established the existence of a solution to the hyperelliptic Riemann-Hilbert Problem 4.3.1. We now pursue the construction of these two functions.

4.3.2 Formulae for the Baker-Akhiezer Functions

To establish the existence part of the argument, we now provide formulae for the two Baker-Akhiezer functions. There are several ingredients we need to define. See Dubrovin's paper [D81] for any details we do not give here. The first is a *homology basis* on X. One starts with the system of equivalence classes of closed, noncontractable oriented contours on X, with two contours being considered equivalent if their difference (the union with the orientation of one contour reversed) forms the oriented boundary of a surface in X. The equivalence classes are referred to by representatives. Two contour representatives of the same class are called *homologous cycles*; the integral of any meromorphic differential without residues gives the same value over any two homologous cycles. The system of homology classes may be viewed as a linear space with integer coefficients. The zero element of this space is the equivalence class of contractable oriented loops on X. It is a fundamental topological result that this space has dimension $2G$. A homology basis is a basis $\{a_1, \ldots, a_G, b_1, \ldots, b_G\}$ of this linear space that has certain properties with respect to contour intersection. Let C_1 and C_2 be two closed oriented contours on X. The *intersection number* $C_1 \circ C_2$ is defined as the number of times C_2 crosses C_1 from the right of C_1 minus the number of times C_2 crosses C_1 from the left. The intersection number is a skew-symmetric bilinear class function. A homology basis is required to have the following properties:

$$a_j \circ a_k = b_j \circ b_k = 0, \qquad a_j \circ b_k = \delta_{jk}. \tag{4.57}$$

This does not make the basis unique, even up to homology equivalence of class representatives. Any linear transformation of the basis elements in the matrix group $Sp(2G, \mathbb{Z})$ will preserve the intersection number but modify the particular basis. Ultimately, we select a particular homology basis in order to simplify the *appearance* of the formulae we write down, but of course the results themselves, by uniqueness, are independent of this choice. Once the homology cycles a_j are fixed, the dual cycles b_j are determined by the intersection relations up to transformations of the form

$$b_k \rightarrow b_k + \sum_{j=1}^{G} s_{kj} a_j, \tag{4.58}$$

where s_{kj} are integers and $s_{kj} = s_{jk}$. Of course, for $G = 0$ there are no homology cycles at all.

The next ingredient we need are the *normalized holomorphic differentials*. On X there is a complex G-dimensional linear space of holomorphic differentials, with basis elements $\nu_k(P)$ for $k = 1, \ldots, G$ that can be written in the form

$$\nu_k(P) = \frac{\sum_{j=0}^{G-1} c_{kj} \lambda(P)^j}{R_X(P)} \, d\lambda(P), \tag{4.59}$$

where $R_X(P)$ is a "lifting" of the function $R(\lambda)$ from the cut plane to X: if P is on the first sheet of X, then $R_X(P) = R(\lambda(P))$, and if P is on the second sheet of X, then $R_X(P) = -R(\lambda(P))$. The coefficients c_{kj} are uniquely determined by the constraint that the differentials satisfy the normalization conditions

$$\oint_{a_j} \nu_k(P) = 2\pi i \delta_{jk}. \tag{4.60}$$

From the normalized differentials, one defines a $G \times G$ matrix $\mathbf{H}$ (the *period matrix*) by the formula

$$H_{jk} = \oint_{b_j} \nu_k(P). \tag{4.61}$$

It is a consequence of the standard theory of Riemann surfaces that $\mathbf{H}$ is a symmetric matrix whose real part is *negative definite*.

Associated with the matrix $\mathbf{H}$ and therefore with the choice of homology basis on X is the *Riemann theta function* defined for $\mathbf{w} \in \mathbb{C}^G$ by the Fourier series

$$\Theta(\mathbf{w}) := \sum_{\mathbf{n} \in \mathbb{Z}^G} \exp\left(\frac{1}{2}\mathbf{n}^T \mathbf{H} \mathbf{n} + \mathbf{n}^T \mathbf{w}\right). \tag{4.62}$$

It is an entire function on $\mathbb{C}^G$.

Let $\mathbf{e}_k$ denote the standard unit vectors in $\mathbb{C}^G$, and let $\mathbf{h}_k$ denote the kth column of the matrix $\mathbf{H}$, that is, $\mathbf{h}_k := \mathbf{H}\mathbf{e}_k$ for $k = 1, \ldots, G$. Denote by $\Lambda \subset \mathbb{C}^G$ the lattice generated by linear combinations with integer coefficients of the vectors $2\pi i \mathbf{e}_k$ and $\mathbf{h}_k$ for $k = 1, \ldots, G$. That is,

$$\Lambda := 2\pi i \mathbb{Z}\mathbf{e}_1 + \cdots + 2\pi i \mathbb{Z}\mathbf{e}_G + \mathbb{Z}\mathbf{h}_1 + \cdots + \mathbb{Z}\mathbf{h}_G. \tag{4.63}$$

The *Jacobian variety* of X, $\mathrm{Jac}(X)$, is simply the complex torus $\mathbb{C}^G/\Lambda$. We arbitrarily fix a basepoint P_0 on X. The *Abel-Jacobi mapping* $\mathbf{A} : X \to \mathrm{Jac}(X)$ is then defined componentwise as

$$A_k(P; P_0) := \int_{P_0}^{P} \nu_k(P'), \qquad k = 1, \ldots, G, \tag{4.64}$$

where P' is an integration variable. The range of the mapping is in the Jacobian because the path of integration is not specified. The Abel-Jacobi mapping is also defined for integral divisors $\mathcal{D} = P_1 + \cdots + P_M$ by summation,

$$\mathbf{A}(\mathcal{D}; P_0) := \mathbf{A}(P_1; P_0) + \cdots + \mathbf{A}(P_M; P_0), \tag{4.65}$$

and finally extended to nonintegral divisors $\mathcal{D} = \mathcal{D}^+ - \mathcal{D}^-$ for integral divisors $\mathcal{D}^\pm$ by $\mathbf{A}(\mathcal{D}; P_0) := \mathbf{A}(\mathcal{D}^+; P_0) - \mathbf{A}(\mathcal{D}^-; P_0)$. If the degree of the divisor $\mathcal{D}$ is zero, then $\mathbf{A}(\mathcal{D}; P_0)$ is independent of the base point P_0. *Abel's theorem* states that if $\mathcal{D}^+$ catalogs the zeros and $\mathcal{D}^-$ the poles of a meromorphic function on the compact surface X, then with $\mathcal{D} = \mathcal{D}^+ - \mathcal{D}^-$, $\mathbf{A}(\mathcal{D}; P_0) = 0$ in the Jacobian or, equivalently, the integral always yields a lattice vector in $\Lambda \subset \mathbb{C}^G$. Note that Abel's theorem applied to the functions $f_1(P)$ and $f_2(P)$ yields the identities

$$\begin{aligned} \mathbf{A}(\mathcal{D}_1; P_0) &= \mathbf{A}(\mathcal{D}^0; P_0) - \mathbf{A}(\infty_2; P_0) \qquad (\mathrm{mod}\ \Lambda), \\ \mathbf{A}(\mathcal{D}_2; P_0) &= \mathbf{A}(\mathcal{D}^0; P_0) - \mathbf{A}(\infty_1; P_0) \qquad (\mathrm{mod}\ \Lambda). \end{aligned} \tag{4.66}$$

Finally, a particularly important element of the Jacobian is the *Riemann constant vector* $\mathbf{K}$, which is defined, modulo the lattice Λ, componentwise by

$$K_k := \pi i + \frac{H_{kk}}{2} - \frac{1}{2\pi i} \sum_{\substack{j=1 \\ j \neq k}}^{G} \oint_{a_j} \left(\nu_j(P) \int_{P_0}^{P} \nu_k(P') \right), \tag{4.67}$$

where the index k varies between 1 and G.

Next, we need to define a certain meromorphic differential on X. Let $\Omega(P)$ be holomorphic away from the points ∞_1 and ∞_2, where it has the behavior

$$\begin{aligned} \Omega(P) &= dp(\lambda(P)) + O\left(\frac{d\lambda(P)}{\lambda(P)^2}\right), \qquad P \to \infty_1, \\ \Omega(P) &= -dp(\lambda(P)) + O\left(\frac{d\lambda(P)}{\lambda(P)^2}\right), \qquad P \to \infty_2, \end{aligned} \tag{4.68}$$

and made unique by the normalization conditions

$$\oint_{a_j} \Omega(P) = 0, \qquad j = 1, \ldots, G. \tag{4.69}$$

Let the vector $\mathbf{U} \in \mathbb{C}^G$ be defined componentwise by

$$U_j := \oint_{b_j} \Omega(P). \tag{4.70}$$

Note that $\Omega(P)$ has no residues.

With these ingredients, we may now give formulae for the functions $z_1(P)$ and $z_2(P)$. First, define $y_1(P)$ and $y_2(P)$ by choosing particular vectors $\mathbf{V}_m \in \mathbb{C}^G$ corresponding to the points $\mathbf{A}(\mathcal{D}_m; P_0) + \mathbf{K} \in \text{Jac}(X)$ and then setting

$$y_m(P) := \frac{\Theta(\mathbf{A}(P; P_0) - \mathbf{V}_m + i\mathbf{U}/\hbar)}{\Theta(\mathbf{A}(P; P_0) - \mathbf{V}_m)} \exp\left(\frac{i}{\hbar}\int_{P_0}^{P} \Omega(P')\right), \tag{4.71}$$

where $m = 1$ or $m = 2$. The path of integration in the exponent is the same path as in the Abel-Jacobi mapping but is otherwise unspecified. The fact that these formulae actually define functions that do not depend on that path follows from the transformation laws for the theta function:

$$\Theta(\mathbf{w} + 2\pi i \mathbf{e}_k) = \Theta(\mathbf{w}), \quad \Theta(\mathbf{w} + \mathbf{h}_k) = \exp(-H_{kk}/2 - w_k)\Theta(\mathbf{w}). \tag{4.72}$$

So, if the path of integration is augmented by adding one of the homology cycles a_k, then the exponent is invariant by the normalization condition for $\Omega(P)$. At the same time, we have $\mathbf{A}(P; P_0) \to \mathbf{A}(P; P_0) + 2\pi i \mathbf{e}_k$, and by the first transformation law we see that $y_m(P)$ is invariant. Similarly, if the homology cycle b_k is added to the path of integration, then the exponential transforms by producing a factor of $\exp(iU_k/\hbar)$. And at the same time $\mathbf{A}(P; P_0) \to \mathbf{A}(P; P_0) + \mathbf{h}_k$, and by the second transformation law we again deduce that $y_m(P)$ is invariant. This means that the functions $y_m(P)$ are well defined given the choices of homology, basepoint P_0, and representatives $\mathbf{V}_m$.

Now, since each divisor $\mathcal{D}_m$ is nonspecial, the zeros of the denominator are exactly the points $P_1^m, \dots, P_G^m$. Since the theta function is entire and the differential $\Omega(P)$ is holomorphic away from the points ∞_1 and ∞_2, it follows that $y_m(P)$ is meromorphic on $X \setminus \{\infty_1, \infty_2\}$, with poles exactly at the points of the divisor $\mathcal{D}_m$. Near the points ∞_1 and ∞_2, we have

$$\begin{aligned} y_m(P) &= \exp(ip(\lambda(P))/\hbar)O(1), \quad P \to \infty_1, \\ y_m(P) &= \exp(-ip(\lambda(P))/\hbar)O(1), \quad P \to \infty_2, \end{aligned} \tag{4.73}$$

where the leading-order term in each expansion depends on m. To obtain formulae for $z_1(P)$ and $z_2(P)$, it then suffices to appropriately normalize the functions $y_1(P)$ and $y_2(P)$. So, let

$$\begin{aligned} N_1 &:= \lim_{P \to \infty_1} y_1(P)\exp(-ip(\lambda(P))/\hbar), \\ N_2 &:= \lim_{P \to \infty_2} y_2(P)\exp(ip(\lambda(P))/\hbar). \end{aligned} \tag{4.74}$$

Then we set

$$z_1(P) := \frac{y_1(P)}{N_1}, \qquad z_2(P) := \frac{y_2(P)}{N_2}. \tag{4.75}$$

These functions at last satisfy all the required conditions, and by the Riemann-Roch argument or the equivalent uniqueness argument for the hyperelliptic Riemann-Hilbert Problem 4.3.1 we summarized earlier, *they are the only such functions.* In particular, $z_1(P)$ and $z_2(P)$ do not depend on the choice of homology cycles, the choice of basepoint P_0, or the choice of representatives $\mathbf{V}_m$. While this is true, certain properties of the functions can be elucidated by making particular convenient choices of these arbitrary "gauge" parameters.

Thus, we have established the existence of the two Baker-Akhiezer functions $z_1(P)$ and $z_2(P)$, which amounts to the solution of the hyperelliptic Riemann-Hilbert Problem 4.3.1 or, equivalently, the solution of the model Riemann-Hilbert Problem 4.2. But these formulae become more effective if we break the gauge symmetry by specifying all paths of integration concretely in the cut plane. We carry out this program now.

4.3.3 Making the Formulae Concrete

We now develop these formulae in more detail. First observe that for the functions $f_1(P)$ and $f_2(P)$ we have the explicit representations

$$\begin{aligned} f_1(P) &= \frac{R_X(P) - (\lambda(P) - \lambda_0^*)(\lambda(P) - \lambda_1) \cdots (\lambda(P) - \lambda_G^*)}{2R_X(P)}, \\ f_2(P) &= \frac{R_X(P) + (\lambda(P) - \lambda_0^*)(\lambda(P) - \lambda_1) \cdots (\lambda(P) - \lambda_G^*)}{2R_X(P)}. \end{aligned} \tag{4.76}$$

Next, we make the observation that

$$\begin{aligned} \Omega(P) &= dh(\lambda(P)), && P \text{ on the first sheet of } X, \\ \Omega(P) &= -dh(\lambda(P)), && P \text{ on the second sheet of } X. \end{aligned} \tag{4.77}$$

To see this, one *defines* an abelian differential $\widehat{\Omega}(P)$ on X by the right-hand side of (4.77); it is not difficult to see from the jump relations for the scalar function h that this indeed defines a meromorphic differential on the whole of the compact surface X. Next, it follows from the definition of $\Omega(P)$ near the points ∞_1 and ∞_2 that the difference $\Omega(P) - \widehat{\Omega}(P)$ is a holomorphic differential on X because the singularities cancel. The difference will therefore be identically zero if it can be shown that

$$\oint_{a_j} (\Omega(P) - \widehat{\Omega}(P)) = 0 \tag{4.78}$$

for all $j = 1, \ldots, G$. But the first term vanishes by definition of $\Omega(P)$, and it can be shown from the jump relations for $h(\lambda)$ that the same is true of $\widehat{\Omega}(P)$, regardless of the choice of homology basis.

Let us specify a particular homology basis. Although somewhat nonstandard, the basis we specify is chosen to behave nicely under complex conjugation. For topological purposes, we can deform each sheet of X so that the contour becomes a straight line along which the endpoints occur from left to right in order: $\lambda_G^*, \ldots, \lambda_1^*, \lambda_0^*, \lambda_0, \lambda_1, \ldots, \lambda_G$. The basis we choose is then illustrated in figure 4.5. Although this illustration is for genus $G = 4$, it should be obvious how the pattern generalizes for other genera.

The point of our choice of homology cycles is that it simplifies certain integrals on the Riemann surface X.

Definition 4.3.1 *By an antisymmetric differential $\omega(P)$ on X, we mean one for which whenever λ is not a branch point and $P^+(\lambda)$ and $P^-(\lambda)$ are the distinct preimages on X of λ under the sheet-projection mapping, then $\omega(P^-(\lambda)) = -\omega(P^+(\lambda))$.*

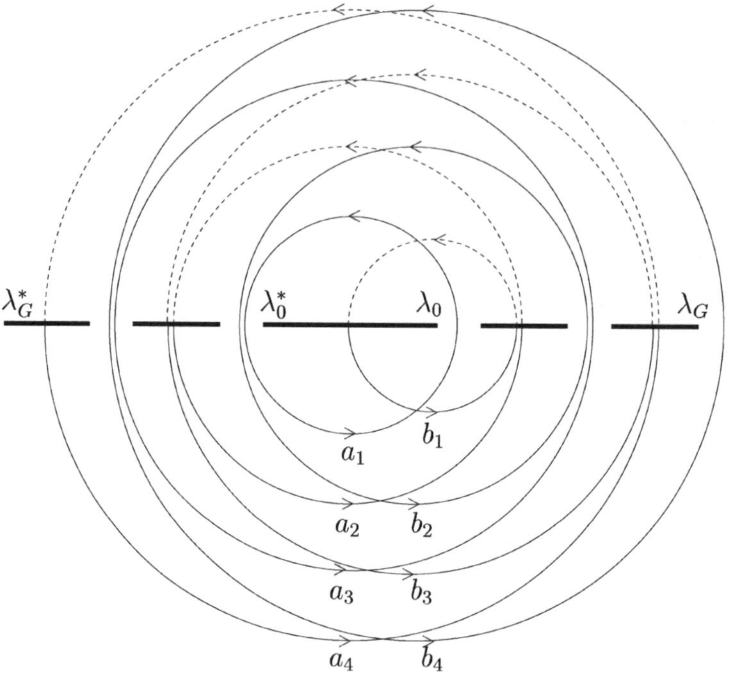

Figure 4.5 Our particular choice of homology cycles, illustrated for $G = 4$, on a surface that is smoothly deformed so that the cuts lie along a straight line. The cuts are illustrated in bold, with endpoints labeled from left to right: $\lambda_G^*, \ldots, \lambda_0^*, \lambda_0, \ldots, \lambda_G$. Paths on the first (second) sheet are indicated with solid (dashed) lines.

Such antisymmetric differentials include the holomorphic differentials $\nu_k(P)$ and the meromorphic differential $\Omega(P)$. With the particular choice of homology basis illustrated in figure 4.5, it is easy to express loop integrals of any antisymmetric differential $\omega(P)$ in terms of integrals only on the first sheet of X, along the jump contour of the model Riemann-Hilbert Problem 4.2 for $\widetilde{\mathbf{O}}(\lambda)$. The paths of integration on the first sheet of X corresponding to the cycles making up the homology basis are illustrated in figure 4.6. The first consequence of this choice is that the integral of any antisymmetric differential over an a-cycle is equal to twice the value of an integral over a concrete path on the first sheet of the cut plane that has a symmetry under complex conjugation of the plane. Given an oriented path c on the cut plane, denote by c^* the path obtained by complex conjugation followed by reversal of orientation. From figure 4.6, keeping in mind that in the figure complex conjugation of the plane corresponds to left/right reflection, one then sees that

$$\frac{1}{2}a_j - \frac{1}{2}a_j^* = 0 \tag{4.79}$$

for all $j = 1, \ldots, G$. For the homology cycles b_j, we can find similar relations,

$$\frac{1}{2}b_j + \frac{1}{2}b_j^* \equiv 0, \quad \text{modulo } \frac{1}{2}\{a_1, \ldots, a_G\}. \tag{4.80}$$

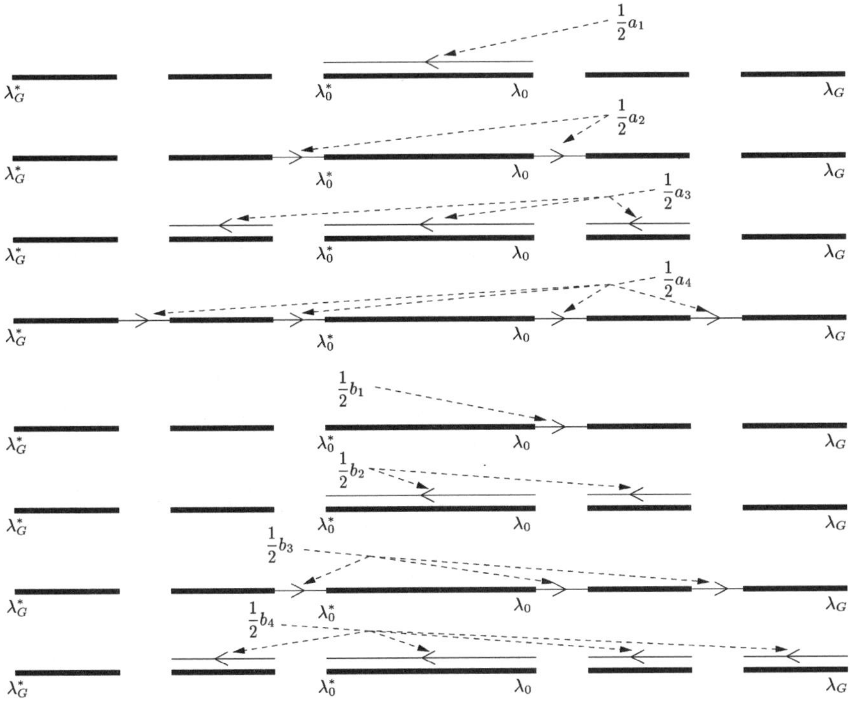

Figure 4.6 Evaluating homology cycle integrals of any antisymmetric differential by integrating on the first sheet along the immediate left of the jump contour of the model Riemann-Hilbert Problem 4.2. In each line of this figure, the jump contour is imagined as a straight line oriented from left to right and the bands (cuts) are shown in bold. It should be clear how this picture generalizes to other genera.

For example, $b_1 + b_1^* = a_2$, $b_2 + b_2^* = a_1 + a_3$, and so on. It is easy to see that with such a choice of homology cycles, the constants c_{kj} in the holomorphic differentials are all made manifestly real by the normalization condition (4.60). Indeed, the linear equations implied by (4.60) for the constants c_{kj} all have real coefficients, and the system is invertible. Once it is known that these constants are all real, the symmetries (4.80) can be used to show that

$$\Im(H_{jk}) = 2\pi n_{jk}, \qquad \text{where } n_{jk} \in \mathbb{Z}, \qquad j,k = 1,\ldots,G. \tag{4.81}$$

Next, note that since the coefficients of the polynomial $p(\lambda)$ are real, it follows from the symmetry of the cycles a_k in this special homology basis on X that for any path c on the cut plane,

$$\int_{c^*} \Omega(P) = -\left(\int_c \Omega(P)\right)^*, \tag{4.82}$$

where $\Omega(P)$ is the meromorphic differential defined by (4.68) and (4.69). Using this relation together with the symmetry relations (4.80) and the normalization conditions

(4.69) defining the differential $\Omega(P)$, we find

$$\begin{aligned}\oint_{b_j} \Omega(P) &= 2\int_{\frac{1}{2}b_j} \Omega(P) = -2\int_{\frac{1}{2}b_j^*} \Omega(P)\\ &= 2\left(\int_{\frac{1}{2}b_j} \Omega(P)\right)^* = \left(\oint_{b_j} \Omega(P)\right)^*,\end{aligned} \tag{4.83}$$

where the first and last integrals in the chain of equalities are loop integrals on the Riemann surface X and the intermediate integrals are all taken on concrete paths in the cut plane according to figure 4.6. This calculation shows that the components of the vector $\mathbf{U}$ are all real.

The next gauge symmetry we break is the invariance with respect to choice of basepoint P_0. For several reasons, it is convenient in the hyperelliptic context to choose P_0 to be a branch point; here we take $P_0 = \lambda_0$. One advantage of this choice is that the Riemann constant vector takes a particularly simple form. Using the fact that with our choice of homology basis, the hyperelliptic (sheet-exchanging) involution of X takes each a cycle into its opposite (i.e., into the same loop with opposite orientation), one finds that

$$\mathbf{K} \equiv \widetilde{\mathbf{K}} \qquad (\text{mod } \Lambda), \tag{4.84}$$

where the components of $\widetilde{\mathbf{K}}$ are given by

$$\widetilde{K}_k = \pi i + \frac{H_{kk}}{2}, \quad k = 1, \ldots, G. \tag{4.85}$$

By the observation (4.81) about the Riemann matrix $\mathbf{H}$, we see that the imaginary parts of the components of $\widetilde{\mathbf{K}}$ are all integer multiples of π.

For concreteness we now choose vectors $\mathbf{V}_m \in \mathbb{C}^G$ for $m = 1, 2$ so that $\mathbf{A}(\mathcal{D}_m; \lambda_0) + \mathbf{K} = \mathbf{V}_m$ (mod Λ). Before doing this, however, we first select a specific path of integration used to define the Abel-Jacobi mapping itself. Given a point $P \in X$, this is done by specifying a path from $P_0 = \lambda_0$ to P modulo homotopy.

Definition 4.3.2 *Let $P \in X$. By C_P we mean any element of the homotopy equivalence class of paths from λ_0 to P on X such that the following three things are true:*

1. *Each point on C_P lies on the same sheet as P (being a branch point, the basepoint is considered to lie on both sheets);*
2. *C_P completely avoids the whole portion of the contour from λ_G^* through to λ_G;*
3. *C_P begins on the "+" side of the basepoint λ_0 on the contour.*

Note that to define the path C_P, it is essential to view X as two copies of the plane cut along the jump contour of the model Riemann-Hilbert Problem 4.2.

Using the path C_P in the Abel mapping and recalling that the same path is used in the integration of the differential Ω, we see that

$$\begin{aligned}\int_{C_P} \Omega(P') &= h(\lambda(P)) - h_+(\lambda_0), && P \text{ on the first sheet of } X,\\ \int_{C_P} \Omega(P') &= -h(\lambda(P)) + h_+(\lambda_0), && P \text{ on the second sheet of } X.\end{aligned} \tag{4.86}$$

Note that it is sufficient here for C_P to be defined modulo homotopy because the meromorphic differential $\Omega(P)$ has no residues. To give representatives $\mathbf{V}_m$ for $\mathbf{A}(\mathcal{D}_m;\lambda_0)+\mathbf{K}$, we first choose to represent the Riemann constant vector $\mathbf{K}$ in $\mathbb{C}^G$ *exactly* by the vector $\widetilde{\mathbf{K}}$ defined by (4.85). To represent $\mathbf{A}(\mathcal{D}_1;\lambda_0)$ and $\mathbf{A}(\mathcal{D}_2;\lambda_0)$ in $\mathbb{C}^G$, it suffices by Abel's theorem to represent, respectively, $\mathbf{A}(\mathcal{D}^0;\lambda_0)-\mathbf{A}(\infty_2;\lambda_0)$ and $\mathbf{A}(\mathcal{D}^0;\lambda_0)-\mathbf{A}(\infty_1;\lambda_0)$ (cf. (4.66)). To begin with, we define $\mathbf{A}(\infty_m;\lambda_0)$ for $m=1$ and $m=2$ by setting

$$A_k(\infty_m,\lambda_0)=\int_{C_{\infty m}}\nu_k(P), \tag{4.87}$$

with the path C_{∞_m} on X chosen according to Definition 4.3.2.

It then follows that for these representatives, $\mathbf{A}(\infty_2;\lambda_0)=-\mathbf{A}(\infty_1;\lambda_0)$. Finally, to represent $\mathbf{A}(\mathcal{D}^0;\lambda_0)$, we associate with each branch point in $\mathcal{D}^0$ a point on the first sheet of X immediately on the "+" side of the contour, and we compute $\mathbf{A}(\mathcal{D}^0;\lambda_0)$ as a sum of integrals with paths determined according to Definition 4.3.2. Each such integral may be realized as lying on the "+" side of the contour and, by the scheme described in figure 4.6, can be identified with a specific half-period in $\Lambda/2$, where the lattice Λ is defined by (4.63). To be quite precise, let $\mathbf{A}^{\text{cut}}(\lambda)$ denote the Abel mapping with basepoint and contour C_P chosen according to Definition 4.3.2 for the point P on the first sheet of X for which $\lambda(P)=\lambda$. Then

$$\begin{aligned} V_{1,k}&=(A_k^{\text{cut}}(\lambda_{1+}^*)+A_k^{\text{cut}}(\lambda_{2+})+A_k^{\text{cut}}(\lambda_{3+}^*)+\cdots+A_k^{\text{cut}}(\lambda_{G+}))\\ &\quad+A_k^{\text{cut}}(\infty)+\pi i+\frac{H_{kk}}{2},\\ V_{2,k}&=(A_k^{\text{cut}}(\lambda_{1+}^*)+A_k^{\text{cut}}(\lambda_{2+})+A_k^{\text{cut}}(\lambda_{3+}^*)+\cdots+A_k^{\text{cut}}(\lambda_{G+}))\\ &\quad-A_k^{\text{cut}}(\infty)+\pi i+\frac{H_{kk}}{2}, \end{aligned} \tag{4.88}$$

where k varies between 1 and G.

In terms of these gauge choices, we find

$$y_m(P)=\begin{cases}\dfrac{\Theta(\mathbf{A}^{\text{cut}}(\lambda(P))-\mathbf{V}_m+i\mathbf{U}/\hbar)}{\Theta(\mathbf{A}^{\text{cut}}(\lambda(P))-\mathbf{V}_m)}e^{ih(\lambda(P))/\hbar}e^{-ih_+(\lambda_0)/\hbar},\\ \qquad P\text{ on first sheet},\\ \dfrac{\Theta(-\mathbf{A}^{\text{cut}}(\lambda(P))-\mathbf{V}_m+i\mathbf{U}/\hbar)}{\Theta(-\mathbf{A}^{\text{cut}}(\lambda(P))-\mathbf{V}_m)}e^{-ih(\lambda(P))/\hbar}e^{ih_+(\lambda_0)/\hbar},\\ \qquad P\text{ on second sheet.}\end{cases} \tag{4.89}$$

The normalizing constants N_m defined by (4.74) are easily obtained from (4.89), since $p(\lambda)=h(\lambda)+O(1/\lambda)$ as λ tends to infinity. Thus,

$$\begin{aligned} N_1&=\frac{\Theta(\mathbf{A}^{\text{cut}}(\infty)-\mathbf{V}_1+i\mathbf{U}/\hbar)}{\Theta(\mathbf{A}^{\text{cut}}(\infty)-\mathbf{V}_1)}e^{-ih_+(\lambda_0)/\hbar},\\ N_2&=\frac{\Theta(-\mathbf{A}^{\text{cut}}(\infty)-\mathbf{V}_2+i\mathbf{U}/\hbar)}{\Theta(-\mathbf{A}^{\text{cut}}(\infty)-\mathbf{V}_2)}e^{ih_+(\lambda_0)/\hbar}. \end{aligned} \tag{4.90}$$

Combining these with the sheetwise formula (4.89) for $y_m(P)$ gives a similar sheetwise formula for the functions $z_m(P)$. From these, one obtains sheetwise formulae for the functions $v_m(P)$ defined originally in (4.52) by introducing

$$b^{\pm}(\lambda) = \frac{R(\lambda) \pm (\lambda - \lambda_0^*)(\lambda - \lambda_1) \cdots (\lambda - \lambda_G^*)}{2R(\lambda)} \tag{4.91}$$

and then observing that for P on the first sheet of X, $f_1(P) = b^-(\lambda(P))$ and $f_2(P) = b^+(\lambda(P))$, while for P on the second sheet of X, $f_1(P) = b^+(\lambda(P))$ and $f_2(P) = b^-(\lambda(P))$. Since, according to (4.52), the first (respectively, second) column of $\mathbf{Q}$ is simply the vector (v_1, v_2) restricted to the first (respectively, second) sheet of X, we have thus obtained an explicit representation of the matrix $\mathbf{Q}$. By the elementary transformations at the beginning of this section, we see that we have proved the following theorem.

Theorem 4.3.1 *For even $G > 0$, the unique solution of the outer model Riemann-Hilbert Problem* 4.2 *is given by the formulae*

$$\begin{aligned}
\tilde{O}_{11}(\lambda) &= \frac{b^-(\lambda)}{\beta(\lambda)} \frac{\Theta(\mathbf{A}^{\text{cut}}(\infty) - \mathbf{V}_1)}{\Theta(\mathbf{A}^{\text{cut}}(\lambda) - \mathbf{V}_1)} \cdot \frac{\Theta(\mathbf{A}^{\text{cut}}(\lambda) - \mathbf{V}_1 + i\mathbf{U}/\hbar)}{\Theta(\mathbf{A}^{\text{cut}}(\infty) - \mathbf{V}_1 + i\mathbf{U}/\hbar)}, \\
\tilde{O}_{12}(\lambda) &= \frac{b^+(\lambda)}{\beta(\lambda)} e^{2ih_+(\lambda_0)/\hbar} \frac{\Theta(\mathbf{A}^{\text{cut}}(\infty) - \mathbf{V}_1)}{\Theta(-\mathbf{A}^{\text{cut}}(\lambda) - \mathbf{V}_1)} \\
&\quad \cdot \frac{\Theta(-\mathbf{A}^{\text{cut}}(\lambda) - \mathbf{V}_1 + i\mathbf{U}/\hbar)}{\Theta(\mathbf{A}^{\text{cut}}(\infty) - \mathbf{V}_1 + i\mathbf{U}/\hbar)}, \\
\tilde{O}_{21}(\lambda) &= \frac{b^+(\lambda)}{\beta(\lambda)} e^{-2ih_+(\lambda_0)/\hbar} \frac{\Theta(-\mathbf{A}^{\text{cut}}(\infty) - \mathbf{V}_2)}{\Theta(\mathbf{A}^{\text{cut}}(\lambda) - \mathbf{V}_2)} \\
&\quad \cdot \frac{\Theta(\mathbf{A}^{\text{cut}}(\lambda) - \mathbf{V}_2 + i\mathbf{U}/\hbar)}{\Theta(-\mathbf{A}^{\text{cut}}(\infty) - \mathbf{V}_2 + i\mathbf{U}/\hbar)}, \\
\tilde{O}_{22}(\lambda) &= \frac{b^-(\lambda)}{\beta(\lambda)} \frac{\Theta(-\mathbf{A}^{\text{cut}}(\infty) - \mathbf{V}_2)}{\Theta(-\mathbf{A}^{\text{cut}}(\lambda) - \mathbf{V}_2)} \cdot \frac{\Theta(-\mathbf{A}^{\text{cut}}(\lambda) - \mathbf{V}_2 + i\mathbf{U}/\hbar)}{\Theta(-\mathbf{A}^{\text{cut}}(\infty) - \mathbf{V}_2 + i\mathbf{U}/\hbar)}.
\end{aligned} \tag{4.92}$$

Note that from the jump relations for $h(\lambda)$,

$$2h_+(\lambda_0) = -\theta_1 - \alpha_0. \tag{4.93}$$

The matrix $\tilde{\mathbf{O}}(\lambda)$ therefore has the property that it is uniformly bounded as $\hbar$ tends to zero in any fixed closed set that does not contain an endpoint λ_k or λ_k^*. This kind of behavior is crucial for controlling the error of these approximations in §4.5. Moreover, away from the endpoints, all derivatives with respect to λ are uniformly bounded as $\hbar$ tends to zero. The $\hbar$ dependence is totally explicit and contributes only global phase oscillations.

4.3.4 Properties of the Semiclassical Solution of the Nonlinear Schrödinger Equation

Consider the function $\tilde{\psi}$ defined from the solution of the outer model Riemann-Hilbert Problem 4.2 by

$$\tilde{\psi} := 2i \lim_{\lambda \to \infty} \lambda \tilde{O}_{12}(\lambda). \tag{4.94}$$

Since by direct computation,

$$b^+(\lambda) = \lambda^{-1} \sum_{k=0}^{G} \frac{i}{2}(-1)^{k+1}\Im(\lambda_k) + O(\lambda^{-2}), \quad \text{as } \lambda \to \infty, \tag{4.95}$$

we find

$$\tilde{\psi} = ae^{iU_0/\hbar}\frac{\Theta(\mathbf{Y}+i\mathbf{U}/\hbar)}{\Theta(\mathbf{Z}+i\mathbf{U}/\hbar)}, \tag{4.96}$$

where

$$a = \frac{\Theta(\mathbf{Z})}{\Theta(\mathbf{Y})}\sum_{k=0}^{G}(-1)^k\Im(\lambda_k) \tag{4.97}$$

and

$$U_0 = -(\theta_1 + \alpha_0), \tag{4.98}$$

with

$$\mathbf{Y} = -\mathbf{A}^{\text{cut}}(\infty) - \mathbf{V}_1, \qquad \mathbf{Z} = \mathbf{A}^{\text{cut}}(\infty) - \mathbf{V}_1. \tag{4.99}$$

Subject to finding an appropriate complex phase function $g^\sigma(\lambda)$ as described in §4.2, we prove in §4.5 that the function $\tilde{\psi}$ captures the leading-order behavior of the true solution ψ of the nonlinear Schrödinger equation as $\hbar$ tends to zero. Here, we show simply that this asymptotic solution is locally a slowly modulated $G+1$-phase wavetrain. To see this, set $x = x_0 + \hbar\hat{x}$ and $t = t_0 + \hbar\hat{t}$, and expand $\tilde{\psi}$ in Taylor series for small $\hbar$, recalling that all quantities depend parametrically on x and t:

$$\tilde{\psi} = a^0 e^{iU_0^0/\hbar}e^{i(k_0^0\hat{x}-w_0^0\hat{t})}\frac{\Theta(\mathbf{Y}^0 + i\mathbf{U}^0/\hbar + i(\mathbf{k}^0\hat{x} - \mathbf{w}^0\hat{t}))}{\Theta(\mathbf{Z}^0 + i\mathbf{U}^0/\hbar + i(\mathbf{k}^0\hat{x} - \mathbf{w}^0\hat{t}))}\cdot(1+O(\hbar)), \tag{4.100}$$

where

$$k_n = \partial_x U_n \quad \text{and} \quad w_n = -\partial_t U_n \quad \text{for } n = 0, \ldots, G \tag{4.101}$$

and the superscript "0" indicates evaluation for $x = x_0$ and $t = t_0$. As a generalization of the exponential function, the theta function is 2π periodic in each imaginary direction in $\mathbb{C}^G$. Therefore, for fixed x_0 and t_0, we see that the leading-order approximation is a multiphase wavetrain with wavenumbers $k_0^0, \ldots, k_G^0$ and frequencies $w_0^0, \ldots, w_G^0$ with respect to the variables $\hat{x}$ and $\hat{t}$. It is a simple consequence of the definition of the wavenumbers and frequencies that the modulations of the waveform due to variations in x_0 and t_0 are constrained by *conservation of waves*:

$$\partial_t k_n + \partial_x w_n = 0. \tag{4.102}$$

4.3.5 Genus Zero

For $G = 0$ the whole construction given in this section degenerates somewhat, and theta functions are not required. For completeness we give all details here for this special case. The function $k(\lambda)$ can be evaluated explicitly by residues:

$$k(\lambda) = -\frac{\alpha_0}{2R(\lambda)}. \tag{4.103}$$

Therefore

$$h(\lambda) = -\frac{1}{2}\alpha_0 = h_0, \tag{4.104}$$

and $p(\lambda)$ is simply a constant (with respect to λ) function h_0. It follows that the functions $v_1(P)$ and $v_2(P)$ that we seek on the compact Riemann surface X of genus zero are in fact meromorphic functions on the whole of X. These functions each have a single simple pole at the point λ_0 and then $v_1(\infty_2) = 0$ while $v_2(\infty_1) = 0$. To give explicit formulae for $v_1(P)$ and $v_2(P)$, we use the "lifting" $R_X(P)$ of the function $R(\lambda)$ to X. This meromorphic function satisfies $R_X(P) \sim -\lambda(P)$ as $P \to \infty_1$ and $R_X(P) \sim \lambda(P)$ as $P \to \infty_2$. We then have

$$\begin{aligned} v_1(P) &= -\frac{1}{2}e^{ih_0/\hbar}\left[\frac{\lambda(P) - \lambda_0^*}{R_X(P)} - 1\right], \\ v_2(P) &= \frac{1}{2}e^{-ih_0/\hbar}\left[\frac{\lambda(P) - \lambda_0^*}{R_X(P)} + 1\right]. \end{aligned} \tag{4.105}$$

Restricting, respectively, to the first and second sheets of X gives an explicit formula for the matrix $\mathbf{Q}(\lambda)$ in the cut plane:

$$\mathbf{Q}(\lambda) = \frac{1}{2R(\lambda)}\exp(i\sigma_3 h_0/\hbar)\begin{bmatrix} -\lambda + \lambda_0^* + R(\lambda) & \lambda - \lambda_0^* + R(\lambda) \\ \lambda - \lambda_0^* + R(\lambda) & -\lambda + \lambda_0^* + R(\lambda) \end{bmatrix}. \tag{4.106}$$

Finally, in terms of the function $\beta(\lambda)$, we see that we have proved the following theorem.

Theorem 4.3.2 *For $G = 0$, the unique solution of the outer model Riemann-Hilbert Problem* 4.2 *is given explicitly by*

$$\widetilde{\mathbf{O}}(\lambda) = \frac{1}{2R(\lambda)\beta(\lambda)} \cdot \begin{bmatrix} -\lambda + \lambda_0^* + R(\lambda) & (\lambda - \lambda_0^* + R(\lambda))\exp(-i\alpha_0/\hbar) \\ (\lambda - \lambda_0^* + R(\lambda))\exp(i\alpha_0/\hbar) & -\lambda + \lambda_0^* + R(\lambda) \end{bmatrix}. \tag{4.107}$$

It is then a direct matter to compute the corresponding semiclassical asymptotic description of the solution of the nonlinear Schrödinger equation:

$$\tilde{\psi} := 2i \lim_{\lambda\to\infty} \lambda \widetilde{O}_{12}(\lambda) = \Im(\lambda_0)e^{-i\alpha_0/\hbar}, \tag{4.108}$$

where we recall that generally λ_0 and α_0 depend on x and t.

4.3.6 The Outer Approximation for $\mathbf{N}^\sigma(\lambda)$

As a final step in this section, we use the solution of the outer model problem to construct an approximation of $\mathbf{N}^\sigma$. This approximation is obtained from $\widetilde{\mathbf{O}}$ by redefining the matrix within the lenses on either side of the bands and is given explicitly by

$$\widehat{\mathbf{N}}^\sigma_{\text{out}}(\lambda) := \widetilde{\mathbf{O}}(\lambda)\mathbf{D}^\sigma(\lambda)^{-1}, \tag{4.109}$$

where $\mathbf{D}^\sigma(\lambda)$ is the explicit piecewise-analytic "lens transformation" relating $\widetilde{\mathbf{N}}^\sigma(\lambda)$ and $\mathbf{O}^\sigma(\lambda)$:

$$\mathbf{O}^\sigma(\lambda) = \widetilde{\mathbf{N}}^\sigma(\lambda)\mathbf{D}^\sigma(\lambda). \tag{4.110}$$

Recall that the matrix $\mathbf{D}^\sigma(\lambda)$ is equal to the identity outside of all lenses. In between the contours C_{k+}^+ and I_k^+,

$$\mathbf{D}^\sigma(\lambda) := \sigma_1^{\frac{1-J}{2}} \begin{bmatrix} 1 & i\exp(-iJ\alpha_k/\hbar)\exp(-ir_k(\lambda)/\hbar) \\ 0 & 1 \end{bmatrix} \sigma_1^{\frac{1-J}{2}}, \tag{4.111}$$

while in between the contours I_k^+ and C_{k-}^+,

$$\mathbf{D}^\sigma(\lambda) := \sigma_1^{\frac{1-J}{2}} \begin{bmatrix} 1 & -i\exp(-iJ\alpha_k/\hbar)\exp(ir_k(\lambda)/\hbar) \\ 0 & 1 \end{bmatrix} \sigma_1^{\frac{1-J}{2}}, \tag{4.112}$$

and for all λ in the lower half-plane, $\mathbf{D}^\sigma(\lambda) = \sigma_2\mathbf{D}^\sigma(\lambda^*)^*\sigma_2$. Here, the functions $r_k(\lambda)$ are defined by Definition 4.2.4. It is easy to check that from the properties of $\widetilde{\mathbf{O}}(\lambda)$, the approximation of $\mathbf{N}^\sigma(\lambda)$ defined above is analytic on the real axis. Also, we have the following useful fact, whose proof is immediate.

Lemma 4.3.1 *The outer approximation $\widehat{\mathbf{N}}^\sigma_{\mathrm{out}}(\lambda)$ is analytic in the complex λ-plane except for $\lambda \in C \cup C^*$ and the boundaries of the lenses $C_{k\pm}^+$ and $C_{k\pm}^-$. In each closed subinterval of the interior of any one band or gap of C_σ or $[C^*]_\sigma$, $\widehat{\mathbf{N}}^\sigma_{\mathrm{out}}(\lambda)$ takes on continuous boundary values. In each band these boundary values satisfy exactly the jump relation*

$$\widehat{\mathbf{N}}^\sigma_{\mathrm{out},+}(\lambda) = \widehat{\mathbf{N}}^\sigma_{\mathrm{out},-}(\lambda)\mathbf{v}^\sigma_{\widetilde{\mathbf{N}}}(\lambda), \tag{4.113}$$

whereas in the gaps,

$$\widehat{\mathbf{N}}^\sigma_{\mathrm{out},+}(\lambda) = \widehat{\mathbf{N}}^\sigma_{\mathrm{out},-}(\lambda)\exp(iJ\theta^\sigma(\lambda)\sigma_3/\hbar), \tag{4.114}$$

where we recall that in each gap the function $\theta^\sigma(\lambda)$ is a real constant. On the lens boundaries, the jump relation is

$$\begin{aligned} \widehat{\mathbf{N}}^\sigma_{\mathrm{out},+}(\lambda) &= \widehat{\mathbf{N}}^\sigma_{\mathrm{out},-}(\lambda)\mathbf{D}^\sigma_-(\lambda), \qquad && \lambda \in C_{k+}^\pm, \\ \widehat{\mathbf{N}}^\sigma_{\mathrm{out},+}(\lambda) &= \widehat{\mathbf{N}}^\sigma_{\mathrm{out},-}(\lambda)\mathbf{D}^\sigma_+(\lambda)^{-1}, \qquad && \lambda \in C_{k-}^\pm. \end{aligned} \tag{4.115}$$

Finally, for λ in any $\hbar$-independent closed set that does not contain $\lambda_0, \ldots, \lambda_G$ or $\lambda_0^*, \ldots, \lambda_G^*$, $\widehat{\mathbf{N}}^\sigma_{\mathrm{out}}(\lambda)$ is uniformly bounded as $\hbar$ tends to zero.

4.4 INNER APPROXIMATIONS

As pointed out in §4.2, the ad hoc approximations made in obtaining the outer model problem from the original Riemann-Hilbert problem for $\mathbf{N}^\sigma(\lambda)$, given a complex phase function $g^\sigma(\lambda)$, clearly break down in the neighborhood of each endpoint $\lambda_0, \ldots, \lambda_G$, its complex conjugate, and also near $\lambda = 0$. Therefore, we now turn our attention to these troublesome neighborhoods and develop inner model problems to approximate $\mathbf{N}^\sigma(\lambda)$ locally in each case. It suffices to construct approximations of $\mathbf{N}^\sigma(\lambda)$ near $\lambda_0, \lambda_1, \ldots, \lambda_G$ and near $\lambda = 0$, because we may then use complex-conjugation symmetry to obtain approximations near the conjugate points.

In this section, we work under one further assumption about $g^\sigma(\lambda)$ that is justified generically (with respect to the parameters x and t) in chapter 5. Thus we have the following:

Working assumption: The density $\rho^\sigma(\lambda)$ vanishes exactly like a square root, and not to higher order, at each endpoint λ_k for $k = 0, \ldots, G$. (4.116)

The generic nature of this assumption is clarified somewhat in Lemma 5.1.3. Higher-order vanishing of $\rho^\sigma(\lambda)$ at an endpoint corresponds to the intersection of two curves in the real (x, t)-plane: a curve where the inequality in a gap fails and a curve where the inequality in an adjacent band fails. So, in making this assumption we are omitting from consideration a set of isolated points in the (x, t)-plane. The nature of the local approximations near the endpoints depends crucially on the degree of vanishing of $\rho^\sigma(\lambda)$, and we want to consider here only the most likely case. For details on the analogous construction necessary for less generic cases, see section 5 of [DKMVZ99B].

4.4.1 Local Analysis for λ near the Endpoint λ_{2k} for $k = 0, \ldots, G/2$

Near an endpoint λ_{2k} for $k = 0, \ldots, G/2$, the approximation of replacing $\mathbf{N}^\sigma(\lambda)$ by $\widetilde{\mathbf{N}}^\sigma(\lambda)$, the continuum limit, is expected to be valid; the trouble is with the approximation of the matrix $\mathbf{O}^\sigma(\lambda)$ by the matrix $\widetilde{\mathbf{O}}(\lambda)$. Locally, the contour of the Riemann-Hilbert problem for $\mathbf{O}^\sigma(\lambda)$ looks like that shown in figure 4.7. Recall the jump relations for $\mathbf{O}^\sigma(\lambda)$. For $\lambda \in \Gamma_{k+1}^+$,

$$\mathbf{O}_+^\sigma(\lambda) = \mathbf{O}_-^\sigma(\lambda)\sigma_1^{\frac{1-J}{2}} \begin{bmatrix} \exp(iJ\theta_{k+1}/\hbar) & 0 \\ i\exp(\tilde{\phi}^\sigma(\lambda)/\hbar) & \exp(-iJ\theta_{k+1}/\hbar) \end{bmatrix} \sigma_1^{\frac{1-J}{2}}, \tag{4.117}$$

for $\lambda \in I_k^+$,

$$\mathbf{O}_+^\sigma(\lambda) = \mathbf{O}_-^\sigma(\lambda)\sigma_1^{\frac{1-J}{2}} \begin{bmatrix} 0 & i\exp(-iJ\alpha_k/\hbar) \\ i\exp(iJ\alpha_k/\hbar) & 0 \end{bmatrix} \sigma_1^{\frac{1-J}{2}}, \tag{4.118}$$

and for $\lambda \in C_{k\pm}^+$,

$$\mathbf{O}_+^\sigma(\lambda) = \mathbf{O}_-^\sigma(\lambda)\sigma_1^{\frac{1-J}{2}} \begin{bmatrix} 1 & -i\exp(-iJ\alpha_k/\hbar)\exp(\mp ir_k(\lambda)/\hbar) \\ 0 & 1 \end{bmatrix} \sigma_1^{\frac{1-J}{2}}. \tag{4.119}$$

The constants α_k and θ_{k+1} are related to the functions $\tilde{\phi}^\sigma(\lambda)$ and $r_k(\lambda)$ (recall that the latter is the analytic continuation of $\theta^\sigma(\lambda)$ off of I_k^+ according to Definition 4.2.4) by

$$r_k(\lambda_{2k}) = J\theta_{k+1}, \qquad \tilde{\phi}^\sigma(\lambda_{2k}) = iJ\alpha_k. \tag{4.120}$$

Recall also that by the conditions imposed in §4.2 on the complex phase function $g^\sigma(\lambda)$ via its density function $\rho^\sigma(\lambda)$ (cf. Definition 4.2.5), we have that $\Re(\tilde{\phi}^\sigma(\lambda)) < 0$ for $\lambda \in \Gamma_{k+1}^+ \backslash \{\lambda_{2k}\}$ and similarly that $\Re(-ir_k(\lambda)) < 0$ for $\lambda \in C_{k+}^+ \backslash \{\lambda_{2k}\}$ and

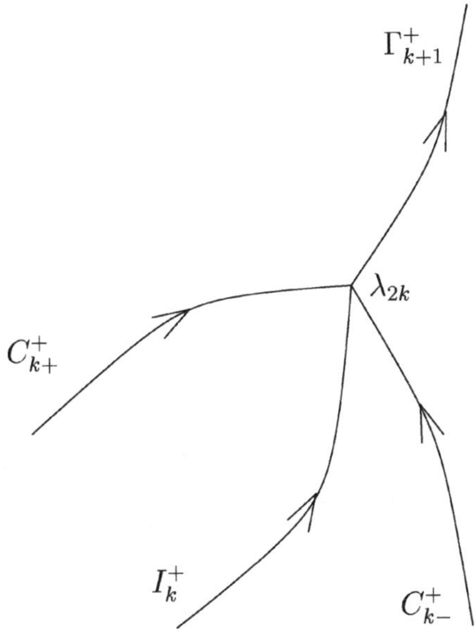

Figure 4.7 The jump matrix near λ_{2k} differs from the identity on a self-intersecting contour with λ_{2k} at the intersection point.

$\Re(ir_k(\lambda)) < 0$ for $\lambda \in C^+_{k-}\backslash\{\lambda_{2k}\}$. Also, the working assumption (4.116) that $\rho^\sigma(\lambda)$ vanishes like a square root at $\lambda = \lambda_{2k}$ implies that the function $r_k(\lambda)$ differs from $J\theta_{k+1}$ by a quantity that vanishes like $(\lambda - \lambda_{2k})^{3/2}$. In fact, the analytic continuation formulae (4.32) and (4.33) imply that for $\lambda \in C^+_{k+}$ we have

$$\tilde{\phi}^\sigma(\lambda) - iJ\alpha_k = i(r_k(\lambda) - J\theta_{k+1}), \tag{4.121}$$

where $r_k(\lambda)$ is the continuation of $\theta^\sigma(\lambda)$ from I^+_k to the left and where $\tilde{\phi}^\sigma(\lambda)$ is the continuation of the function with the same name from Γ^+_{k+1} to the left. Similarly, for $\lambda \in C^+_{k-}$ we have

$$\tilde{\phi}^\sigma(\lambda) - iJ\alpha_k = -i(r_k(\lambda) - J\theta_{k+1}), \tag{4.122}$$

where here $r_k(\lambda)$ is the continuation of $\theta^\sigma(\lambda)$ from I^+_k to the right and $\tilde{\phi}^\sigma(\lambda)$ is the continuation of the function with the same name from Γ^+_{k+1} to the right.

These facts suggest a local change of variables. Let $\zeta = \zeta(\lambda)$ be defined by

$$\zeta(\lambda) := \left(\frac{r_k(\lambda) - J\theta_{k+1}}{\hbar}\right)^{2/3}, \tag{4.123}$$

and note that $\zeta \in \mathbb{R}_+$ when $\lambda \in I^+_k$. Since by assumption $r_k(\lambda)$ approaches its value at λ_{2k} like $(\lambda - \lambda_{2k})^{3/2}$, this change of variables is an *invertible analytic map* of some sufficiently small (but with size independent of $\hbar$) neighborhood of λ_k containing no other endpoints into the ζ-plane. In terms of this change of variables, the discussion preceding (4.32) and (4.33) implies that for $\lambda \in \Gamma^+_{k+1}$,

$$\frac{\tilde{\phi}^\sigma(\lambda) - iJ\alpha_k}{\hbar} = -(-\zeta)^{3/2}. \tag{4.124}$$

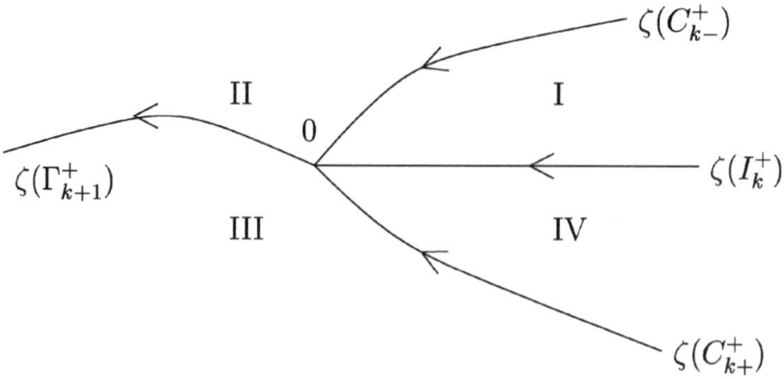

Figure 4.8 The image of the local jump contours in the ζ-plane. The point $\lambda = \lambda_{2k}$ is mapped to $\zeta = 0$, and the contour I_k^+ is mapped to the positive real ζ-axis.

The transformation $\lambda \mapsto \zeta(\lambda)$ maps the local contour diagram shown in figure 4.7 into the ζ-plane as shown in figure 4.8. We now center a disk D_{2k} in the λ-plane at $\lambda = \lambda_{2k}$, and we choose the radius of the disk to be sufficiently small so that D_{2k} contains no other endpoints and so that the map $\zeta(\lambda)$ is a biholomorphic map of D_{2k} to the ζ-plane. We consider this radius to be independent of $\hbar$. We also exploit the fact that, as remarked in §4.2, the contours Γ_{k+1}^+ and $C_{k\pm}^+$ are not specifically determined, to *choose* them within the disk D_{2k} (taken here to be sufficiently small independent of $\hbar$) so that $\zeta(\Gamma_{k+1}^+ \cap D_{2k})$ lies on the negative real ζ-axis and $\zeta(C_{k\pm}^+ \cap D_{2k})$ lies on the straight ray on which $\arg(\zeta) = \mp\pi/3$. This choice straightens out the contours shown in figure 4.8. The image $\zeta(D_{2k})$ is a domain containing $\zeta = 0$ and *expanding* as $\hbar$ tends to zero.

For expressing the exact jump conditions of the matrix $\mathbf{O}^\sigma(\lambda)$ in terms of the new variable ζ, it is convenient to introduce a matrix $\mathbf{S}_{2k}(\lambda)$ defined by

$$\mathbf{S}_{2k}(\zeta) := \begin{cases} \mathbf{O}^\sigma(\lambda(\zeta))\sigma_1^{\frac{1-J}{2}} \\ \qquad \times \exp(iJ\sigma_3(\theta_{k+1} - \alpha_k)/(2\hbar)), & \zeta \in \mathrm{I} \cup \mathrm{II}, \\ \mathbf{O}^\sigma(\lambda(\zeta))\sigma_1^{\frac{1-J}{2}} \\ \qquad \times \exp(-iJ\sigma_3(\theta_{k+1} + \alpha_k)/(2\hbar)), & \zeta \in \mathrm{III} \cup \mathrm{IV}. \end{cases} \tag{4.125}$$

Then, the exact jump relations for $\mathbf{O}^\sigma(\lambda)$ become quite simple. For $\zeta \in \zeta(\Gamma_{k+1}^+)$,

$$\mathbf{S}_{2k+}(\zeta) = \mathbf{S}_{2k-}(\zeta) \begin{bmatrix} 1 & 0 \\ i\exp(-(-\zeta)^{3/2}) & 1 \end{bmatrix}, \tag{4.126}$$

for $\zeta \in \zeta(I_k^+)$,

$$\mathbf{S}_{2k+}(\zeta) = \mathbf{S}_{2k-}(\zeta) \cdot i\sigma_1, \tag{4.127}$$

and for $\zeta \in \zeta(C_{k\pm}^+)$,

$$\mathbf{S}_{2k+}(\zeta) = \mathbf{S}_{2k-}(\zeta) \begin{bmatrix} 1 & -i\exp(\mp i\zeta^{3/2}) \\ 0 & 1 \end{bmatrix}. \tag{4.128}$$

We want to view this as a Riemann-Hilbert problem to be solved exactly in $\zeta(D_{2k})$, but to pose this problem correctly we need to include auxiliary conditions to ensure that the local solution matches well onto that of the outer model problem.

In order to compare with the solution $\widetilde{\mathbf{O}}(\lambda)$ of the outer model problem obtained in §4.3, we may introduce an analogous local representation of $\widetilde{\mathbf{O}}(\lambda)$ in terms of the variable ζ. Define the matrix $\widetilde{\mathbf{S}}_{2k}(\zeta)$ by

$$\widetilde{\mathbf{S}}_{2k}(\zeta) := \begin{cases} \widetilde{\mathbf{O}}(\lambda(\zeta))\sigma_1^{\frac{1-J}{2}} \exp(iJ\sigma_3(\theta_{k+1} - \alpha_k)/(2\hbar)), & \zeta \in \mathrm{I} \cup \mathrm{II}, \\ \widetilde{\mathbf{O}}(\lambda(\zeta))\sigma_1^{\frac{1-J}{2}} \exp(-iJ\sigma_3(\theta_{k+1} + \alpha_k)/(2\hbar)), & \zeta \in \mathrm{III} \cup \mathrm{IV}. \end{cases} \tag{4.129}$$

Clearly, this matrix is analytic in $\zeta(D_{2k})$ except for positive real ζ, where it has continuous boundary values for $\zeta \neq 0$ that satisfy the jump relation $\widetilde{\mathbf{S}}_{2k+}(\zeta) = \widetilde{\mathbf{S}}_{2k-}(\zeta) \cdot i\sigma_1$.

Lemma 4.4.1 *The matrix $\widetilde{\mathbf{S}}_{2k}(\zeta)$ determined from the solution of the outer model problem has a unique representation*

$$\widetilde{\mathbf{S}}_{2k}(\zeta) = \widetilde{\mathbf{S}}_{2k}^{\mathrm{hol}}(\zeta)\widetilde{\mathbf{S}}^{\mathrm{loc,\,even}}(\zeta), \tag{4.130}$$

where

$$\widetilde{\mathbf{S}}^{\mathrm{loc,\,even}}(\zeta) := \frac{(-\zeta)^{\sigma_3/4}}{\sqrt{2}} \begin{bmatrix} 1 & 1 \\ -1 & 1 \end{bmatrix} \tag{4.131}$$

and where $\widetilde{\mathbf{S}}_{2k}^{\mathrm{hol}}(\zeta)$ is holomorphic in the interior of $\zeta(D_{2k})$.

Proof. Observe that by direct calculation, the matrix $\widetilde{\mathbf{S}}^{\mathrm{loc,\,even}}(\zeta)$ is analytic for all ζ except on the positive real axis, where it satisfies the jump relation $\widetilde{\mathbf{S}}_{+}^{\mathrm{loc,\,even}}(\zeta) = \widetilde{\mathbf{S}}_{-}^{\mathrm{loc,\,even}}(\zeta) \cdot i\sigma_1$. The factor of $1/\sqrt{2}$ is included in (4.131) so that $\widetilde{\mathbf{S}}^{\mathrm{loc,\,even}}(\zeta)$ transforms nicely under matrix inversion. Since both $\widetilde{\mathbf{S}}_{2k}(\zeta)$ and $\widetilde{\mathbf{S}}^{\mathrm{loc,\,even}}(\zeta)$ have determinant one and have smooth boundary values except at $\zeta = 0$, it follows that the quotient $\widetilde{\mathbf{S}}_{2k}(\zeta)\widetilde{\mathbf{S}}^{\mathrm{loc,\,even}}(\zeta)^{-1}$ is analytic in $\zeta(D_{2k})\setminus\{0\}$. But by construction, the matrix $\widetilde{\mathbf{O}}(\lambda)$ obtained in §4.3 is $O((\lambda - \lambda_{2k})^{-1/4})$, and consequently $\widetilde{\mathbf{S}}(\zeta)$ is $O(|\zeta|^{-1/4})$ at the origin since $\zeta(\lambda)$ is an analytic mapping. Therefore, the quotient is bounded at $\zeta = 0$ and hence is analytic throughout the interior of $\zeta(D_{2k})$. □

The main idea of this result is that the matrix $\widetilde{\mathbf{S}}_{2k}(\zeta)$ has a representation in terms of an analytic piece, which contains all of the complicated global information and is defined only in $\zeta(D_{2k})$, and a local piece, which is actually defined for almost all $\zeta \in \mathbb{C}$ and is of a canonical form, satisfying very simple jump relations. In particular, the local piece $\widetilde{\mathbf{S}}^{\mathrm{loc,\,even}}(\zeta)$ does not depend on $\hbar$, even though $\widetilde{\mathbf{S}}_{2k}(\zeta)$ does. We now seek a similar decomposition of the matrix $\mathbf{S}_{2k}(\zeta)$.

Let Σ_I denote the positive real axis in the ζ-plane, oriented from infinity into the origin, let Σ_Γ denote the negative real axis in the ζ-plane, oriented from the origin to infinity, and finally let $\Sigma^{\pm}$ denote the rays with angles $\arg(\zeta) = \mp\pi/3$, both oriented from infinity to the origin. Denote the union of these contours by Σ^{loc}. See figure 4.9. Consider the following Riemann-Hilbert problem.

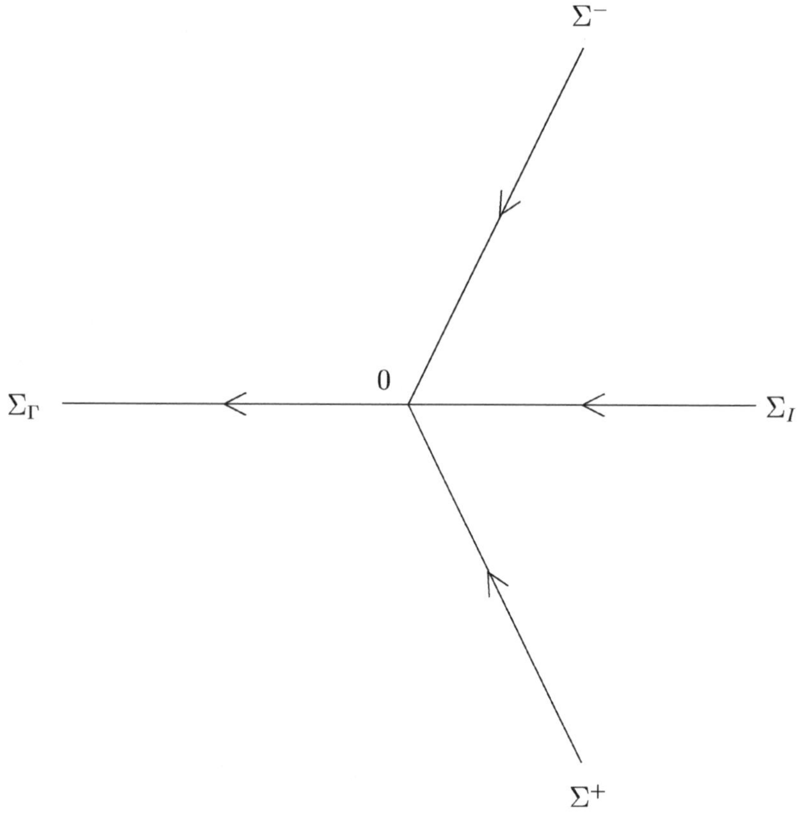

Figure 4.9 The oriented contour Σ^{loc}. All rays extend to $\zeta = \infty$.

RIEMANN-HILBERT PROBLEM 4.4.1 (Local model for even endpoints). *Find a matrix* $\mathbf{S}^{\mathrm{loc,\,even}}(\zeta)$ *satisfying the following:*

1. **Analyticity:** $\mathbf{S}^{\mathrm{loc,\,even}}(\zeta)$ *is analytic for* $\zeta \in \mathbb{C}\backslash\Sigma^{\mathrm{loc}}$.
2. **Boundary behavior:** $\mathbf{S}^{\mathrm{loc,\,even}}(\zeta)$ *assumes continuous boundary values from within each sector of* $\mathbb{C}\backslash\Sigma^{\mathrm{loc}}$, *with continuity holding also at the point of self-intersection.*
3. **Jump conditions:** *The boundary values taken on* Σ^{loc} *satisfy*

$$\mathbf{S}_+^{\mathrm{loc,\,even}}(\zeta)=\mathbf{S}_-^{\mathrm{loc,\,even}}(\zeta)\begin{bmatrix}1 & 0\\ i\exp(-(-\zeta)^{3/2}) & 1\end{bmatrix},\quad \zeta\in\Sigma_\Gamma,$$
$$\mathbf{S}_+^{\mathrm{loc,\,even}}(\zeta)=\mathbf{S}_-^{\mathrm{loc,\,even}}(\zeta)\begin{bmatrix}1 & -i\exp(\mp i\zeta^{3/2})\\ 0 & 1\end{bmatrix},\quad \zeta\in\Sigma^{\pm},\quad (4.132)$$
$$\mathbf{S}_+^{\mathrm{loc,\,even}}(\zeta)=\mathbf{S}_-^{\mathrm{loc,\,even}}(\zeta)\cdot i\sigma_1,\quad \zeta\in\Sigma_I.$$

4. **Normalization:** $\mathbf{S}^{\mathrm{loc,\,even}}(\zeta)$ *is similar to* $\widetilde{\mathbf{S}}^{\mathrm{loc,\,even}}(\zeta)$ *at* $\zeta = \infty$, *where* $\widetilde{\mathbf{S}}^{\mathrm{loc,\,even}}(\zeta)$ *is defined by* (4.131). *Precisely,*

$$\lim_{\zeta\to\infty}\mathbf{S}^{\mathrm{loc,\,even}}(\zeta)\widetilde{\mathbf{S}}^{\mathrm{loc,\,even}}(\zeta)^{-1}=\mathbb{I}\qquad (4.133)$$

with the limit being uniform with respect to direction.

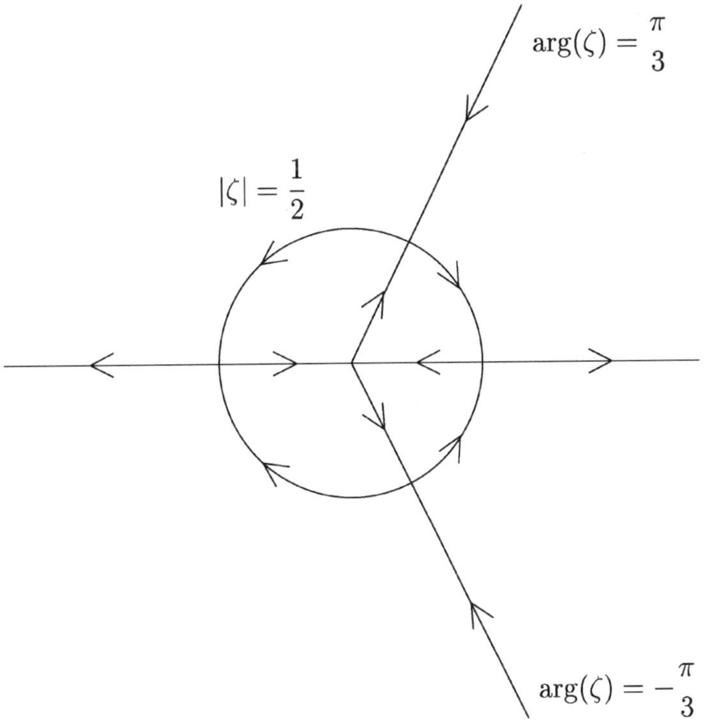

Figure 4.10 The contour Σ_L is the union of a circle of radius $R = 1/2$, the real axis, and the rays $\arg(\zeta) = \pm(\pi/3)$, oriented as shown.

Lemma 4.4.2 *Riemann-Hilbert Problem* 4.4.1 *has a unique solution, with the additional property that there exists a constant* $M > 0$ *such that the estimate*

$$\|\mathbf{S}^{\text{loc, even}}(\zeta)\widetilde{\mathbf{S}}^{\text{loc, even}}(\zeta)^{-1} - \mathbb{I}\| \leq M|\zeta|^{-1} \tag{4.134}$$

holds for all sufficiently large $|\zeta|$. *The solution* $\mathbf{S}^{\text{loc, even}}(\zeta)$ *is universal in the sense that it does not depend on* $\hbar$.

Proof. We first introduce an auxiliary Riemann-Hilbert problem. Let Σ_L be the oriented contour illustrated in figure 4.10. For $\zeta \in \Sigma_L\backslash\{0, 1/2, -1/2, \exp(i\pi/3)/2, \exp(-i\pi/3)/2\}$, we define a jump matrix $\mathbf{v}_L(\zeta)$ as follows. For $0 < |\zeta| < 1/2$, set

$$\mathbf{v}_L(\zeta) := \begin{cases} i\sigma_1, & \arg(\zeta) = 0, \\ \begin{bmatrix} 1 & i\exp(\pm i\zeta^{3/2}) \\ 0 & 1 \end{bmatrix}, & \arg(\zeta) = \pm\pi/3, \\ \begin{bmatrix} 1 & 0 \\ -i\exp(-(-\zeta)^{3/2}) & 1 \end{bmatrix}, & \arg(\zeta) = \pi, \end{cases} \tag{4.135}$$

for $|\zeta| > 1/2$, set

$$\mathbf{v_L}(\zeta) := \begin{cases} \mathbb{I}, & \arg(\zeta) = 0, \\ \widetilde{\mathbf{S}}^{\text{loc, even}}(\zeta) \begin{bmatrix} 1 & -i\exp(\pm i\zeta^{3/2}) \\ 0 & 1 \end{bmatrix} \widetilde{\mathbf{S}}^{\text{loc, even}}(\zeta)^{-1}, & \arg(\zeta) = \pm\pi/3, \\ \widetilde{\mathbf{S}}^{\text{loc, even}}(\zeta) \begin{bmatrix} 1 & 0 \\ i\exp(-(-\zeta)^{3/2}) & 1 \end{bmatrix} \widetilde{\mathbf{S}}^{\text{loc, even}}(\zeta)^{-1}, & \arg(\zeta) = \pi, \end{cases} \tag{4.136}$$

and for $|\zeta| = 1/2$, set

$$\mathbf{v_L}(\zeta) := \begin{cases} \widetilde{\mathbf{S}}^{\text{loc, even}}(\zeta)^{-1}, & 0 < \arg(\zeta) < \pi/3 \text{ and } -\pi < \arg(\zeta) < -\pi/3, \\ \widetilde{\mathbf{S}}^{\text{loc, even}}(\zeta), & \pi/3 < \arg(\zeta) < \pi \text{ and } -\pi/3 < \arg(\zeta) < 0. \end{cases} \tag{4.137}$$

□

Consider the following problem.

Riemann-Hilbert Problem 4.4.2 (Auxiliary local problem). *Find a matrix* $\mathbf{L}(\zeta)$ *satisfying the following:*

1. **Analyticity:** $\mathbf{L}(\zeta)$ *is analytic for* $\zeta \in \mathbb{C} \setminus \Sigma_\mathbf{L}$ *and takes continuous boundary values on* Σ_L*, including self-intersection points;*
2. **Boundary behavior:** $\mathbf{L}(\zeta)$ *takes continuous boundary values from each connected component of* $\mathbb{C} \setminus \Sigma_\mathbf{L}$*, with continuity holding also at corner points corresponding to self-intersections of* $\Sigma_\mathbf{L}$*;*
3. **Jump conditions:** *The boundary values taken on* $\Sigma_\mathbf{L} \setminus$ {self-intersection points} *satisfy*

$$\mathbf{L}_+(\zeta) = \mathbf{L}_-(\zeta)\mathbf{v_L}(\zeta), \tag{4.138}$$

   with the jump matrix $\mathbf{v_L}(\zeta)$ *defined by* (4.135), (4.136), *and* (4.137)*;*
4. **Normalization:** $\mathbf{L}(\zeta)$ *is normalized at infinity:*

$$\mathbf{L}(\zeta) \to \mathbb{I} \quad \text{as } \zeta \to \infty, \tag{4.139}$$

   uniformly with respect to direction.

Observe that the jump matrix $\mathbf{v_L}(\zeta)$ has the following properties:

1. $\mathbf{v_L}(\zeta)$ has determinant one for all $\zeta \in \Sigma_\mathbf{L}$.
2. $\mathbf{v_L}(\zeta)$ is smooth on each open arc and in particular is Lipschitz.
3. At each point ζ_0 of self-intersection of $\Sigma_\mathbf{L}$, let the intersecting arcs be enumerated in counterclockwise order (beginning with any arc) as $\Sigma_\mathbf{L}^{(1)}, \dots, \Sigma_\mathbf{L}^{(n)}$, where n is even. The limits $\mathbf{v}_\mathbf{L}^{(k)} := \lim_{\zeta\to\zeta_0,\, \zeta\in\Sigma_\mathbf{L}^{(k)}} \mathbf{v_L}(\zeta)$ exist and satisfy

$$\mathbf{v}_\mathbf{L}^{(1)}\mathbf{v}_\mathbf{L}^{(2)-1}\mathbf{v}_\mathbf{L}^{(3)} \dots \mathbf{v}_\mathbf{L}^{(n-1)}\mathbf{v}_\mathbf{L}^{(n)-1} = \mathbb{I}. \tag{4.140}$$

4. $\mathbf{v_L}(\zeta) - \mathbb{I} = O(|\zeta|^{-1})$ as $|\zeta| \to \infty$. In fact, the decay is exponentially fast in $|\zeta|$.

The first two conditions are obvious. Checking the third condition is a direct computation that we omit, and the fourth condition follows from the fact that the $|\zeta|^{1/4}$ growth of the conjugating factors $\tilde{\mathbf{S}}^{\text{loc, even}}(\zeta)$ and $\tilde{\mathbf{S}}^{\text{loc, even}}(\zeta)^{-1}$ is controlled easily by the exponential decay of $\exp(\pm i\zeta^{3/2})$ for $\arg(\zeta) = \pm\pi/3$ and of $\exp(-(-\zeta)^{3/2})$ for $\arg(\zeta) = \pi$.

It then follows from Theorem A.1.1 proved in appendix A that there will exist a unique solution $\mathbf{L}(\zeta)$ of Riemann-Hilbert Problem 4.4.2 that additionally satisfies the following conditions:

1. $\mathbf{L}(\zeta)$ is uniformly bounded and satisfies $\|\mathbf{L}(\zeta) - \mathbb{I}\| = O(|\zeta|^{-\mu})$ as $|\zeta| \to \infty$ for all $\mu < 1$, and
2. the boundary values $\mathbf{L}_{\pm}(\zeta)$ taken on each component of $\mathbb{C}\backslash\Sigma_{\mathbf{L}}$ are Hölder-continuous for all exponents strictly less than 1,

if and only if the corresponding homogeneous Riemann-Hilbert problem has only the trivial solution, that is, the Fredholm alternative applies. If there exists a solution $\mathbf{L}(\zeta)$, then define $\mathbf{S}^{\text{loc, even}}(\zeta)$ by

$$\mathbf{S}^{\text{loc, even}}(\zeta) := \mathbf{L}(\zeta) \cdot \begin{cases} \mathbb{I}, & |\zeta| < 1/2, \\ \tilde{\mathbf{S}}^{\text{loc, even}}(\zeta), & |\zeta| > 1/2. \end{cases} \tag{4.141}$$

The function so defined has a holomorphic extension through the circle $|\zeta| = 1/2$, since it takes boundary values there from both sides that are continuous and equal. It is easy to check that it solves Riemann-Hilbert Problem 4.4.1. Uniqueness follows from the analogous property of $\mathbf{L}(\zeta)$.

So, we must now show that such a matrix $\mathbf{L}(\zeta)$ exists by proving that all solutions of the homogeneous problem are trivial. Let us define this homogeneous problem.

RIEMANN-HILBERT PROBLEM 4.4.3 (Homogeneous auxiliary local problem). *Let $\mu \in (0, 1)$ be given. Find a matrix $\mathbf{L}_0(\zeta)$ satisfying the following:*

1. **Analyticity:** $\mathbf{L}_0(\zeta)$ *is analytic for* $\zeta \in \mathbb{C}\backslash\Sigma_{\mathbf{L}}$*;*
2. **Boundary behavior:** $\mathbf{L}_0(\zeta)$ *takes boundary values from each connected component of its domain of analyticity that are Hölder-continuous with exponent* μ*, including at self-intersection (corner) points;*
3. **Jump conditions:** *The boundary values* $\mathbf{L}_{0\pm}(\zeta)$ *that* $\mathbf{L}_0(\zeta)$ *assumes on any smooth oriented component of* $\Sigma_{\mathbf{L}}\backslash\{\text{self-intersection points}\}$ *satisfy*
$$\mathbf{L}_{0+}(\zeta) = \mathbf{L}_{0-}(\zeta)\mathbf{v}_{\mathbf{L}}(\zeta); \tag{4.142}$$
4. **Homogeneous normalization:** *The matrix function* $\mathbf{L}_0(\zeta)$ *vanishes for large* ζ*, satisfying the precise estimate*
$$\|\mathbf{L}_0(\zeta)\| \le M|\zeta|^{-\mu}, \tag{4.143}$$
holding for some $M > 0$ *and all sufficiently large* $|\zeta|$.

Thus, a solution of the homogeneous Riemann-Hilbert Problem 4.4.1 is similar to $\mathbf{L}(\zeta)$ but vanishes for large ζ. The identity matrix in the normalization condition for $\mathbf{L}(\zeta)$ is replaced with the zero matrix. Note that, according to the discussion following the statement of Theorem A.1.1 in appendix A, it suffices to find a $\mu_0 < 1$ such that all nontrivial solutions of the homogeneous Riemann-Hilbert Problem 4.4.1 with exponents $\mu > \mu_0$ can be ruled out. Unfortunately, the jump matrix $\mathbf{v_L}(\zeta)$ lacks the symmetry needed to apply the general theory described in appendix A, so we must construct a specific argument. We suppose that the Hölder exponent satisfies $\mu > \mu_0 = 3/4$. Let $\mathbf{L}_0(\zeta)$ be a corresponding solution of the homogeneous Riemann-Hilbert Problem 4.4.1. First, set

$$\mathbf{S}_0^{\text{loc, even}}(\zeta) := \mathbf{L}_0(\zeta) \cdot \begin{cases} \mathbb{I}, & |\zeta| < 1/2, \\ \widetilde{\mathbf{S}}^{\text{loc, even}}(\zeta), & |\zeta| > 1/2. \end{cases} \tag{4.144}$$

This matrix is analytic in each sector for $|\zeta| = 1/2$, and on the real axis and the rays $\arg(\zeta) = \pm\pi/3$ it satisfies the same jump conditions as $\mathbf{S}^{\text{loc, even}}(\zeta)$. As $\zeta \to \infty$, we have for some $M > 0$ the estimate

$$\|\mathbf{S}_0^{\text{loc, even}}(\zeta)\| \le \|\mathbf{L}_0(\zeta)\| \cdot \|\widetilde{\mathbf{S}}^{\text{loc, even}}(\zeta)\| \le M|\zeta|^{1/4-\mu}. \tag{4.145}$$

Next, set

$$\mathbf{A}(\zeta) = \mathbf{S}_0^{\text{loc, even}}(\zeta) \cdot \begin{cases} \mathbb{I}, & -\pi < \arg(\zeta) < -\pi/3, \\ \sigma_1, & \pi/3 < \arg(\zeta) < \pi, \\ \begin{bmatrix} 1 & i\exp(i\zeta^{3/2}) \\ 0 & 1 \end{bmatrix} \sigma_1, & 0 < \arg(\zeta) < \pi/3, \\ \begin{bmatrix} 1 & -i\exp(-i\zeta^{3/2}) \\ 0 & 1 \end{bmatrix}, & -\pi/3 < \arg(\zeta) < 0. \end{cases} \tag{4.146}$$

Since the matrices multiplying $\mathbf{S}_0^{\text{loc, even}}(\zeta)$ above are uniformly bounded, $\mathbf{A}(\zeta)$ retains the decay properties of $\mathbf{S}_0^{\text{loc, even}}(\zeta)$. Also, $\mathbf{A}(\zeta)$ is analytic for $\zeta \in \mathbb{C}\backslash\mathbb{R}$. On the real axis, oriented from right to left, there is the jump condition

$$\mathbf{A}_+(\zeta) = \mathbf{A}_-(\zeta) \begin{cases} \begin{bmatrix} i\exp(-(-\zeta)^{3/2}) & 1 \\ 1 & 0 \end{bmatrix}, & \zeta \in \mathbb{R}_-, \\ \begin{bmatrix} i & \exp(-i\zeta^{3/2}) \\ \exp(i\zeta^{3/2}) & 0 \end{bmatrix}, & \zeta \in \mathbb{R}_+. \end{cases} \tag{4.147}$$

Now, the matrix function $\mathbf{Q}(\zeta) := \mathbf{A}(\zeta^*)^\dagger \mathbf{A}(\zeta)$ is also analytic for $\zeta \in \mathbb{C}\backslash\mathbb{R}$, and since $\|\mathbf{A}(\zeta)\| = O(|\zeta|^{1/4-\mu})$ for large $|\zeta|$, we can apply Cauchy's theorem for all $\mu > \mu_0 = 3/4$ to deduce that

$$i\int_{-\infty}^{\infty} \mathbf{Q}_+(\zeta)\, d\zeta = \mathbf{0}. \tag{4.148}$$

Since for ζ real, $\mathbf{Q}_+(\zeta) = \mathbf{A}_-(\zeta)^\dagger \mathbf{A}_+(\zeta)$, (4.148) becomes, using the relations (4.147),

$$\int_{-\infty}^{0} \mathbf{A}_-(\zeta)^\dagger \mathbf{A}_-(\zeta) \begin{bmatrix} -\exp(-(-\zeta)^{3/2}) & i \\ i & 0 \end{bmatrix} d\zeta + \int_0^{\infty} \mathbf{A}_-(\zeta)^\dagger \mathbf{A}_-(\zeta) \begin{bmatrix} -1 & i\exp(-i\zeta^{3/2}) \\ i\exp(i\zeta^{3/2}) & 0 \end{bmatrix} d\zeta = \mathbf{0}. \tag{4.149}$$

Adding this equation to its conjugate transpose and looking at the (1, 1) entry of the resulting matrix equation, one finds

$$\int_{-\infty}^{0} \|\mathbf{A}_-^{(1)}(\zeta)\|_2^2 \exp(-(-\zeta)^{3/2})\, d\zeta + \int_0^{\infty} \|\mathbf{A}_-^{(1)}(\zeta)\|_2^2\, d\zeta = 0, \tag{4.150}$$

where $\mathbf{A}^{(k)}(\zeta)$ is the kth column of $\mathbf{A}(\zeta)$ and consequently $\mathbf{A}^{(1)}(\zeta) \equiv \mathbf{0}$ for $\Im(\zeta) \geq 0$. From (4.147), it then follows immediately that $\mathbf{A}^{(2)}(\zeta) \equiv \mathbf{0}$ for $\Im(\zeta) < 0$. The jump relations (4.147) then relate the boundary values of the remaining, possibly nonzero, entries of $\mathbf{A}(\zeta)$ by

$$\mathbf{A}_+^{(1)}(\zeta) = \begin{cases} \mathbf{A}_-^{(2)}, & \zeta \in \mathbb{R}_-, \\ \exp(i\zeta^{3/2})\mathbf{A}_-^{(2)}(\zeta), & \zeta \in \mathbb{R}_+. \end{cases} \tag{4.151}$$

So, defining scalar functions $a_k(\zeta)$ for $\zeta \in \mathbb{C} \setminus \mathbb{R}$ by

$$a_k(\zeta) := \begin{cases} \mathbf{A}_k^{(1)}(\zeta), & \Im(\zeta) < 0, \\ \mathbf{A}_k^{(2)}(\zeta), & \Im(\zeta) > 0, \end{cases} \tag{4.152}$$

we see that both functions are analytic for $\zeta \in \mathbb{C} \setminus \mathbb{R}_+$, both are $O(|\zeta|^{1/4-\mu})$ for large $|\zeta|$, and both take continuous boundary values on $\mathbb{R}_+$, where they satisfy

$$a_{k+}(\zeta) = \exp(i\zeta^{3/2})a_{k-}(\zeta), \tag{4.153}$$

with the ray considered oriented from infinity to the origin.

We now show that necessarily $a_k(\zeta) \equiv 0$. Given $a(\zeta) := a_k(\zeta)$ satisfying the above properties, define a scalar function $b(\zeta)$ that is analytic in the *extended* plane $-\pi/3 < \arg(\zeta) < 2\pi + \pi/3$ by setting

$$b(\zeta) := \begin{cases} a(\zeta), & 0 \leq \arg(\zeta) \leq 2\pi, \\ a([\zeta]) \exp(i[\zeta]^{3/2}), & 2\pi \leq \arg(\zeta) \leq 2\pi + \pi/3, \\ & \quad [\zeta] := |\zeta| \exp(i(\arg(\zeta) - 2\pi)), \\ a([\zeta]) \exp(-i[\zeta]^{3/2}), & -\pi/3 \leq \arg(\zeta) \leq 0, \\ & \quad [\zeta] := |\zeta| \exp(i(\arg(\zeta) + 2\pi)). \end{cases} \tag{4.154}$$

We are using the notation $[\zeta]$ for the class representative of ζ with $0 < \arg([\zeta]) < 2\pi$. From the jump relation for $a(\zeta)$, it follows that $b(\zeta)$ is analytic for $\arg(\zeta) = 0$ and $\arg(\zeta) = 2\pi$. In the extended plane where $b(\zeta) \not\equiv a(\zeta)$, we have $|b(\zeta)| < |a([\zeta])|$; moreover, from the mere algebraic decay of $a(\zeta)$ for large $|\zeta|$, we find that $b(\zeta)$ decays *exponentially* for large $|\zeta|$ in these regions and in particular on the boundaries $\arg(\zeta) = -\pi/3$ and $\arg(\zeta) = 2\pi + \pi/3$. Finally, define an analytic function of w for $\Re(w) \geq 0$ by

$$c(w) := b(-w^{8/3}). \tag{4.155}$$

From the exponential decay of $b(\zeta)$ for $\arg(\zeta) = -\pi/3$ and $\arg(\zeta) = 2\pi + \pi/3$, it follows that $|c(iy)| \leq M \exp(-y^4) \leq M' \exp(-|y|)$ for all $y \in \mathbb{R}$. Also $c(w)$ is uniformly bounded for all $\Re(w) \geq 0$. Now, we recall *Carlson's theorem* [RS78]:

> Suppose that $f(z)$ is a complex-valued function defined and continuous for $\Re(z) \geq 0$ and analytic for $\Re(z) > 0$. Suppose that $|f(z)| \leq M \exp(A|z|)$ for $\Re(z) \geq 0$ and $|f(iy)| \leq M \exp(-B|y|)$ for all $y \in \mathbb{R}$, where $B > 0$. Then $f(z)$ is identically zero.

Applying this result of complex analysis for $A = 0$ and $B = 1$, we deduce that $c(w) \equiv 0$. This in turn implies that $a_k(\zeta) \equiv 0$ and, in conjunction with our earlier results, that $\mathbf{A}(\zeta) \equiv \mathbf{0}$. Consequently, we find that $\mathbf{L}_0(\zeta) \equiv \mathbf{0}$. Therefore, all solutions of the homogeneous Riemann-Hilbert Problem 4.4.1 with Hölder exponents $\mu > \mu_0 = 3/4$ are trivial, and the required function $\mathbf{L}(\zeta)$ exists by the Fredholm alternative (cf. Theorem A.1.1). Because $\mathbf{L}(\zeta)$ has Hölder-continuous boundary values, the matrix $\mathbf{S}^{\mathrm{loc,\,even}}(\zeta)$ defined by (4.141) is a solution of the Riemann-Hilbert Problem 4.4.1, taking uniformly continuous boundary values on Σ^{loc}.

Finally, we notice that since the jump matrix $\mathbf{v}_{\mathbf{L}}(\zeta)$ is analytic on each ray of $\Sigma_{\mathbf{L}}$ and decays exponentially to the identity as $\zeta \to \infty$, all of the order-ζ^{-1} moments vanish, and it follows from Theorem A.1.3 that $\|\mathbf{L}(\zeta) - \mathbb{I}\|$ is uniformly of order $|\zeta|^{-1}$ for large ζ. Using the formula (4.141), we see that this in turn implies the decay estimate (4.134), which completes the proof of Lemma 4.4.2.

Remark. While it is sufficient for our purposes to present an argument for the existence of the matrix function $\mathbf{S}^{\mathrm{loc,\,even}}(\zeta)$ based on abstract Fredholm theory as done here, the solution to the Riemann-Hilbert Problem 4.4.1 can even be given *explicitly* in terms of Airy functions. See [D99, DKMVZ99A, DKMVZ99B] for these formulae. In those papers, it was essential to have an explicit accurate description of the local behavior near the endpoint, whereas we require only qualitative properties sufficient to establish ultimately that the explicit approximation afforded by the solution of the outer model Riemann-Hilbert Problem 4.2 is part of a uniformly valid approximation to $\mathbf{N}^{\sigma}(\lambda)$. Indeed, it is only the expansion of $\mathbf{N}^{\sigma}(\lambda)$ for sufficiently large λ that we need to compute asymptotics for the focusing nonlinear Schrödinger equation.

As was the case with the matrix $\widetilde{\mathbf{S}}^{\mathrm{loc,\,even}}(\zeta)$, the matrix $\mathbf{S}^{\mathrm{loc,\,even}}(\zeta)$ solving Riemann-Hilbert Problem 4.4.1 is independent of all parameters $\hbar$, x, and t of our asymptotic analysis. We now propose a factorized representation of an approximation to $\mathbf{S}_{2k}(\zeta)$ by setting for $\zeta \in \zeta(D_{2k})$,

$$\widehat{\mathbf{S}}_{2k}(\zeta) := \widetilde{\mathbf{S}}_{2k}^{\mathrm{hol}}(\zeta)\mathbf{S}^{\mathrm{loc,\,even}}(\zeta). \tag{4.156}$$

This matrix depends on x, t, and $\hbar$ through the holomorphic prefactor. It satisfies *exactly* the same jump relations within $\zeta(D_{2k})$ as does $\mathbf{S}_{2k}(\zeta)$.

Finally, we use the matrix $\widehat{\mathbf{S}}_{2k}(\zeta)$ to construct a local approximation of $\mathbf{N}^{\sigma}(\lambda)$ valid within D_{2k}. First, we apply to the matrix $\widehat{\mathbf{S}}_{2k}(\zeta)$ the change of variables (4.123) and (4.125) connecting $\mathbf{S}_{2k}(\zeta)$ to $\mathbf{O}^{\sigma}(\lambda)$. This yields a matrix that exactly satisfies the jump relations for $\mathbf{O}^{\sigma}(\lambda)$ and that by construction matches well onto the matrix $\widetilde{\mathbf{O}}(\lambda)$ at the boundary of D_{2k}. To recover the approximation for $\mathbf{N}^{\sigma}(\lambda)$, one multiplies by the explicit triangular factors $\mathbf{D}^{\sigma}(\lambda)$, relating, by definition, the matrices $\mathbf{N}^{\sigma}(\lambda)$ and $\mathbf{O}^{\sigma}(\lambda)$ in the lens halves, when $\zeta(\lambda)$ is in regions I and IV of the ζ-plane. Thus, for $\lambda \in D_{2k}$ the local approximation is defined as follows. For $\zeta(\lambda)$ in region I of the ζ-plane,

$$\begin{aligned}\widehat{\mathbf{N}}_{2k}^{\sigma}(\lambda) :=& \widetilde{\mathbf{S}}_{2k}^{\mathrm{hol}}(\zeta(\lambda))\mathbf{S}^{\mathrm{loc,\,even}}(\zeta(\lambda))\exp(-iJ\sigma_3(\theta_{k+1} - \alpha_k)/(2\hbar)) \\ &\times \begin{bmatrix} 1 & i\exp(-iJ\alpha_k/\hbar)\exp(ir_k(\lambda)/\hbar) \\ 0 & 1 \end{bmatrix} \sigma_1^{\frac{1-J}{2}},\end{aligned} \tag{4.157}$$

for $\zeta(\lambda)$ in region II of the ζ-plane,

$$\widehat{\mathbf{N}}^{\sigma}_{2k}(\lambda) := \widetilde{\mathbf{S}}^{\text{hol}}_{2k}(\zeta(\lambda))\mathbf{S}^{\text{loc, even}}(\zeta(\lambda))\exp(-iJ\sigma_3(\theta_{k+1} - \alpha_k)/(2\hbar))\sigma_1^{\frac{1-J}{2}}, \tag{4.158}$$

for $\zeta(\lambda)$ in region III of the ζ-plane,

$$\widehat{\mathbf{N}}^{\sigma}_{2k}(\lambda) := \widetilde{\mathbf{S}}^{\text{hol}}_{2k}(\zeta(\lambda))\mathbf{S}^{\text{loc, even}}(\zeta(\lambda))\exp(iJ\sigma_3(\theta_{k+1} + \alpha_k)/(2\hbar))\sigma_1^{\frac{1-J}{2}}, \tag{4.159}$$

and for $\zeta(\lambda)$ in region IV of the ζ-plane,

$$\widehat{\mathbf{N}}^{\sigma}_{2k}(\lambda) := \widetilde{\mathbf{S}}^{\text{hol}}_{2k}(\zeta(\lambda))\mathbf{S}^{\text{loc, even}}(\zeta(\lambda))\exp(iJ\sigma_3(\theta_{k+1} + \alpha_k)/(2\hbar)) \times \begin{bmatrix} 1 & -i\exp(-iJ\alpha_k/\hbar)\exp(-ir_k(\lambda)/\hbar) \\ 0 & 1 \end{bmatrix}\sigma_1^{\frac{1-J}{2}}. \tag{4.160}$$

Here the function $r_k(\lambda)$ is defined in Definition 4.2.4.

We finish our local analysis near the endpoint λ_{2k} by recording several crucial properties of this local approximation.

Lemma 4.4.3 *The local approximation $\widehat{\mathbf{N}}^{\sigma}_{2k}(\lambda)$ is analytic for $\lambda \in D_{2k}\setminus(D_{2k} \cap C)$ and takes continuous boundary values on C_σ which satisfy exactly $\widehat{\mathbf{N}}^{\sigma}_{2k+}(\lambda) = \widehat{\mathbf{N}}^{\sigma}_{2k-}(\lambda)\mathbf{v}^{\sigma}_{\widetilde{\mathbf{N}}}(\lambda)$.*

Proof. This is an elementary consequence of the fact that for all $\zeta \in \zeta(D_{2k})$, $\widehat{\mathbf{S}}_{2k}(\zeta)$ satisfies the exact same jump relations as $\mathbf{S}_{2k}(\zeta)$. □

Lemma 4.4.4 *There exists some $M > 0$ such that for all $\lambda \in D_{2k}$ and all sufficiently small $\hbar$,*

$$\|\widehat{\mathbf{N}}^{\sigma}_{2k}(\lambda)\| \le M\hbar^{-1/3}. \tag{4.161}$$

The same estimate holds for the inverse, since the local approximation has determinant one.

Proof. Recall the exact representation (4.130) of the function $\widetilde{\mathbf{S}}_{2k}(\zeta)$ related to the outer solution $\widetilde{\mathbf{O}}(\lambda)$ obtained in §4.3 by a change of variables. It follows from the construction in §4.3 that $\widetilde{\mathbf{S}}_{2k}(\zeta(\lambda))$ blows up like $(\lambda - \lambda_{2k})^{-1/4}$ with a leading coefficient that is uniformly bounded as $\hbar$ tends to zero. Now for $\lambda - \lambda_{2k}$ small, the Taylor expansion for the analytic map $\zeta(\lambda)$ gives $\zeta(\lambda) = M'\hbar^{-2/3}(\lambda-\lambda_{2k})+O((\lambda-\lambda_k)^2)$, where M' is a constant that is bounded as $\hbar$ goes to zero. Approximating $\zeta(\lambda)$ on the right-hand side of (4.130) by such a formula, one sees that the holomorphic prefactor $\widetilde{\mathbf{S}}^{\text{hol}}_{2k}(\zeta(\lambda))$ must be of the form

$$\widetilde{\mathbf{S}}^{\text{hol}}_{2k}(\zeta(\lambda)) = \mathbf{T}(\lambda)\hbar^{-\sigma_3/6}, \tag{4.162}$$

with $\mathbf{T}(\lambda)$ being a matrix analytic in D_{2k} that is uniformly bounded as $\hbar$ tends to zero. Now, since $\mathbf{S}^{\text{loc, even}}(\zeta)$ is bounded only by $|\zeta|^{1/4}$ for large ζ, we get a uniform estimate for all $\lambda \in D_{2k}$ of the form $\|\mathbf{S}^{\text{loc, even}}(\zeta(\lambda))\| = O(\hbar^{-1/6})$ as well. These bounds, along with the definition of $\widehat{\mathbf{N}}^{\sigma}_{2k}(\lambda)$, yield the desired estimate. □

Lemma 4.4.5 *There exists some $M > 0$ such that for all $\lambda \in \partial D_{2k}$ and for all sufficiently small $\hbar$,*

$$\|\widehat{\mathbf{N}}^{\sigma}_{2k}(\lambda)\widehat{\mathbf{N}}^{\sigma}_{\text{out}}(\lambda)^{-1} - \mathbb{I}\| \le M\hbar^{1/3}, \tag{4.163}$$

where $\widehat{\mathbf{N}}^{\sigma}_{\text{out}}(\lambda)$ is defined by (4.109) in §4.3.6.

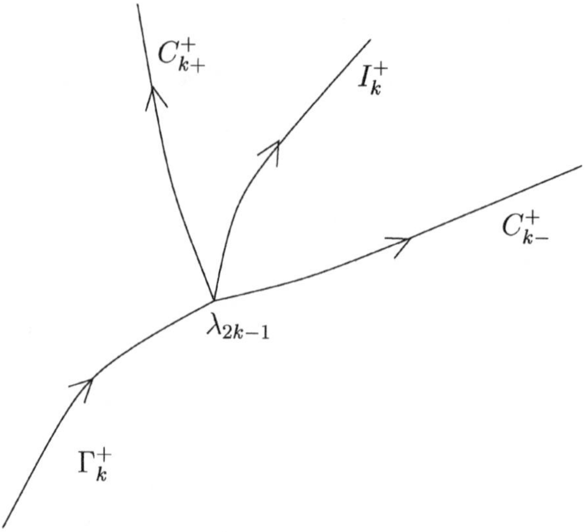

Figure 4.11 The jump matrix for $\mathbf{O}^\sigma(\lambda)$ near λ_{2k-1} differs from the identity on a self-intersecting contour with λ_{2k-1} at the intersection point.

Proof. By definition, we have for all $\lambda \in D_{2k}$,

$$\widehat{\mathbf{N}}_{2k}^\sigma(\lambda)\widehat{\mathbf{N}}_{\text{out}}^\sigma(\lambda)^{-1} = \widetilde{\mathbf{S}}_{2k}^{\text{hol}}(\zeta(\lambda))\mathbf{S}^{\text{loc, even}}(\zeta(\lambda))\widetilde{\mathbf{S}}^{\text{loc, even}}(\zeta(\lambda))^{-1}\widetilde{\mathbf{S}}_{2k}^{\text{hol}}(\zeta(\lambda))^{-1}. \tag{4.164}$$

Now, as $\hbar$ tends to zero, $\zeta(\lambda) \to \infty$ for all $\lambda \in \partial D_{2k}$; in particular $|\zeta| \sim \hbar^{-2/3}$ for all $\lambda \in \partial D_{2k}$. Therefore, directly from the large-ζ asymptotic properties of the matrix $\mathbf{S}^{\text{loc, even}}(\zeta)$ in the estimate (4.134), we have for $\lambda \in \partial D_{2k}$,

$$\widehat{\mathbf{N}}_{2k}^\sigma(\lambda)\widehat{\mathbf{N}}_{\text{out}}^\sigma(\lambda)^{-1} = \mathbb{I} + \widetilde{\mathbf{S}}_{2k}^{\text{hol}}(\zeta(\lambda))[O(|\zeta(\lambda)|^{-1})]\widetilde{\mathbf{S}}_{2k}^{\text{hol}}(\zeta(\lambda))^{-1}. \tag{4.165}$$

From the proof of Lemma 4.4.4, the conjugating factors are each uniformly bounded for $\lambda \in D_{2k}$ by $O(\hbar^{-1/6})$. Using this fact in (4.165) yields the desired bound. □

4.4.2 Local Analysis for λnear the Endpoint λ_{2k-1} for $k = 1, \ldots, G/2$

The analysis near λ_{2k-1} proceeds in a similar manner, beginning with the exact jump relations for the matrix $\mathbf{O}^\sigma(\lambda)$. The local contour structure is illustrated in figure 4.11. The exact jump relations for $\mathbf{O}^\sigma(\lambda)$ in a neighborhood of λ_{2k-1} are for $\lambda \in \Gamma_k^+$,

$$\mathbf{O}_+^\sigma(\lambda) = \mathbf{O}_-^\sigma(\lambda)\sigma_1^{\frac{1-J}{2}}\begin{bmatrix} \exp(iJ\theta_k/\hbar) & 0 \\ i\exp(\tilde{\phi}^{J,\omega}(\lambda)/\hbar) & \exp(-iJ\theta_k/\hbar)\end{bmatrix}\sigma_1^{\frac{1-J}{2}}, \tag{4.166}$$

for $\lambda \in I_k^+$,

$$\mathbf{O}_+^\sigma(\lambda) = \mathbf{O}_-^\sigma(\lambda)\sigma_1^{\frac{1-J}{2}}\begin{bmatrix} 0 & i\exp(-iJ\alpha_k/\hbar) \\ i\exp(iJ\alpha_k/\hbar) & 0\end{bmatrix}\sigma_1^{\frac{1-J}{2}}, \tag{4.167}$$

and for $\lambda \in C_{k\pm}^+$,

$$\mathbf{O}_+^\sigma(\lambda) = \mathbf{O}_-^\sigma(\lambda)\sigma_1^{\frac{1-J}{2}} \begin{bmatrix} 1 & -i\exp(-iJ\alpha_k/\hbar)\exp(\mp i r_k(\lambda)/\hbar) \\ 0 & 1 \end{bmatrix} \sigma_1^{\frac{1-J}{2}}. \quad (4.168)$$

Here, we have the identifications

$$r_k(\lambda_{2k-1}) = J\theta_k, \qquad \tilde{\phi}^\sigma(\lambda_{2k-1}) = iJ\alpha_k. \quad (4.169)$$

Once again, analytic continuation arguments using the complex phase function $g^\sigma(\lambda)$ yield useful relations between the function $\tilde{\phi}^\sigma(\lambda)$ continued from Γ_k^+ and the function $r_k(\lambda)$ continued from I_k^+. One finds that for $\lambda \in C_{k+}^+$,

$$\tilde{\phi}^\sigma(\lambda) - iJ\alpha_k = i(r_k(\lambda) - J\theta_k), \quad (4.170)$$

where $\tilde{\phi}^\sigma(\lambda)$ and $r_k(\lambda)$ are continued, respectively, from Γ_k^+ and I_k^+ to the left, and for $\lambda \in C_{k-}^+$,

$$\tilde{\phi}^\sigma(\lambda) - iJ\alpha_k = -i(r_k(\lambda) - J\theta_k), \quad (4.171)$$

where here $\tilde{\phi}^\sigma(\lambda)$ and $r_k(\lambda)$ are continued, respectively, from Γ_k^+ and I_k^+ to the right.

The appropriate analytic change of variables $\lambda \mapsto \zeta(\lambda)$ suggested by these continuation facts and the degree of vanishing of $r_k(\lambda) - J\theta_k$ at $\lambda = \lambda_{2k-1}$ now is specified by

$$\zeta(\lambda) := \left(-\frac{r_k(\lambda) - J\theta_k}{\hbar}\right)^{2/3}, \quad (4.172)$$

and we note that $\zeta \in \mathbb{R}_+$ when $\lambda \in I_k^+$. It follows from the analytic continuation properties described above (see (4.170) and (4.171)) that

$$\frac{\tilde{\phi}^\sigma(\lambda) - iJ\alpha_k}{\hbar} = -(-\zeta)^{3/2}. \quad (4.173)$$

The transformation $\zeta(\lambda)$ takes the local contours illustrated in figure 4.11 into the ζ-plane as shown in figure 4.12. We fix a disk D_{2k-1} centered at λ_{2k-1} of sufficiently small radius independent of $\hbar$. As before, we choose the contours Γ_k^+, C_{k+}^+, and C_{k-}^+ within D_{2k-1} so that their images in $\zeta(D_{2k-1})$ lie, respectively, on the straight

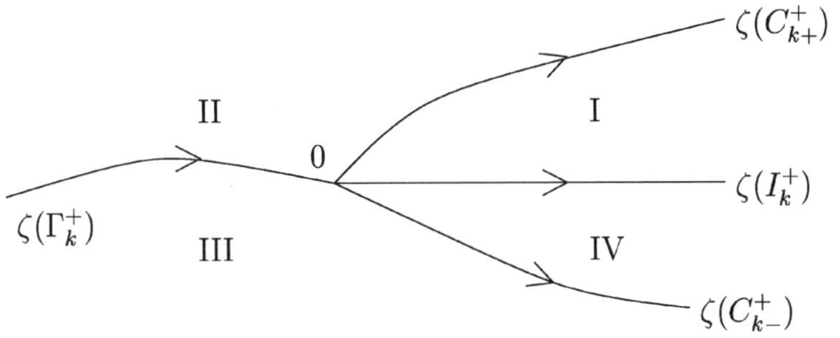

Figure 4.12 The image of the local contours in the ζ-plane. The intersection point is $\zeta = 0$, and the image of I_k^+ is the positive real ζ-axis.

rays $\arg(\zeta) = \pi$, $\arg(\zeta) = \pi/3$, and $\arg(\zeta) = -\pi/3$. These choices straighten out the contours in figure 4.12 within the expanding neighborhood $\zeta(D_{2k-1})$.

We make the change of dependent variable:

$$\mathbf{S}_{2k-1}(\zeta) := \begin{cases} \mathbf{O}^{\sigma}(\lambda(\zeta))\sigma_1^{\frac{1-J}{2}} \exp(iJ\sigma_3(\theta_k - \alpha_k)/(2\hbar)), & \zeta \in \mathrm{I} \cup \mathrm{II}, \\ \mathbf{O}^{\sigma}(\lambda(\zeta))\sigma_1^{\frac{1-J}{2}} \exp(-iJ\sigma_3(\theta_k + \alpha_k)/(2\hbar)), & \zeta \in \mathrm{III} \cup \mathrm{IV}. \end{cases} \tag{4.174}$$

Consequently, the matrix $\mathbf{S}_{2k-1}(\zeta)$ satisfies locally simple jump relations. For $\zeta \in \zeta(\Gamma_k^+)$,

$$\mathbf{S}_{2k-1,+}(\zeta) = \mathbf{S}_{2k-1,-}(\zeta) \begin{bmatrix} 1 & 0 \\ i\exp(-(-\zeta)^{3/2}) & 1 \end{bmatrix}, \tag{4.175}$$

for $\zeta \in \zeta(I_k^+)$,

$$\mathbf{S}_{2k-1,+}(\zeta) = \mathbf{S}_{2k-1,-}(\zeta) \cdot i\sigma_1, \tag{4.176}$$

and for $\zeta \in \zeta(C_{k\pm}^+)$,

$$\mathbf{S}_{2k-1,+}(\zeta) = \mathbf{S}_{2k-1,-}(\zeta) \begin{bmatrix} 1 & -i\exp(\pm i\zeta^{3/2}) \\ 0 & 1 \end{bmatrix}. \tag{4.177}$$

Along with this, we consider the matrix $\widetilde{\mathbf{S}}_{2k-1}(\zeta)$ defined for $\zeta \in \zeta(D_{2k-1})$ in terms of the solution $\widetilde{\mathbf{O}}(\lambda)$, obtained in §4.3, of the outer model problem by

$$\widetilde{\mathbf{S}}_{2k-1}(\zeta) := \begin{cases} \widetilde{\mathbf{O}}(\lambda(\zeta))\sigma_1^{\frac{1-J}{2}} \exp(iJ\sigma_3(\theta_k - \alpha_k)/(2\hbar)), & \zeta \in \mathrm{I} \cup \mathrm{II}, \\ \widetilde{\mathbf{O}}(\lambda(\zeta))\sigma_1^{\frac{1-J}{2}} \exp(-iJ\sigma_3(\theta_k + \alpha_k)/(2\hbar)), & \zeta \in \mathrm{III} \cup \mathrm{IV}. \end{cases} \tag{4.178}$$

As before, this matrix is analytic in $\zeta(D_{2k-1})$ except for $\zeta \in \mathbb{R}_+$, where it takes continuous boundary values for all $\zeta \neq 0$ that satisfy $\widetilde{\mathbf{S}}_{2k-1,+}(\zeta) = \widetilde{\mathbf{S}}_{2k-1,-}(\zeta) \cdot (-i\sigma_1)$. Although the jump relation for $\widetilde{\mathbf{S}}_{2k-1}(\zeta)$ is formally the same as for $\widetilde{\mathbf{S}}_{2k}(\zeta)$ in §4.4.1, one should keep in mind here that according to figure 4.12 the orientation of $\mathbb{R}_+$ has been reversed and is oriented here from the origin to $\zeta = \infty$. As before, one can prove the following decomposition result.

Lemma 4.4.6 *The matrix $\widetilde{\mathbf{S}}_{2k-1}(\zeta)$ determined from the solution of the outer model problem has a unique representation*

$$\widetilde{\mathbf{S}}_{2k-1}(\zeta) = \widetilde{\mathbf{S}}_{2k-1}^{\mathrm{hol}}(\zeta)\widetilde{\mathbf{S}}^{\mathrm{loc,\,odd}}(\zeta), \tag{4.179}$$

where

$$\widetilde{\mathbf{S}}^{\mathrm{loc,\,odd}}(\zeta) := \frac{(-\zeta)^{-\sigma_3/4}}{\sqrt{2}} \begin{bmatrix} 1 & 1 \\ -1 & 1 \end{bmatrix} \tag{4.180}$$

and where $\widetilde{\mathbf{S}}_{2k-1}^{\mathrm{hol}}(\zeta)$ is holomorphic in the interior of $\zeta(D_{2k-1})$.

This exact local representation of the outer model problem solution is thus written in terms of a complicated analytic part and a simple explicit local part that is a function of ζ alone (and is in particular independent of $\hbar$).

We obtain a similar factorization of $\mathbf{S}_{2k-1}(\zeta)$ as follows. Recall the oriented contour Σ^{loc} defined in §4.4.1 and illustrated in figure 4.9. Consider the following Riemann-Hilbert problem.

RIEMANN-HILBERT PROBLEM 4.4.4 (Local model for odd endpoints). *Find a matrix $\mathbf{S}^{\text{loc, odd}}(\zeta)$ satisfying the following:*

1. **Analyticity:** *$\mathbf{S}^{\text{loc, odd}}(\zeta)$ is analytic for $\zeta \in \mathbb{C}\setminus\Sigma^{\text{loc}}$.*
2. **Boundary behavior:** *$\mathbf{S}^{\text{loc, odd}}(\zeta)$ assumes continuous boundary values from within each sector of $\mathbb{C} \setminus \Sigma^{\text{loc}}$, with continuity holding also at the point of self-intersection.*
3. **Jump conditions:** *The boundary values taken on Σ^{loc} satisfy*

$$\begin{aligned}
\mathbf{S}_+^{\text{loc, odd}}(\zeta) &= \mathbf{S}_-^{\text{loc, odd}}(\zeta)\begin{bmatrix} 1 & 0 \\ -i\exp(-(-\zeta)^{3/2}) & 1 \end{bmatrix}, & \zeta \in \Sigma_\Gamma, \\
\mathbf{S}_+^{\text{loc, odd}}(\zeta) &= \mathbf{S}_-^{\text{loc, odd}}(\zeta)\begin{bmatrix} 1 & i\exp(\mp i\zeta^{3/2}) \\ 0 & 1 \end{bmatrix}, & \zeta \in \Sigma^\pm, \\
\mathbf{S}_+^{\text{loc, odd}}(\zeta) &= \mathbf{S}_-^{\text{loc, odd}}(\zeta)(-i\sigma_1), & \zeta \in \Sigma_I.
\end{aligned} \tag{4.181}$$

4. **Normalization:** *$\mathbf{S}^{\text{loc, odd}}(\zeta)$ is similar to $\widetilde{\mathbf{S}}^{\text{loc, odd}}(\zeta)$ at $\zeta = \infty$, where $\widetilde{\mathbf{S}}^{\text{loc, odd}}(\zeta)$ is defined by* (4.180). *Precisely,*

$$\lim_{\zeta\to\infty} \mathbf{S}^{\text{loc, odd}}(\zeta)\widetilde{\mathbf{S}}^{\text{loc, odd}}(\zeta)^{-1} = \mathbb{I}, \tag{4.182}$$

with the limit being uniform with respect to direction.

LEMMA 4.4.7 *Riemann-Hilbert Problem* 4.4.2 *has a unique solution, with the additional property that there exists some $M > 0$ such that the estimate*

$$\|\mathbf{S}^{\text{loc, odd}}(\zeta)\widetilde{\mathbf{S}}^{\text{loc, odd}}(\zeta)^{-1} - \mathbb{I}\| \leq M|\zeta|^{-1} \tag{4.183}$$

holds for all sufficiently large $|\zeta|$. The solution $\mathbf{S}^{\text{loc, odd}}(\zeta)$ is universal in the sense that it does not depend on $\hbar$.

Proof. Rather than repeating similar arguments to those used in the proof of Lemma 4.4.2, we simply use the matrix $\mathbf{S}^{\text{loc, even}}(\zeta)$ whose existence is guaranteed by that same lemma to construct a solution $\mathbf{S}^{\text{loc, odd}}(\zeta)$ of Riemann-Hilbert Problem 4.4.2. For $\zeta \in \mathbb{C} \setminus \Sigma^{\text{loc}}$, set

$$\mathbf{S}^{\text{loc, odd}}(\zeta) := (-i\sigma_1) \cdot \mathbf{S}^{\text{loc, even}}(\zeta) \cdot (i\sigma_3). \tag{4.184}$$

It is a direct matter to check that the jump relations and normalization condition for $\mathbf{S}^{\text{loc, even}}(\zeta)$, along with the smoothness and decay of the boundary values given in Lemma 4.4.2, imply that the matrix thus defined is a solution of Riemann-Hilbert Problem 4.4.2 with the desired properties. Uniqueness follows from the uniform boundedness for finite ζ, continuity of the boundary values, and Liouville's theorem. □

We now propose an approximation to $\mathbf{S}_{2k-1}(\zeta)$ defined for $\zeta \in \zeta(D_{2k-1})$ by

$$\widehat{\mathbf{S}}_{2k-1}(\zeta) := \widetilde{\mathbf{S}}^{\text{hol}}_{2k-1}(\zeta)\mathbf{S}^{\text{loc, odd}}(\zeta). \tag{4.185}$$

When we take into account the fact that the contour Σ^{loc} is the union of $\zeta(I_k^+)$, $\zeta(\Gamma_k^+)$, $\zeta(C_{k+}^+)$, and $\zeta(C_{k-}^+)$ *with the orientation reversed*, we see that $\widehat{\mathbf{S}}_{2k-1}(\zeta)$ satisfies exactly the same jump relations as $\mathbf{S}_{2k-1}(\zeta)$.

As before, we may use this matrix to define a local approximation of $\mathbf{N}^\sigma(\lambda)$ valid for $\lambda \in D_{2k-1}$. We define this approximation as follows. For $\zeta(\lambda)$ in region I, set

$$\widehat{\mathbf{N}}^\sigma_{2k-1}(\lambda) := \widetilde{\mathbf{S}}^{\text{hol}}_{2k-1}(\zeta(\lambda))\mathbf{S}^{\text{loc, odd}}(\zeta(\lambda)) \exp(-iJ\sigma_3(\theta_k - \alpha_k)/(2\hbar)) \times \begin{bmatrix} 1 & -i\exp(-iJ\alpha_k/\hbar)\exp(-ir_k(\lambda)/\hbar) \\ 0 & 1 \end{bmatrix} \sigma_1^{\frac{1-J}{2}}, \quad (4.186)$$

for $\zeta(\lambda)$ in region II, set

$$\widehat{\mathbf{N}}^\sigma_{2k-1}(\lambda) := \widetilde{\mathbf{S}}^{\text{hol}}_{2k-1}(\zeta(\lambda))\mathbf{S}^{\text{loc, odd}}(\zeta(\lambda)) \exp(-iJ\sigma_3(\theta_k - \alpha_k)/(2\hbar))\sigma_1^{\frac{1-J}{2}}, \quad (4.187)$$

for $\zeta(\lambda)$ in region III, set

$$\widehat{\mathbf{N}}^\sigma_{2k-1}(\lambda) := \widetilde{\mathbf{S}}^{\text{hol}}_{2k-1}(\zeta(\lambda))\mathbf{S}^{\text{loc, odd}}(\zeta(\lambda)) \exp(iJ\sigma_3(\theta_k + \alpha_k)/(2\hbar))\sigma_1^{\frac{1-J}{2}}, \quad (4.188)$$

and for $\zeta(\lambda)$ in region IV, set

$$\widehat{\mathbf{N}}^\sigma_{2k-1}(\lambda) := \widetilde{\mathbf{S}}^{\text{hol}}_{2k-1}(\zeta(\lambda))\mathbf{S}^{\text{loc, odd}}(\zeta(\lambda)) \exp(iJ\sigma_3(\theta_k + \alpha_k)/(2\hbar)) \times \begin{bmatrix} 1 & i\exp(-iJ\alpha_k/\hbar)\exp(ir_k(\lambda)/\hbar) \\ 0 & 1 \end{bmatrix} \sigma_1^{\frac{1-J}{2}}. \quad (4.189)$$

As in §4.4.1, we can characterize the local approximation of $\mathbf{N}^\sigma(\lambda)$ near the endpoint λ_{2k-1} by the following results, all of which are proved in exactly the same manner as their analogues for the corresponding approximations valid near λ_{2k}.

Lemma 4.4.8 *The local approximation $\widehat{\mathbf{N}}^\sigma_{2k-1}(\lambda)$ is analytic for $\lambda \in D_{2k-1}\setminus(D_{2k-1}\cap C)$ and takes continuous boundary values on C_σ that satisfy exactly $\widehat{\mathbf{N}}^\sigma_{2k-1,+}(\lambda) = \widehat{\mathbf{N}}^\sigma_{2k-1,-}(\lambda)\mathbf{v}^\sigma_{\tilde{\mathbf{N}}}(\lambda)$.*

Lemma 4.4.9 *There exists some $M > 0$ such that for all $\lambda \in D_{2k-1}$ and all sufficiently small $\hbar$,*

$$\|\widehat{\mathbf{N}}^\sigma_{2k-1}(\lambda)\| \le M\hbar^{-1/3}. \quad (4.190)$$

The same estimate holds for the inverse matrix, since the local approximation has determinant one.

Lemma 4.4.10 *There exists some $M > 0$ such that for all $\lambda \in \partial D_{2k-1}$ and for all sufficiently small $\hbar$,*

$$\left\|\widehat{\mathbf{N}}^\sigma_{2k-1}(\lambda)\widehat{\mathbf{N}}^\sigma_{\text{out}}(\lambda)^{-1} - \mathbb{I}\right\| \le M\hbar^{1/3}. \quad (4.191)$$

4.4.3 Local Analysis for λ near the Origin

Near the origin, the ad hoc replacement of $\mathbf{N}^\sigma(\lambda)$ with the "continuum limit" approximation $\widehat{\mathbf{N}}^\sigma(\lambda)$ breaks down for two reasons. First, at the level of the Riemann-Hilbert problem for the matrix $\mathbf{O}^\sigma(\lambda)$, the function $\theta^\sigma(\lambda)$ is not analytic at $\lambda = 0$ and therefore the origin must lie at the junction of two lenses, one corresponding to the band I_0^+ connecting the origin to λ_0 and the second being

I_0^-, the reflection of I_0^+ in the real axis. Furthermore, the terminal portion of the loop contour C_σ, namely, the gap $\Gamma_{G/2+1}^+$, and its complex conjugate $\Gamma_{G/2+1}^-$ meet at the origin. Although $\Re(\tilde{\phi}^\sigma(\lambda))$ is negative by assumption on the interior of this gap, it always vanishes at the origin, which means that significant errors may be introduced by simply replacing the jump matrix on $\Gamma_{G/2+1}^+$ by the identity on a neighborhood of the origin. The breakdown of the approximations leading to the outer model Riemann-Hilbert Problem 4.2 by these mechanisms is thus similar to the corresponding breakdown near the endpoints $\lambda_0, \ldots, \lambda_G$.

On the other hand, a second mechanism for failure of our formal approximations at the origin is unlike what happens at the nonzero endpoints. There is additional difficulty at the origin entering at the level of the "discrete" (referring to a discrete WKB eigenvalue measure $d\mu_{\hbar_N}^{\mathrm{WKB}}$ in the logarithmic integral) Riemann-Hilbert Problem 4.1 for $\mathbf{N}^\sigma(\lambda)$. Namely, the replacement of the function $\phi^\sigma(\lambda)$ by $\tilde{\phi}^\sigma(\lambda)$ is not valid in any neighborhood of the origin. Here, an additional contribution coming from the function $W(w)$ defined by (3.64) must be included in any uniformly valid approximation.

The situation near the origin is more complicated than near the endpoints $\lambda_0, \ldots,$ λ_G because analytic continuation properties of the functions appearing in the jump matrix do not favor the sort of convenient change of variables that yields a model Riemann-Hilbert problem that does not involve $\hbar$ and yet captures the asymptotic behavior of the solution in a local neighborhood of fixed size independent of $\hbar$. Thus, we are led to work in a shrinking neighborhood of the origin and to introduce a less elegant local change of variables.

The procedure we use is to consider the local error between the matrix $\mathbf{N}^\sigma(\lambda)$ satisfying the phase-conjugated Riemann-Hilbert Problem 4.1 and its outer approximation $\widehat{\mathbf{N}}_{\mathrm{out}}^\sigma(\lambda)$ defined in §4.3 by (4.109). Thus, near the origin set

$$\mathbf{E}^{\sigma,\,\mathrm{loc}}(\lambda) := \mathbf{N}^\sigma(\lambda)\widehat{\mathbf{N}}_{\mathrm{out}}^\sigma(\lambda)^{-1}. \tag{4.192}$$

Near the origin, this matrix is analytic except on the contours shown in figure 4.13. On these contours, we have jump relations of the form

$$\mathbf{E}_+^{\sigma,\,\mathrm{loc}}(\lambda) = \mathbf{E}_-^{\sigma,\,\mathrm{loc}}(\lambda)\big(\widetilde{\mathbf{O}}_-(\lambda)\mathbf{D}_-^\sigma(\lambda)^{-1}\mathbf{v}_{\mathbf{N}}^\sigma(\lambda)\mathbf{D}_+^\sigma(\lambda)\widetilde{\mathbf{O}}_+(\lambda)^{-1}\big). \tag{4.193}$$

Recall that $\widetilde{\mathbf{O}}(\lambda)$ is the matrix obtained as the solution of the outer model Riemann-Hilbert Problem 4.2 in §4.3, and $\mathbf{D}^\sigma(\lambda)$ is defined to be the identity outside all "lenses" while inside the lenses is given by (4.111) and (4.112). Therefore, the boundary values of $\mathbf{D}^\sigma(\lambda)$ are equal for $\lambda \in \Gamma_{G/2+1}^\pm$ and the boundary values of $\widetilde{\mathbf{O}}(\lambda)$ only differ for $\lambda \in I_0^\pm$. Also recall that $\mathbf{v}_{\mathbf{N}}^\sigma(\lambda)$ differs from the identity matrix only for $\lambda \in I_0^\pm$ and $\lambda \in \Gamma_{G/2+1}^\pm$. Finally, the jump matrix in the lower half-plane is determined from that in the upper half-plane by the symmetry that $\mathbf{E}^{\sigma,\mathrm{loc}}(\lambda)$ satisfies by construction: $\mathbf{E}^{\sigma,\mathrm{loc}}(\lambda^*) = \sigma_2\mathbf{E}^{\sigma,\mathrm{loc}}(\lambda)^*\sigma_2$.

Let $U_\hbar$ be a sufficiently small disk neighborhood of the origin, whose radius we will specify later. For $\lambda \in U_\hbar$, we introduce a change of variables of the form

$$\mathbf{F}^\sigma(\lambda) := \mathbf{C}^\sigma(\lambda)^{-1}\mathbf{E}^{\sigma,\mathrm{loc}}(\lambda)\mathbf{C}^\sigma(\lambda). \tag{4.194}$$

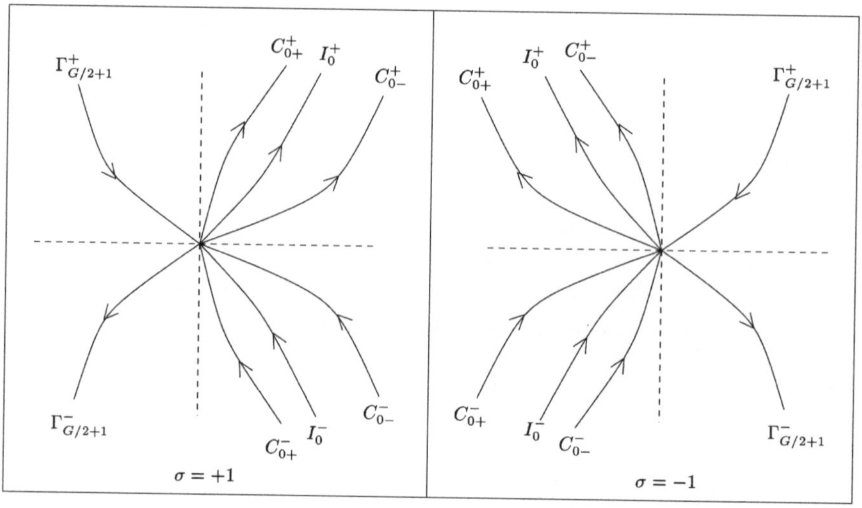

Figure 4.13 The support of the jump matrix for $\mathbf{E}^{\sigma,\mathrm{loc}}(\lambda)$ near $\lambda = 0$. The picture depends on the index σ. On the left, $\sigma = +1$. On the right, $\sigma = -1$. The real and imaginary axes of the λ-plane are shown with dashed lines.

The conjugating factors are specified as follows. Let $U_\hbar^+$ (respectively, $U_\hbar^-$) denote the part of $U_\hbar$ lying to the left (respectively, right) of $I_0^+ \cup I_0^-$. Then we set

$$\mathbf{C}^\sigma(\lambda) := \widetilde{\mathbf{O}}(\lambda)\sigma_1^{\frac{1-J}{2}} \cdot \left\{ \begin{array}{ll} \mathbb{I}, & \lambda \in U_\hbar^\sigma \\ \begin{bmatrix} 0 & i\sigma\exp(-iJ\alpha_0/\hbar) \\ i\sigma\exp(iJ\alpha_0/\hbar) & 0 \end{bmatrix}, & \lambda \in U_\hbar^{-\sigma} \end{array} \right\}$$
$$\times \exp(-i(J\alpha_0 + \sigma r_0(0))\sigma_3/(2\hbar)). \tag{4.195}$$

Recall that $r_0(0)$ and α_0 are purely real constants and that the function $r(\lambda)$ is defined in Definition 4.2.4. This, along with the properties of the solution $\widetilde{\mathbf{O}}(\lambda)$ of the outer model problem developed in §4.3, implies that *the matrices* $\mathbf{C}^\sigma(\lambda)$ *and their inverses are analytic and uniformly bounded in any sufficiently small neighborhood of the origin.*

The exact jump relations for $\mathbf{F}^+(\lambda)$ (by which we mean $\mathbf{F}^\sigma(\lambda)$ for $\sigma = +1$) on the contours in the upper half-plane are as follows. For $\lambda \in \Gamma^+_{G/2+1}$,

$$\mathbf{F}^+_+(\lambda) = \mathbf{F}^+_-(\lambda)\begin{bmatrix} 1 & 0 \\ i\exp(\delta^+/\hbar)\exp((\tilde{\phi}^+(\lambda) - \tilde{\phi}^+(0))/\hbar)(1 - d^+(\lambda)) & 1 \end{bmatrix}, \tag{4.196}$$

for $\lambda \in C^+_{0+}$,

$$\mathbf{F}^+_+(\lambda) = \mathbf{F}^+_-(\lambda)\begin{bmatrix} 1 & -i\exp(-i(r_0(\lambda) - r_0(0))/\hbar) \\ 0 & 1 \end{bmatrix}, \tag{4.197}$$

for $\lambda \in C^+_{0-}$,

$$\mathbf{F}^+_+(\lambda) = \mathbf{F}^+_-(\lambda)\begin{bmatrix} 1 & 0 \\ -i\exp(i(r_0(\lambda) - r_0(0))/\hbar) & 1 \end{bmatrix}, \tag{4.198}$$

and for $\lambda \in I_0^+$,

$$\mathbf{F}_+^+(\lambda) = \mathbf{F}_-^+(\lambda) \begin{bmatrix} 1 - d^+(\lambda) & -i\exp(-i(r_0(\lambda) - r_0(0))/\hbar)d^+(\lambda) \\ -i\exp(i(r_0(\lambda) - r_0(0))/\hbar)d^+(\lambda) & 1 + d^+(\lambda) \end{bmatrix}. \quad (4.199)$$

These are expressed in terms of the quantities

$$d^\sigma(\lambda) := 1 - \exp((\phi^\sigma(\lambda) - \tilde{\phi}^\sigma(\lambda))/\hbar) \quad (4.200)$$

and

$$\delta^\sigma := \tilde{\phi}^\sigma(0) - iJ\alpha_0 - i\sigma r_0(0), \quad (4.201)$$

where by $\tilde{\phi}^\sigma(0)$ we mean the limit as $\lambda \to 0$ in the gap $\Gamma_{G/2+1}^+$ and where we recall that $r_0(\lambda)$ is defined as the analytic continuation of $\theta^\sigma(\lambda)$ from the band I_0^+.

Similarly, the jump relations for $\mathbf{F}^-(\lambda)$ (i.e., for $\mathbf{F}^\sigma(\lambda)$ in the case when $\sigma = -1$) in the upper half-plane are as follows. For $\lambda \in \Gamma_{G/2+1}^+$,

$$\mathbf{F}_+^-(\lambda) = \mathbf{F}_-^-(\lambda) \begin{bmatrix} 1 & 0 \\ i\exp(\delta^-/\hbar)\exp((\tilde{\phi}^-(\lambda) - \tilde{\phi}^-(0))/\hbar)(1 - d^-(\lambda)) & 1 \end{bmatrix}, \quad (4.202)$$

for $\lambda \in C_{0+}^+$,

$$\mathbf{F}_+^-(\lambda) = \mathbf{F}_+^-(\lambda) \begin{bmatrix} 1 & 0 \\ -i\exp(-i(r_0(\lambda) - r_0(0))/\hbar) & 1 \end{bmatrix}, \quad (4.203)$$

for $\lambda \in C_{0-}^+$,

$$\mathbf{F}_+^-(\lambda) = \mathbf{F}_+^-(\lambda) \begin{bmatrix} 1 & -i\exp(i(r_0(\lambda) - r_0(0))/\hbar) \\ 0 & 1 \end{bmatrix}, \quad (4.204)$$

and for $\lambda \in I_0^+$,

$$\mathbf{F}_+^-(\lambda) = \mathbf{F}_+^-(\lambda) \begin{bmatrix} 1 + d^-(\lambda) & -i\exp(i(r_0(\lambda) - r_0(0))/\hbar)d^-(\lambda) \\ -i\exp(-i(r_0(\lambda) - r_0(0))/\hbar)d^-(\lambda) & 1 - d^-(\lambda) \end{bmatrix}. \quad (4.205)$$

In both cases, the jump relations on the corresponding contours in the lower half-plane are obtained by the symmetry $\mathbf{F}^\sigma(\lambda^*) = \sigma_2\mathbf{F}^\sigma(\lambda)^*\sigma_2$.

We now observe a consequence of the fact that we are considering only values of $\hbar$ in the "quantum" sequence $\hbar = \hbar_N$ for $N = 1, 2, 3, \ldots$ (cf. definition (3.9) of $\hbar_N$). Consider first the case $\sigma = +1$. We know that in this case the function $\theta^\sigma(\lambda) - i\tilde{\phi}^\sigma(\lambda)$ is analytic on the bounded interior of the loop C, except on the support of the asymptotic eigenvalue measure $\rho^0(\eta)\,d\eta$, namely, the imaginary interval $[0, iA]$. If we orient this interval from the origin to iA, then we can calculate the explicit jump relation

$$(\theta^\sigma(\lambda) - i\tilde{\phi}^\sigma(\lambda))_+ - (\theta^\sigma(\lambda) - i\tilde{\phi}^\sigma(\lambda))_- = 2\pi \int_\lambda^{iA} \rho^0(\eta)\, d\eta. \tag{4.206}$$

Applying this relation to the limiting values of $\theta^\sigma(\lambda)$ and $\tilde{\phi}^\sigma(\lambda)$ taken as $\lambda \to 0$ along the boundary in either I_0^+ or $\Gamma_{G/2+1}^+$, we take advantage of the fact that throughout I_0^+ we have the identity $\tilde{\phi}^\sigma(\lambda) \equiv iJ\alpha_0$ and throughout $\Gamma_{G/2+1}^+$ we have the identity $\theta^\sigma(\lambda) \equiv 0$. Along with similar reasoning for the case $\sigma = -1$, we finally obtain the formula

$$\delta^\sigma = 2\pi i\sigma \int_0^{iA} \rho^0(\eta)\, d\eta = -2\pi i\sigma N\hbar_N, \tag{4.207}$$

where the second equality follows from the definition (3.9) of the quantum sequence of values of $\hbar$. Consequently, whenever $\hbar = \hbar_N$ for any $N = 1, 2, 3, \ldots$, we conclude that $\exp(\delta^\sigma/\hbar) \equiv 1$.

As in the local analysis near the nonzero endpoints, we again use the freedom of placement of the contours $\Gamma_{G/2+1}^+$ and $C_{0\pm}^+$ to ensure that in some *fixed* disk neighborhood U of the origin, these contours are radial straight lines in the λ-plane with slopes independent of $\hbar$. Let $I_0^{+\prime}$ and $\Gamma_{G/2+1}^{+\prime}$, respectively, denote the tangent lines to I_0^+ and $\Gamma_{G/2+1}^+$ at the origin. Note that the tangent line $\Gamma_{G/2+1}^{+\prime}$ is confined to some sector for the inequality $\Re(\tilde{\phi}^\sigma(\lambda)) < 0$ to be satisfied but is otherwise arbitrary, while the tangent line $I_0^{+\prime}$ is not free, being fixed by the measure reality condition. For concreteness, we choose the contours so that for $\sigma = +1$, $C_{0+}^+ \cap U$ bisects the sector between $I_0^{+\prime}$ at the origin and the positive imaginary axis while $C_{0-}^+ \cap U$ bisects the sector between the positive real axis and $I_0^{+\prime}$ at the origin. For $\sigma = -1$, we arrange that $C_{0+}^+ \cap U$ bisects the sector between the negative real axis and $I_0^{+\prime}$, while $C_{0-}^+ \cap U$ bisects the sector between $I_0^{+\prime}$ and the positive imaginary axis. Let κ denote $\arg(I_0^{+\prime})$ and ξ denote $\arg(\Gamma_{G/2+1}^{+\prime})$. For $\sigma = +1$, we have $0 < \kappa < \pi/2$ and $\pi/2 < \xi < \pi$, while for $\sigma = -1$, we have $0 < \xi < \pi/2$ and $\pi/2 < \kappa < \pi$.

Our strategy is to approximate the exact jump relations for the matrices $\mathbf{F}^\sigma(\lambda)$ in terms of a crude rescaled local variable $\zeta = -i\rho^0(0)\lambda/\hbar$, combining careful asymptotic analysis of $d^\sigma(\lambda)$ with elementary Taylor approximations of $r_0(\lambda) - r_0(0)$ and $\tilde{\phi}^\sigma(\lambda) - \tilde{\phi}^\sigma(0)$. First, note that the definitions of $\phi^\sigma(\lambda)$ and $\tilde{\phi}^\sigma(\lambda)$ imply

$$1 - d^\sigma(\lambda) = \left(\prod_{n=0}^{N-1} \frac{\lambda - \lambda_{\hbar_N,n}^{\mathrm{WKB}*}}{\lambda - \lambda_{\hbar_N,n}^{\mathrm{WKB}}}\right) \exp\left(-\frac{1}{\hbar_N}\left[\int_0^{iA} L_\eta^0(\lambda)\rho^0(\eta)\, d\eta \right.\right.$$
$$\left.\left. + \int_{-iA}^0 L_\eta^0(\lambda)\rho^0(\eta^*)^*\, d\eta\right]\right), \tag{4.208}$$

so that in particular we see that $d^\sigma(\lambda)$ is independent of σ. Recall now Theorem 3.2.1, which gives

$$1 - d^\sigma(\lambda) = \frac{1}{W(-i\zeta)}\left(1 + O(\hbar_N^{1/3})\right) \tag{4.209}$$

uniformly for bounded λ outside any sector including the imaginary axis or equivalently for $\zeta = O(\hbar_N^{-1})$. Now, the function $W(w)$ defined by (3.64) has a cut on the positive real w-axis, which corresponds to the positive imaginary ζ-axis. Thus, we find that

$$1 - d^\sigma(\lambda) = \frac{\Gamma(1/2 + i\zeta)}{\Gamma(1/2 - i\zeta)}(-i\zeta)^{-2i\zeta}\exp(2i\zeta)\big(1 + O\big(\hbar_N^{1/3}\big)\big) \times \begin{cases} \exp(\pi\zeta), & 0 < \arg(\zeta) < \dfrac{\pi}{2}, \\ \exp(-\pi\zeta), & \dfrac{\pi}{2} < \arg(\zeta) < \pi. \end{cases} \tag{4.210}$$

To compactly express these asymptotics, we define the analytic functions $h^\sigma(\zeta)$ for $\Im(\zeta) \geq 0$ and $\arg(\zeta) \neq \pi/2$ by setting

$$h^\sigma(\zeta) := 1 - \frac{\Gamma(1/2 + i\zeta)}{\Gamma(1/2 - i\zeta)}(-i\zeta)^{-2i\zeta}\exp((2i + \sigma\pi)\zeta). \tag{4.211}$$

These functions are uniformly bounded if ζ is bounded away from a sector containing the positive imaginary axis. From Stirling's formula, we deduce their asymptotic behavior for large ζ in the upper half-plane:

$$h^+(\zeta) = \begin{cases} \dfrac{1}{12i\zeta} + O(|\zeta|^{-2}), & 0 < \arg(\zeta) < \dfrac{\pi}{2}, \\ 1 + O(\exp(2\pi\Re(\zeta))), & \dfrac{\pi}{2} < \arg(\zeta) < \pi, \end{cases}$$
$$h^-(\zeta) = \begin{cases} 1 + O(\exp(-2\pi\Re(\zeta))), & 0 < \arg(\zeta) < \dfrac{\pi}{2}, \\ \dfrac{1}{12i\zeta} + O(|\zeta|^{-2}), & \dfrac{\pi}{2} < \arg(\zeta) < \pi. \end{cases} \tag{4.212}$$

Next, define the constants u and v by

$$\begin{aligned} u &:= \frac{1}{-i\rho^0(0)} \cdot \lim_{\lambda\to 0,\, \lambda\in\Gamma^+_{G/2+1}} \frac{d\tilde{\phi}^\sigma}{d\lambda}(\lambda), \\ v &:= \frac{1}{-i\rho^0(0)} \cdot \lim_{\lambda\to 0,\, \lambda\in I_0^+} \frac{dr_0}{d\lambda}(\lambda) \\ &= \frac{\pi}{-i\rho^0(0)} \lim_{\lambda\to 0,\, \lambda\in I_0^+} \rho^\sigma(\lambda). \end{aligned} \tag{4.213}$$

These constants are of course independent of $\hbar$. For $\lambda \in I_0^{+\prime}$, $v\zeta = -i\rho^0(0)v\lambda/\hbar_N$ is real and negative. Likewise, for $\lambda \in \Gamma^{+\prime}_{G/2+1}$, the real part of the product $u\zeta = -i\rho^0(0)u\lambda/\hbar_N$ is negative. For $\lambda \in \Gamma^+_{G/2+1}$, we therefore have

$$\exp((\tilde{\phi}^\sigma(\lambda) - \tilde{\phi}^\sigma(0))/\hbar_N) = \exp(u\zeta)\exp(O(\lambda^2/\hbar_N)). \tag{4.214}$$

Similarly, for $\lambda \in I_0^+ \cup C_{0+}^+ \cup C_{0-}^+$,

$$\exp(\pm i(r_0(\lambda) - r_0(0))/\hbar_N) = \exp(\pm iv\zeta)\exp(O(\lambda^2/\hbar_N)). \tag{4.215}$$

We are now going to use these results to propose a model Riemann-Hilbert problem for an approximation to $\mathbf{F}^\sigma(\lambda)$. The range of validity of the asymptotic formulae (4.214) and (4.215) places restrictions on the size of the neighborhood $U_\hbar$.

In fact, the radius of $U_\hbar$ must *shrink* as $\hbar \downarrow 0$. If R is the radius $U_\hbar$, we will need to have $R^2/\hbar \ll 1$ in order for the error factors in the Taylor approximations (4.214) and (4.215) to be negligible for $\lambda \in U_\hbar$. On the other hand, to characterize the local behavior in a universal way (so that $\hbar$ enters into the local approximation in the form of a simple scaling), we will need the image in the ζ-plane of the boundary $\partial U_\hbar$ to be *expanding* as $\hbar \downarrow 0$. This requires $R/\hbar \gg 1$. The radius R is therefore asymptotically bounded above and below: $\hbar \ll R \ll \hbar^{1/2}$. Thus, let δ be a number between 1/2 and 1, and fix the size of $U_\hbar$ by setting $R = \hbar^\delta$. We reserve the choice of a particular value of δ for later optimization of our estimates. Note that for all sufficiently small $\hbar$, $U_\hbar$ is contained in the fixed neighborhood U.

By keeping the leading terms of the jump matrices for $\mathbf{F}^\sigma(\lambda)$ in an expansion for ζ held fixed as $\hbar$ tends to zero, we are led to propose a local model Riemann-Hilbert problem. First, we introduce a contour. Let $\Sigma^\sigma_{\widehat{\mathbf{F}}}$ be the oriented contour shown in figure 4.14 for both signs of σ. Next, we define on the contour a jump matrix $\mathbf{v}^\sigma_{\widehat{\mathbf{F}}}(\zeta)$. For $\sigma = +1$ and $\zeta \in \Sigma^+_{\widehat{\mathbf{F}}}$ with $\Im(\zeta) > 0$, set

$$\mathbf{v}^+_{\widehat{\mathbf{F}}}(\zeta) := \begin{cases} \begin{bmatrix} 1 & 0 \\ i(1-h^+(\zeta))\exp((u-2\pi)\zeta) & 1 \end{bmatrix}, & \arg(\zeta) = \xi, \\ \begin{bmatrix} 1 & -i\exp(-iv\zeta) \\ 0 & 1 \end{bmatrix}, & \arg(\zeta) = \kappa/2 + \pi/4, \\ \begin{bmatrix} 1 & 0 \\ -i\exp(iv\zeta) & 1 \end{bmatrix}, & \arg(\zeta) = \kappa/2, \\ \begin{bmatrix} 1-h^+(\zeta) & -ih^+(\zeta)\exp(-iv\zeta) \\ -ih^+(\zeta)\exp(iv\zeta) & 1+h^+(\zeta) \end{bmatrix}, & \arg(\zeta) = \kappa. \end{cases} \tag{4.216}$$

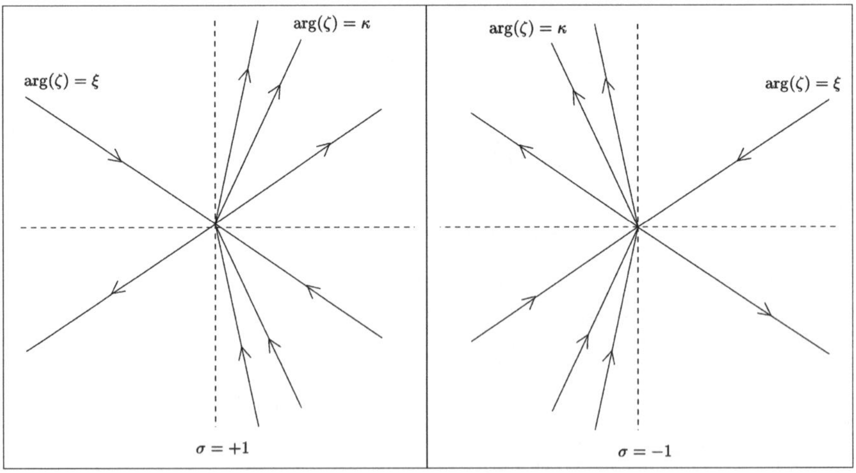

Figure 4.14 The contours $\Sigma^\sigma_{\widehat{\mathbf{F}}}$ for the model Riemann-Hilbert problems near the origin. On the left, $\sigma = +1$. On the right, $\sigma = -1$. In the upper half-plane, the unmarked contours bisect the angles between $\arg(\zeta) = \kappa$ and the boundaries of the quadrant. Although the orientation is as shown, both contours are symmetric with respect to complex conjugation as sets of points. All rays extend to $\zeta = \infty$.

For $\zeta \in \Sigma_{\widehat{\mathbf{F}}}^{+}$ with $\Im(\zeta) < 0$, we set $\mathbf{v}_{\widehat{\mathbf{F}}}^{+}(\zeta) := \sigma_2 \mathbf{v}_{\widehat{\mathbf{F}}}^{+}(\zeta^*)^* \sigma_2$. For the opposite parity, $\sigma = -1$, define for $\zeta \in \Sigma_{\widehat{\mathbf{F}}}^{-}$ and $\Im(\zeta) > 0$,

$$\mathbf{v}_{\widehat{\mathbf{F}}}^{-}(\zeta) := \begin{cases} \begin{bmatrix} 1 & 0 \\ i(1-h^{-}(\zeta))\exp((u+2\pi)\zeta) & 1 \end{bmatrix}, & \arg(\zeta) = \xi, \\ \begin{bmatrix} 1 & 0 \\ -i\exp(-iv\zeta) & 1 \end{bmatrix}, & \arg(\zeta) = \kappa/2 + \pi/2, \\ \begin{bmatrix} 1 & -i\exp(iv\zeta) \\ 0 & 1 \end{bmatrix}, & \arg(\zeta) = \kappa/2 + \pi/4, \\ \begin{bmatrix} 1+h^{-}(\zeta) & -ih^{-}(\zeta)\exp(iv\zeta) \\ -ih^{-}(\zeta)\exp(-iv\zeta) & 1-h^{-}(\zeta) \end{bmatrix}, & \arg(\zeta) = \kappa. \end{cases} \tag{4.217}$$

Again, for $\zeta \in \Sigma_{\widehat{\mathbf{F}}}^{-}$ with $\Im(\zeta) < 0$, we set $\mathbf{v}_{\widehat{\mathbf{F}}}^{-}(\zeta) := \sigma_2 \mathbf{v}_{\widehat{\mathbf{F}}}^{-}(\zeta^*)^* \sigma_2$.

RIEMANN-HILBERT PROBLEM 4.4.5 (Local model for the origin). *Find a matrix function $\widehat{\mathbf{F}}^{\sigma}(\zeta)$ satisfying the following:*

1. **Analyticity:** *$\widehat{\mathbf{F}}^{\sigma}(\zeta)$ is analytic for $\zeta \in \mathbb{C} \setminus \Sigma_{\widehat{\mathbf{F}}}^{\sigma}$;*
2. **Boundary behavior:** *$\widehat{\mathbf{F}}^{\sigma}(\zeta)$ assumes continuous boundary values on $\Sigma_{\widehat{\mathbf{F}}}^{\sigma}$ from each sector of the complement, with continuity also at the self-intersection point;*
3. **Jump condition:** *On the oriented contour $\Sigma_{\widehat{\mathbf{F}}}^{\sigma}$ minus the origin, the boundary values satisfy*
$$\widehat{\mathbf{F}}_{+}^{\sigma}(\zeta) = \widehat{\mathbf{F}}_{-}^{\sigma}(\zeta)\mathbf{v}_{\widehat{\mathbf{F}}}^{\sigma}(\zeta); \tag{4.218}$$
4. **Normalization:** *$\widehat{\mathbf{F}}^{\sigma}(\zeta)$ is normalized at infinity:*
$$\widehat{\mathbf{F}}^{\sigma}(\zeta) \to \mathbb{I} \quad as \quad \zeta \to \infty. \tag{4.219}$$

LEMMA 4.4.11 *Let the parameters u, v, κ, ξ, and σ be fixed. Then Riemann-Hilbert Problem* 4.4.5 *has a unique solution with the additional property that there exists a constant $M > 0$ such that the estimate*

$$\|\widehat{\mathbf{F}}^{\sigma}(\zeta) - \mathbb{I}\| \leq M|\zeta|^{-1} \tag{4.220}$$

holds for all sufficiently large $|\zeta|$. The solution is universal in the sense that it is independent of $\hbar$.

Proof. We begin by introducing an auxiliary Riemann-Hilbert problem. Let $\Sigma_{\mathbf{L}}^{\sigma}$ be the contour illustrated in figure 4.15 for both choices of the parity σ. We define a jump matrix for $\zeta \in \Sigma_{\mathbf{L}}^{\sigma} \setminus \{\text{self-intersection points}\}$ as follows:

$$\mathbf{v}_{\mathbf{L}}^{\sigma}(\zeta) := \begin{cases} \mathbb{I}, & \zeta \in \mathbb{R} \text{ or } |\zeta| = 1/2, \\ \mathbf{v}_{\widehat{\mathbf{F}}}^{\sigma}(\zeta), & \arg(\zeta) = \kappa \text{ and } |\zeta| > 1/2, \\ \mathbf{v}_{\widehat{\mathbf{F}}}^{\sigma}(\zeta)^{-1}, & \arg(\zeta) \neq \kappa \text{ and } |\zeta| > 1/2 \text{ and } \Im(\zeta) > 0, \\ \mathbf{v}_{\widehat{\mathbf{F}}}^{\sigma}(\zeta)^{-1}, & \arg(\zeta) = \kappa \text{ and } |\zeta| < 1/2, \\ \mathbf{v}_{\widehat{\mathbf{F}}}^{\sigma}(\zeta), & \arg(\zeta) \neq \kappa \text{ and } |\zeta| < 1/2 \text{ and } \Im(\zeta) > 0, \\ [\sigma_2 \mathbf{v}_{\mathbf{L}}^{\sigma}(\zeta^*)^* \sigma_2]^{-1}, & \Im(\zeta) < 0. \end{cases} \tag{4.221}$$

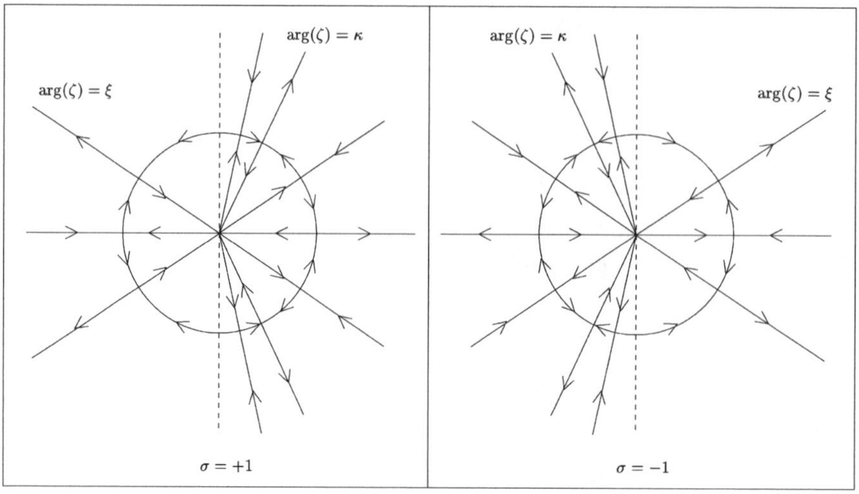

Figure 4.15 The contour $\Sigma_{\mathbf{L}}^{\sigma}$ for $\sigma = +1$ (left) and $\sigma = -1$ (right). In each case, the contour contains a circle of radius 1/2 as well as the real axis. Outside the circle, the rays extend to $\zeta = \infty$ and the orientation is as shown. Inside and on the circle, the orientation is determined so that the contour forms the positively oriented boundary of a multiply connected open region and the negatively oriented boundary of the closure of its complement in $\mathbb{C}$.

RIEMANN-HILBERT PROBLEM 4.4.6 (Reoriented local model for the origin). *Find a matrix $\mathbf{L}^{\sigma}(\zeta)$ satisfying the following:*

1. **Analyticity:** $\mathbf{L}^{\sigma}(\zeta)$ *is analytic for* $\zeta \in \mathbb{C}\setminus\Sigma_{\mathbf{L}}^{\sigma}$;
2. **Boundary behavior:** $\mathbf{L}^{\sigma}(\zeta)$ *assumes continuous boundary values from each connected component of* $\mathbb{C}\setminus\Sigma_{\mathbf{L}}^{\sigma}$ *that are continuous, including corner points corresponding to self-intersections;*
3. **Jump condition:** *On the oriented contour* $\Sigma_{\mathbf{L}}^{\sigma}$ *minus the origin, the boundary values satisfy*

$$\mathbf{L}_{+}^{\sigma}(\zeta) = \mathbf{L}_{-}^{\sigma}(\zeta)\mathbf{v}_{\mathbf{L}}^{\sigma}(\zeta); \tag{4.222}$$

4. **Normalization:** $\mathbf{L}^{\sigma}(\zeta)$ *is normalized at infinity:*

$$\mathbf{L}^{\sigma}(\zeta) \to \mathbb{I} \quad as \quad \zeta \to \infty. \tag{4.223}$$

Riemann-Hilbert Problem 4.4.6 differs from Riemann-Hilbert Problem 4.4.5 of interest only in the orientation of the contour and the introduction of some contour components supporting identity jump matrices. Therefore, the solutions of these two problems are in one-to-one correspondence: $\widehat{\mathbf{F}}^{\sigma}(\zeta) \equiv \mathbf{L}^{\sigma}(\zeta)$. The reorientation of the contour and introduction of the real axis and the circle are simply to rewrite the problem in precisely the form to which the general results from appendix A can be applied.

We now proceed to apply the Hölder theory developed in appendix A. First, observe that on each smooth component of $\Sigma_{\mathbf{L}}^{\sigma}$ and for each $\nu < 1$, the jump matrix $\mathbf{v}_{\mathbf{L}}^{\sigma}(\zeta)$ is Hölder-continuous with exponent ν. Indeed, in the interior of each component, the jump matrix is analytic, and it is easy to see that the only obstruction to arbitrary smoothness is in the limiting behavior at the origin. Here, the term that determines the smoothness is the factor $(-i\zeta)^{-2i\zeta}$ in $h^{\sigma}(\zeta)$. But at $\zeta = 0$, this term is in all Hölder classes with exponents ν *strictly* less than one. Next, note that on each ray of $\Sigma_{\mathbf{L}}^{\sigma}$, the jump matrix decays to the identity matrix as $\zeta \to \infty$ at least as fast as $O(|\zeta|^{-1})$. Indeed, from the asymptotic formulae (4.212), we see that for $\arg(\zeta) = \kappa$, the decay to the identity is $O(|\zeta|^{-1})$. On the two rays on either side of $\arg(\zeta) = \kappa$ in the same quadrant, the decay of the jump matrix to the identity is exponential for large $|\zeta|$ since $v\zeta$ is real and negative for $\arg(\zeta) = \kappa$. For $\arg(\zeta) = \xi$, one sees from (4.212) that the jump matrix decays exponentially to the identity like $O(\exp(u\zeta))$. The symmetry that determines the jump matrix in the lower half-plane in terms of that in the upper half-plane ensures that similar decay properties hold in the lower half-plane, and of course on the real axis the jump matrix is exactly the identity.

Next, we observe that the jump matrices are consistent at the origin, in the following sense. If we number the rays in counterclockwise order starting with the positive real axis as $\Sigma^{(1)}, \ldots, \Sigma^{(10)}$ and define $\mathbf{v}^{(k)} := \lim_{\zeta\to 0,\, \zeta\in\Sigma^{(k)}} \mathbf{v}_{\widehat{\mathbf{F}}}^{\sigma}(\zeta)$, then the cyclic relation

$$\mathbf{v}^{(1)}\mathbf{v}^{(2)-1}\mathbf{v}^{(3)}\mathbf{v}^{(4)-1}\cdots\mathbf{v}^{(9)}\mathbf{v}^{(10)-1} = \mathbb{I} \tag{4.224}$$

holds for all values of the parameters u, v, κ, ξ, and σ. It is easy to see that the same property holds at each intersection point ζ_0 of the circle with a ray (i.e., $|\zeta_0| = 1/2$), since by definition

$$\lim_{\epsilon\downarrow 0} \mathbf{v}_{\mathbf{L}}^{\sigma}((1+\epsilon)\zeta_0)\cdot\mathbb{I}^{-1}\mathbf{v}_{\mathbf{L}}^{\sigma}((1-\epsilon)\zeta_0)\mathbb{I}^{-1} = \mathbb{I}. \tag{4.225}$$

Finally, we observe that for all $\zeta \in \Sigma_{\mathbf{L}}^{\sigma}$ with $\Im(\zeta) \neq 0$, the relation $\mathbf{v}_{\mathbf{L}}^{\sigma}(\zeta^*) = [\sigma_2\mathbf{v}_{\mathbf{L}}^{\sigma}(\zeta)^*\sigma_2]^{-1}$ implies that

$$\mathbf{v}_{\mathbf{L}}^{\sigma}(\zeta^*) = \mathbf{v}_{\mathbf{L}}^{\sigma}(\zeta)^{\dagger}. \tag{4.226}$$

This is not a general fact but is a consequence of the special structure of the jump matrices for this Riemann-Hilbert problem. Also, since the jump matrix is the identity on the real axis, $\mathbf{v}_{\mathbf{L}}^{\sigma}(\zeta) + \mathbf{v}_{\mathbf{L}}^{\sigma}(\zeta)^{\dagger}$ is strictly positive definite for real ζ.

These facts allow us to apply Theorem A.1.2 proved in appendix A to deduce the existence of a matrix $\mathbf{L}^{\sigma}(\zeta)$, which is analytic in $\mathbb{C}\setminus\Sigma_{\mathbf{L}}^{\sigma}$, which for each $\mu < \nu$ takes on boundary values on $\Sigma_{\mathbf{L}}^{\sigma}$ that are uniformly bounded and Hölder-continuous with exponent μ, which for each $\mu < \nu$ satisfies $\mathbf{L}^{\sigma}(\zeta) - \mathbb{I} = O(|\zeta|^{-\mu})$ as $\zeta \to \infty$, and whose boundary values satisfy the jump relation $\mathbf{L}_{+}^{\sigma}(\zeta) = \mathbf{L}_{-}^{\sigma}(\zeta)\mathbf{v}_{\mathbf{L}}^{\sigma}(\zeta)$. Note that the meaning of the subscripts "+" and "−" in regard to the boundary values of $\mathbf{L}^{\sigma}(\zeta)$ refer to the contour $\Sigma_{\mathbf{L}}^{\sigma}$, oriented as shown in figure 4.15. Since the boundary values are uniformly continuous and since $\mathbf{v}_{\mathbf{L}}^{\sigma}(\zeta) = \mathbb{I}$ for all real $\zeta \neq 0$ and for all ζ on the circle of radius 1/2 (except at the self-intersection points where the jump matrix is not defined), the matrix function $\mathbf{L}^{\sigma}(\zeta)$ is in fact analytic at these points.

The function defined by $\widehat{\mathbf{F}}^{\sigma}(\zeta) := \mathbf{L}^{\sigma}(\zeta)$ is easily seen to be the unique solution of Riemann-Hilbert Problem 4.4.3, since it is analytic in $\mathbb{C}\setminus\Sigma_{\widehat{\mathbf{F}}}^{\sigma}$ and since it has Hölder-continuous and uniformly bounded boundary values that satisfy

$\widehat{\mathbf{F}}_+^\sigma(\zeta) = \widehat{\mathbf{F}}_-^\sigma(\zeta)\mathbf{v}_{\widehat{\mathbf{F}}}^\sigma(\zeta)$. Note that here the subscripts "+" and "−" of the boundary values refer to the orientation of $\Sigma_{\widehat{\mathbf{F}}}^\sigma$ as shown in figure 4.14.

It remains only to obtain the decay estimate (4.220). For this, we return to the auxiliary problem for $\mathbf{L}^\sigma(\zeta)$, and we compute the averages of $\zeta(\mathbf{v}_\mathbf{L}^\sigma(\zeta) - \mathbb{I})$ along the various rays. First, observe that as ζ tends to infinity along any ray of $\Sigma_\mathbf{L}^\sigma$ except those satisfying $\arg(\zeta) = \pm\kappa$,

$$\mathbf{v}_\mathbf{L}^\sigma(\zeta) = \mathbb{I} + \text{ exponentially small.} \tag{4.227}$$

Now when $\sigma = +1$, we can use (4.212) to obtain

$$\mathbf{v}_\mathbf{L}^+(\zeta) = \mathbb{I} + \frac{1}{12i\zeta}\begin{bmatrix} -1 & i\exp(-iv\zeta) \\ i\exp(iv\zeta) & 1 \end{bmatrix} + O(|\zeta|^{-2}) \tag{4.228}$$

as $\zeta \to \infty$ with $\arg(\zeta) = \kappa$, and since the coefficient of ζ^{-1} in the expansion is an oscillatory function, we can isolate its asymptotic mean value:

$$\langle\zeta(\mathbf{v}_\mathbf{L}^+(\zeta) - \mathbb{I})\rangle = -\frac{1}{12i}\sigma_3. \tag{4.229}$$

Similarly,

$$\mathbf{v}_\mathbf{L}^+(\zeta) = \mathbb{I} + \frac{1}{12i\zeta}\begin{bmatrix} 1 & i\exp(-iv^*\zeta) \\ i\exp(iv^*\zeta) & -1 \end{bmatrix} + O(|\zeta|^{-2}) \tag{4.230}$$

as $\zeta \to \infty$ with $\arg(\zeta) = -\kappa$, and for the mean of the coefficient of ζ^{-1}, we get

$$\langle\zeta(\mathbf{v}_\mathbf{L}^+(\zeta) - \mathbb{I})\rangle = \frac{1}{12i}\sigma_3. \tag{4.231}$$

In both cases, the integral of the error term is $O(|\zeta|^{-1})$, which is clearly bounded by $\log|\zeta|$. On the other hand, when $\sigma = -1$, we can again use (4.212) to find

$$\mathbf{v}_\mathbf{L}^-(\zeta) = \mathbb{I} + \frac{1}{12i\zeta}\begin{bmatrix} 1 & i\exp(iv\zeta) \\ i\exp(-iv\zeta) & -1 \end{bmatrix} + O(|\zeta|^{-2}) \tag{4.232}$$

as $\zeta \to \infty$ with $\arg(\zeta) = \kappa$, and the average of the coefficient of ζ^{-1} is

$$\langle\zeta(\mathbf{v}_\mathbf{L}^-(\zeta) - \mathbb{I})\rangle = \frac{1}{12i}\sigma_3. \tag{4.233}$$

Similarly,

$$\mathbf{v}_\mathbf{L}^-(\zeta) = \mathbb{I} + \frac{1}{12i\zeta}\begin{bmatrix} -1 & i\exp(iv^*\zeta) \\ i\exp(-iv^*\zeta) & 1 \end{bmatrix} + O(|\zeta|^{-2}) \tag{4.234}$$

as $\zeta \to \infty$ with $\arg(\zeta) = -\kappa$, and for the average of the coefficient of ζ^{-1} along this ray, we find

$$\langle\zeta(\mathbf{v}_\mathbf{L}^-(\zeta) - \mathbb{I})\rangle = -\frac{1}{12i}\sigma_3. \tag{4.235}$$

Once again in both cases the integral of the error is bounded by $O(|\zeta|^{-1})$. Since the sum of the means $\langle\zeta(\mathbf{v}_\mathbf{L}^\sigma(\zeta) - \mathbb{I})\rangle$ taken over all rays vanishes, it follows from

Theorem A.1.3 that the matrix $\mathbf{L}^{\sigma}(\zeta)$ decays at least like $|\zeta|^{-1}$ for large ζ. This implies the decay estimate (4.220) and completes the proof. □

Remark. Note that the construction of $\widehat{\mathbf{F}}^{\sigma}(\zeta)$ presumes that the angle κ is not equal to $\pi/2$. Ultimately this is because the asymptotic approximation of the discrepancy between the discrete sum and the integral given in chapter 3 is not uniformly valid in any sector containing $\arg(\zeta) = \pi/2$ (cf. Theorem 3.2.1). A method for analyzing the case of $\kappa = \pi/2$, which occurs for example when $t = 0$, is described in [M02].

Now, we use the matrix $\widehat{\mathbf{F}}^{\sigma}(\zeta)$ to build an approximation $\widehat{\mathbf{N}}^{\sigma}_{\mathrm{origin}}(\lambda)$ of the matrix $\mathbf{N}^{\sigma}(\lambda)$ for $\lambda \in U_{\hbar}$. Since $\widehat{\mathbf{F}}^{\sigma}(\zeta)$ has jumps on the rays $\arg(\zeta) = \pm\kappa$, which do not quite coincide with the contours I_0^+ and I_0^- in the λ-plane, we want to make one final deformation. For sufficiently small $\hbar$, the tangent $I_0^{+\prime}$ meets the contour I_0^+ within $U_{\hbar}$ only at the origin. Denote the wedge-shaped subset of $U_{\hbar}$ between these two curves by δI_0^+. See figure 4.16. The jump matrix $\mathbf{v}^{\sigma}_{\widehat{\mathbf{F}}}(-i\rho^0(0)\lambda/\hbar_N)$ for $\arg(\lambda) = \kappa$ has an analytic continuation to $\lambda \in \delta I_0^+$ for $\hbar$ sufficiently small, which we denote by $\mathbf{u}^{\sigma}(\zeta)$. If I_0^+ lies to the right of the ray $\arg(\lambda) = \kappa$ in $U_{\hbar}$ for $\hbar$ small enough, then we set

$$\widehat{\mathbf{G}}^{\sigma}(\lambda) := \widehat{\mathbf{F}}^{\sigma}(-i\rho^0(0)\lambda/\hbar_N) \cdot \begin{cases} \mathbf{u}^{\sigma}(-i\rho^0(0)\lambda/\hbar_N), & \lambda \in \delta I_0^+, \\ \sigma_2\mathbf{u}^{\sigma}(-i\rho^0(0)\lambda^*/\hbar_N)^*\sigma_2, & \lambda \in \delta I_0^-, \\ \mathbb{I}, & \lambda \in U_{\hbar} \setminus (\delta I_0^+ \cup \delta I_0^-), \end{cases} \tag{4.236}$$

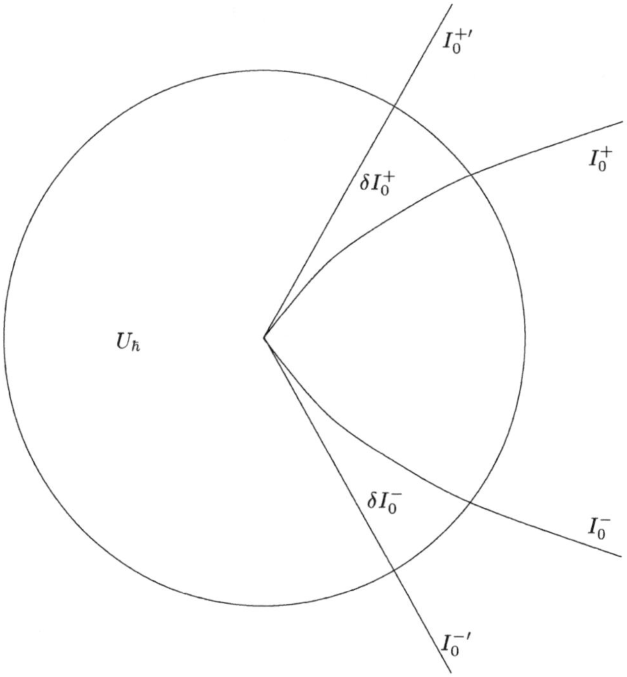

Figure 4.16 The regions δI_0^+ and δI_0^- as defined for sufficiently small $\hbar$.

and if I_0^+ lies to the left of the ray $\arg(\lambda) = \kappa$ in $U_\hbar$ for $\hbar$ small enough, then we set

$$\widehat{\mathbf{G}}^\sigma(\lambda) := \widehat{\mathbf{F}}^\sigma(-i\rho^0(0)\lambda/\hbar_N) \cdot \begin{cases} \mathbf{u}^\sigma(-i\rho^0(0)\lambda/\hbar_N)^{-1}, & \lambda \in \delta I_0^+, \\ [\sigma_2\mathbf{u}^\sigma(-i\rho^0(0)\lambda^*/\hbar_N)^*\sigma_2]^{-1}, & \lambda \in \delta I_0^-, \\ \mathbb{I}, & \lambda \in U_\hbar \backslash (\delta I_0^+ \cup \delta I_0^-). \end{cases} \tag{4.237}$$

Note that the constant $-i\rho^0(0)$ is always manifestly real according to (3.1).

We now record several consequences of this definition.

Lemma 4.4.12 *There exists some $M > 0$ such that for all $\lambda \in U_\hbar$ and all sufficiently small $\hbar$,*

$$\|\widehat{\mathbf{G}}^\sigma(\lambda)\| \leq M. \tag{4.238}$$

Proof. It follows from Lemma 4.4.11 that the factor $\widehat{\mathbf{F}}^\sigma(-i\rho^0(0)\lambda/\hbar)$ is uniformly bounded for *all* λ. Next, we study the behavior of $\mathbf{u}^\sigma(\lambda/\hbar)$ in δI_0^+. Along with its inverse, this matrix is bounded for $\arg(\lambda) = \kappa$. To deform into δI_0^+, we need only consider the behavior of the exponential factors $\exp(\pm iv\zeta)$, since $h^\sigma(\zeta)$ is uniformly bounded. Now, since I_0^+ is smooth and tangent to the ray $\arg(\lambda) = \kappa$, it will be the case for sufficiently small $\hbar$ that $\Im(-i\rho^0(0)v\lambda/\hbar_N) = O(\hbar_N^{2\delta-1})$ uniformly for $\lambda \in \delta I_0^+$ because the radius of $U_\hbar$ is $\hbar_N^\delta$ and $\Im(-i\rho^0(0)v\lambda/\hbar_N) = 0$ for $\arg(\lambda) = \kappa$. Therefore, uniformly for $\lambda \in \delta I_0^+$ we have $\exp(\pm\rho^0(0)v\lambda/\hbar_N) = (1 + O(\hbar_N^{2\delta-1}))\exp(\pm i\Re(-i\rho^0(0)v\lambda/\hbar_N))$. Similar arguments hold for $\lambda \in \delta I_0^-$, and the proof is complete since we always have $\delta > 1/2$. □

Lemma 4.4.13 *The matrix $\widehat{\mathbf{G}}^\sigma(\lambda)$ has the same domain of analyticity for $\lambda \in U_\hbar$ as $\mathbf{F}^\sigma(\lambda)$, bounded by the contours I_0^+, $\Gamma_{G/2+1}^+$, and $C_{0\pm}^+$ and the corresponding contours in the lower half-plane as illustrated in figure 4.13. Define jump matrices for $\lambda \in U_\hbar$ relative to the oriented contour of figure 4.13 by $\mathbf{F}_+^\sigma(\lambda) = \mathbf{F}_-^\sigma(\lambda)\mathbf{v}_\mathbf{F}^\sigma(\lambda)$ and $\widehat{\mathbf{G}}_+^\sigma(\lambda) = \widehat{\mathbf{G}}_-^\sigma(\lambda)\mathbf{v}_{\widehat{\mathbf{G}}}^\sigma(\lambda)$. Then, there exists some $M > 0$ such that for all $\lambda \in U_\hbar$ and all sufficiently small $\hbar$,*

$$\|\mathbf{v}_\mathbf{F}^\sigma(\lambda)\mathbf{v}_{\widehat{\mathbf{G}}}^\sigma(\lambda)^{-1} - \mathbb{I}\| \leq M\hbar_N^{\min(1/3,\,2\delta-1)}. \tag{4.239}$$

Recall that the neighborhood $U_\hbar$ has radius $\hbar_N^\delta$ for $1/2 < \delta < 1$.

Proof. From the definition of $\widehat{\mathbf{G}}^\sigma(\lambda)$, we have $\widehat{\mathbf{G}}_+^\sigma(\lambda) = \widehat{\mathbf{G}}_-^\sigma(\lambda)$ for $\arg(\lambda) = \kappa$. The continuity of the boundary values then implies that $\widehat{\mathbf{G}}^\sigma(\lambda)$ is analytic for $\arg(\lambda) = \kappa$ in $U_\hbar$. This latter statement assumes that I_0^+ does not agree identically with its tangent line, in which case the regions $\delta I_0^\pm$ would be empty and we would have $\widehat{\mathbf{G}}^\sigma(\lambda) \equiv \widehat{\mathbf{F}}^\sigma(-i\rho^0(0)\lambda/\hbar_N)$. The remaining contours of nonanalyticity for $\mathbf{F}^\sigma(\lambda)$ have all been taken to be straight line segments within the fixed disk neighborhood U and therefore agree with the corresponding contours for $\widehat{\mathbf{F}}^\sigma(-i\rho^0(0)\lambda/\hbar_N)$. The statement carries over to $\widehat{\mathbf{G}}^\sigma(\lambda)$, because for sufficiently small $\hbar$ the regions $\delta I_0^\pm$ where $\widehat{\mathbf{G}}^\sigma(\lambda)$ differs from $\widehat{\mathbf{F}}^\sigma(-i\rho^0(0)\lambda/\hbar_N)$ will only meet these contours at the origin. This proves that the domains of analyticity for $\mathbf{F}^\sigma(\lambda)$ and $\widehat{\mathbf{G}}^\sigma(\lambda)$ agree within $U_\hbar$.

The estimate (4.239) follows from our asymptotic analysis of the jump matrix $\mathbf{v}_{\mathbf{F}}^{\sigma}(\lambda)$. On the straight line segments $C_{0\pm}^{+}$ and their images in the lower half-plane, we have $\widehat{\mathbf{G}}^{\sigma}(\lambda) \equiv \widehat{\mathbf{F}}^{\sigma}(-i\rho^0(0)\lambda/\hbar_N)$, and therefore the simple Taylor approximations (4.214) and (4.215) give

$$\begin{aligned}\|\mathbf{v}_{\mathbf{F}}^{\sigma}(\lambda)\mathbf{v}_{\mathbf{G}}^{\sigma}(\lambda)^{-1} - \mathbb{I}\| &= \|\mathbf{v}_{\mathbf{F}}^{\sigma}(\lambda)\mathbf{v}_{\widehat{\mathbf{F}}}^{\sigma}(-i\rho^0(0)\lambda/\hbar_N)^{-1} - \mathbb{I}\| \\ &= \exp(-|\Im(-i\rho^0(0)v\lambda)|/\hbar_N)O(\lambda^2/\hbar_N), \quad (4.240)\end{aligned}$$

which is $O(\hbar^{2\delta-1})$ for all $\lambda \in U_{\hbar}$. On the two straight line segments $\Gamma_{G/2+1}^{\pm}$, we again find that $\widehat{\mathbf{G}}^{\sigma}(\lambda) \equiv \widehat{\mathbf{F}}^{\sigma}(-i\rho^0(0)\lambda/\hbar_N)$, but we have an additional contribution from the asymptotic approximation of $d^{\sigma}(\lambda)$ to take into account. The errors that dominate are determined by the choice of δ, since for $\sigma = +1$ and $\lambda \in \Gamma_{G/2+1}^{+} \cap U_{\hbar}$, we have (using (4.196), (4.210), (4.214), and (4.216))

$$\begin{aligned}\|\mathbf{v}_{\mathbf{F}}^{\sigma}(\lambda)\mathbf{v}_{\mathbf{G}}^{\sigma}(\lambda)^{-1} - \mathbb{I}\| &= (1 - h^{+}(-i\rho^0(0)\lambda/\hbar_N))\exp(-i\rho^0(0)(u - 2\pi)\lambda/\hbar_N) \\ &\quad \times \left(1 + O(\hbar_N^{1/3}) + O(\hbar_N^{2\delta-1})\right), \quad (4.241)\end{aligned}$$

with the first error term coming from Theorem 3.2.1 and the second error term coming from the Taylor approximations (4.214) and (4.215). Similarly, for $\sigma = -1$ and $\lambda \in \Gamma_{G/2+1}^{+} \cap U_{\hbar}$, we have

$$\begin{aligned}\|\mathbf{v}_{\mathbf{F}}^{\sigma}(\lambda)\mathbf{v}_{\mathbf{G}}^{\sigma}(\lambda)^{-1} - \mathbb{I}\| &= (1 - h^{-}(-i\rho^0(0)\lambda/\hbar_N))\exp(-i\rho^0(0)(u + 2\pi)\lambda/\hbar_N) \\ &\quad \times \left(1 + O(\hbar_N^{1/3}) + O(\hbar_N^{2\delta-1})\right). \quad (4.242)\end{aligned}$$

That both of these estimates are uniformly small in $U_{\hbar}$ now follows from the boundedness and asymptotic properties of $h^{\pm}(\zeta)$. By symmetry of the definition of the jump matrices, both of these estimates also hold on $\Gamma_{G/2+1}^{-} \cap U_{\hbar}$. Finally, we note that to verify the estimate (4.239) for $\lambda \in I_0^{\pm} \cap U_{\hbar}$ requires the same sort of analysis as above, with the additional observation that the jump matrix for $\widehat{\mathbf{G}}^{\sigma}(\lambda)$, denoted by $\mathbf{u}^{\sigma}(-i\rho^0(0)\lambda/\hbar_N)$ above, is uniformly bounded in $U_{\hbar}$ according to the arguments in the proof of Lemma 4.4.12. □

Now, we give the local approximation for $\mathbf{N}^{\sigma}$ valid near the origin. We define, for $\lambda \in U_{\hbar}$,

$$\widehat{\mathbf{N}}_{\text{origin}}^{\sigma}(\lambda) := \mathbf{C}^{\sigma}(\lambda)\widehat{\mathbf{G}}^{\sigma}(\lambda)\mathbf{C}^{\sigma}(\lambda)^{-1}\widehat{\mathbf{N}}_{\text{out}}^{\sigma}(\lambda). \quad (4.243)$$

The essential properties of this approximation are the following.

Lemma 4.4.14 *$\widehat{\mathbf{N}}_{\text{origin}}^{\sigma}(\lambda)$ is uniformly bounded independent of $\hbar$ for $\lambda \in U_{\hbar}$. Since $\det(\widehat{\mathbf{N}}_{\text{origin}}^{\sigma}(\lambda)) = 1$, this property is also held by the inverse matrix.*

Proof. The factors $\mathbf{C}^{\sigma}(\lambda)$, $\mathbf{C}^{\sigma}(\lambda)^{-1}$, and $\widehat{\mathbf{N}}_{\text{out}}^{\sigma}(\lambda)$ are clearly all uniformly bounded in $U_{\hbar}$, essentially since the solution $\widetilde{\mathbf{O}}(\lambda)$ of the outer model Riemann-Hilbert Problem 4.2 obtained in §4.3 has bounded boundary values at $\lambda = 0$ on I_0. To analyze $\widehat{\mathbf{G}}^{\sigma}(\lambda)$, we observe that the factor $\widehat{\mathbf{F}}^{\sigma}(-i\rho^0(0)\lambda/\hbar_N)$ is uniformly bounded for all $\zeta = -i\rho^0(0)\lambda/\hbar_N$ and in particular for $\lambda \in U_{\hbar}$. The remaining factor in the definition of $\widehat{\mathbf{G}}^{\sigma}(\lambda)$ is controlled as described above, essentially because of the small size of the neighborhood $U_{\hbar}$. The fact that $\det(\widehat{\mathbf{N}}_{\text{origin}}^{\sigma}(\lambda)) = 1$ follows from the analogous properties of the individual factors and that the determinant is unchanged by conjugation by $\mathbf{C}^{\sigma}(\lambda)$. □

Lemma 4.4.15 *Let the jump matrix for* $\widehat{\mathbf{N}}^{\sigma}_{\mathrm{origin}}(\lambda)$ *for* λ *near the origin on the oriented contour shown in figure 4.13 be denoted* $\mathbf{v}^{\sigma}_{\mathrm{origin}}(\lambda)$. *Then, for some* $M > 0$,

$$\|\mathbf{v}^{\sigma}_{\mathbf{N}}(\lambda)\mathbf{v}^{\sigma}_{\mathrm{origin}}(\lambda)^{-1} - \mathbb{I}\| \leq M\hbar_N^{\min(1/3,\, 2\delta-1)} \tag{4.244}$$

for all λ *on the contour in* $U_\hbar$ *and for all sufficiently small* $\hbar$.

Proof. Using the fact that $\mathbf{C}^{\sigma}(\lambda)$ is analytic in $U_\hbar$, we find

$$\mathbf{v}^{\sigma}_{\mathbf{N}}(\lambda)\mathbf{v}^{\sigma}_{\mathrm{origin}}(\lambda)^{-1} = \mathbf{C}^{\sigma}(\lambda)\widehat{\mathbf{G}}^{\sigma}_{-}(\lambda)\mathbf{v}^{\sigma}_{\mathbf{F}}(\lambda)\mathbf{v}^{\sigma}_{\widehat{\mathbf{G}}}(\lambda)^{-1}\widehat{\mathbf{G}}^{\sigma}_{-}(\lambda)^{-1}\mathbf{C}^{\sigma}(\lambda)^{-1}. \tag{4.245}$$

The estimate (4.244) then follows from Lemma 4.4.13, Lemma 4.4.12, and the uniform boundedness in $U_\hbar$ of $\mathbf{C}^{\sigma}(\lambda)$ and its inverse. □

Lemma 4.4.16 *There exists a constant* M *such that for all* $\lambda \in \partial U_\hbar$ *and for all sufficiently small* $\hbar$,

$$\|\widehat{\mathbf{N}}^{\sigma}_{\mathrm{origin}}(\lambda)\widehat{\mathbf{N}}^{\sigma}_{\mathrm{out}}(\lambda)^{-1} - \mathbb{I}\| \leq M\hbar_N^{1-\delta}. \tag{4.246}$$

Proof. From the definition of the local approximation, we have for $|\lambda| = \hbar_N^{\delta}$,

$$\begin{aligned}\|\widehat{\mathbf{N}}^{\sigma}_{\mathrm{origin}}(\lambda)\widehat{\mathbf{N}}^{\sigma}_{\mathrm{out}}(\lambda)^{-1} - \mathbb{I}\| &= \|\mathbf{C}^{\sigma}(\lambda)\cdot[\widehat{\mathbf{G}}^{\sigma}(\lambda) - \mathbb{I}]\cdot\mathbf{C}^{\sigma}(\lambda)^{-1}\| \\ &\leq \|\mathbf{C}^{\sigma}(\lambda)\|\cdot\|\mathbf{C}^{\sigma}(\lambda)^{-1}\|\cdot\|\widehat{\mathbf{G}}^{\sigma}(\lambda) - \mathbb{I}\| \\ &\leq M'\cdot\|\widehat{\mathbf{G}}^{\sigma}(\lambda) - \mathbb{I}\|\end{aligned} \tag{4.247}$$

for some $M' > 0$, where we have used the uniform boundedness of $\mathbf{C}^{\sigma}(\lambda)$ and its inverse. Now, for all $\lambda \in \partial U_\hbar$ that are not on the boundary of $\delta I_0^{\pm}$, we have $\widehat{\mathbf{G}}^{\sigma}(\lambda) \equiv \widehat{\mathbf{F}}^{\sigma}(-i\rho^0(0)\lambda/\hbar_N)$, and then from the decay property (4.220) of $\widehat{\mathbf{F}}^{\sigma}(\zeta)$ established in Lemma 4.4.11, the estimate (4.246) follows from (4.247). For the other values of $\lambda \in \partial U_\hbar$ where $\widehat{\mathbf{G}}^{\sigma}(\lambda) \not\equiv \widehat{\mathbf{F}}^{\sigma}(-i\rho^0(0)\lambda/\hbar_N)$, we have to take into account an additional factor. But for $|\lambda| = \hbar_N^{\delta}$ on the boundary of $\delta I_0^{\pm}$, it follows from the asymptotic behavior (4.212) of $h^{\sigma}(\zeta)$ and the geometry of the region $\delta I_0^{\pm}$ that this factor is uniformly $\mathbb{I} + O(\hbar_N^{1-\delta})$, and again (4.247) yields the required estimate (4.246). □

4.4.4 Note Added: Exact Solution of Riemann-Hilbert Problem 4.4.5

In fact, Riemann-Hilbert Problem 4.4.5 can also be solved explicitly in terms of special functions for relevant values of the parameters u and v. This is a fact that we have learned recently, and as an explicit solution may be useful in computing higher-order corrections to our semiclassical asymptotics, we want to show here how the solution may be found for the case $\sigma = +1$.

Recall that the problem is posed relative to a self-intersecting contour $\Sigma^{+}_{\widehat{\mathbf{F}}}$ illustrated in figure 4.17. The contour $\Sigma^{+}_{\widehat{\mathbf{F}}}$ is symmetric with respect to complex conjugation (disregarding orientation).

We begin by defining a positive parameter β by the condition that the ray I_0' lie on the line

$$\pi\Re(\zeta) = \beta\Im(\zeta). \tag{4.248}$$

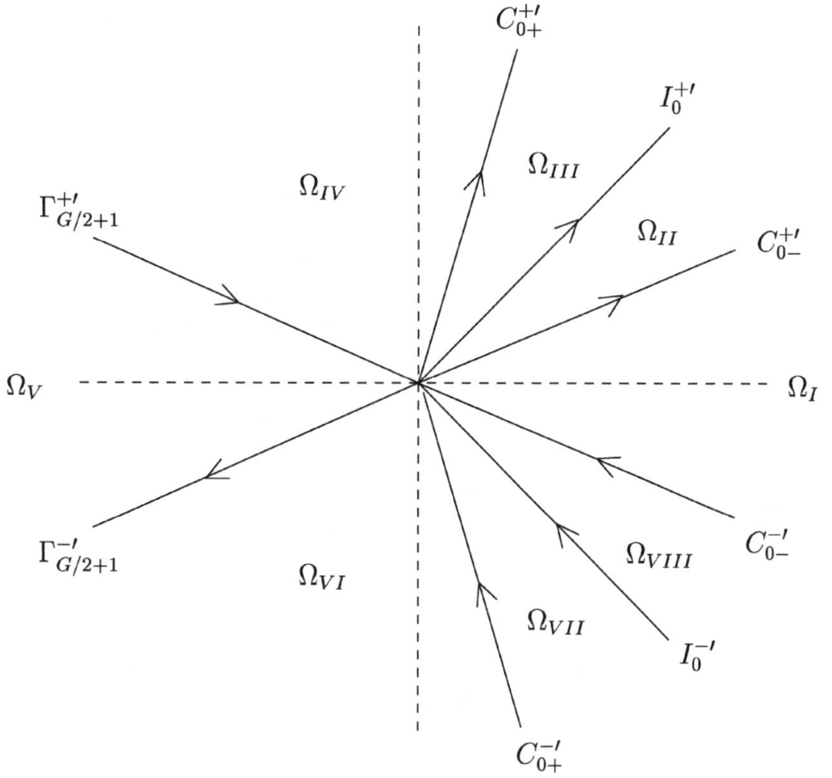

Figure 4.17 The oriented contour $\Sigma_{\widehat{\mathbf{F}}}^{+}$ in the complex ζ-plane. The contour $\Sigma_{\widehat{\mathbf{F}}}^{+}$ consists of rays $C'_{0\pm}$, I'_0, and $\Gamma'_{G/2+1}$ and their complex conjugates, and divides the complex plane into eight sectors denoted $\Omega_{\text{I}} \ldots \Omega_{\text{VIII}}$ as shown.

We need to assume here that the ray C'_{0+} lies in the sector between I'_0 and the positive imaginary axis while the ray C'_{0-} lies in the sector between the real axis and I'_0; also that the ray $\Gamma'_{G/2+1}$ lies in the upper half-plane to the left of the line $\pi\Re(\zeta) + \beta\Im(\zeta) = 0$. These assumptions are consistent with the actual position of the contours near the origin in the case $\sigma = +1$, modulo our freedom of placement of the contours $C_{0\pm}^{+}$ and $\Gamma_{G/2+1}^{+}$, which we are taking to be appropriate straight rays (indicated with primes in figure 4.17) when mapped to the ζ-plane.

Lemma 4.4.17 *The complex constants u and v defined by* (4.213) *are expressed in terms of β by*

$$u = \pi - i\beta \quad \textit{and} \quad v = -\beta + i\pi. \tag{4.249}$$

Proof. This follows from explicit formulae for $\tilde{\phi}^{\sigma}(\lambda)$ and $\rho^{\sigma}(\lambda)$ to be presented in chapter 5. Alternatively, one can work from the description of these two functions in terms of the boundary values of $g^{\sigma}(\lambda)$, a function that is assumed to be analytic for $\Im(\lambda) > 0$ in the vicinity of the origin except for $\lambda \in I_0^{+}$. We leave the details to the interested reader. □

Remark. The conditions (4.249) are essential for the explicit construction of $\widehat{\mathbf{F}}^{+}(\zeta)$ we are about to develop, although the Fredholm-theoretic method of implicit solution presented in §4.4.3 does not require them to hold.

By meromorphic continuation from the sector $\Omega_{\rm V}$ to the positive real axis, we obtain a matrix $\widehat{\mathbf{H}}(\zeta)$ that is meromorphic for $\zeta \in \mathbb{C}\setminus\mathbb{R}_{+}$, having simple poles at the points $\zeta = \pm i(1/2+n)$ for $n = 0, 1, 2, \ldots$. This matrix satisfies $\widehat{\mathbf{H}}(\zeta) \equiv \widehat{\mathbf{F}}^{+}(\zeta)$ for $\zeta \in \Omega_{\rm V}$ and solves a modified Riemann-Hilbert problem, which is obtained from Riemann-Hilbert Problem 4.4.3 by "folding" the contours individually in the upper and lower half-planes in the direction of the positive real axis and making use of the conditions (4.249) to simplify the resulting jump. The Riemann-Hilbert problem satisfied by $\widehat{\mathbf{H}}(\zeta)$ is the following.

Riemann-Hilbert Problem 4.4.7 *Find a 2×2 matrix $\widehat{\mathbf{H}}(\zeta)$ that satisfies the following:*

1. **Analyticity:** *$\widehat{\mathbf{H}}(\zeta)$ is a meromorphic function of ζ for $\zeta \in \mathbb{C}\setminus\mathbb{R}_{+}$.*
2. **Normalization:** *As $\zeta \to \infty$ in $\Omega_{\rm V}$, $\widehat{\mathbf{H}}(\zeta) \to \mathbb{I}$.*
3. **Jump condition:** *$\widehat{\mathbf{H}}(\zeta)$ takes continuous boundary values on $\mathbb{R}_{+}$ and is bounded at the origin. Denoting the boundary value taken from above as $\widehat{\mathbf{H}}_{+}(\zeta)$ and from below as $\widehat{\mathbf{H}}_{-}(\zeta)$ for $\zeta \in \mathbb{R}_{+}$, we have the simple jump condition*
$$\widehat{\mathbf{H}}_{+}(\zeta) = \widehat{\mathbf{H}}_{-}(\zeta)\begin{bmatrix} e^{-2\pi\zeta} & 0 \\ 0 & e^{2\pi\zeta} \end{bmatrix}. \tag{4.250}$$
4. **Singularities:** *$\widehat{\mathbf{H}}(\zeta)$ has simple poles at $\zeta = \pm i(1/2+n)$ for whole numbers n. The singularities of $\widehat{\mathbf{H}}(\zeta)$ are such that near $\zeta = i(1/2+n)$ for $n = 0, 1, 2, 3, \ldots$,*
$$\widehat{\mathbf{H}}(\zeta)\begin{bmatrix} 1 & 0 \\ -i(h^{+}(\zeta)-1)e^{(u-2\pi)\zeta} & 1 \end{bmatrix} = \textit{holomorphic function of } \zeta, \tag{4.251}$$
and near $\zeta = -i(1/2+n)$ for $n = 0, 1, 2, 3, \ldots$,
$$\widehat{\mathbf{H}}(\zeta)\begin{bmatrix} 1 & -i(h^{+}(\zeta^{*})^{*}-1)e^{(u^{*}-2\pi)\zeta} \\ 0 & 1 \end{bmatrix} = \textit{holomorphic function of } \zeta. \tag{4.252}$$

Let the matrix $\widehat{\mathbf{J}}(\zeta)$ be defined explicitly for $\zeta \in \mathbb{C}\setminus\mathbb{R}_{+}$ by
$$\widehat{\mathbf{J}}(\zeta) := \begin{bmatrix} (-\zeta)^{-i\zeta} & 0 \\ 0 & (-\zeta)^{i\zeta} \end{bmatrix}. \tag{4.253}$$
This matrix function is bounded at the origin and resolves the jump condition on the positive real axis in the sense that if $\widehat{\mathbf{H}}(\zeta)$ is the solution of Riemann-Hilbert Problem 4.4.4, then the product
$$\widehat{\mathbf{K}}(\zeta) := \widehat{\mathbf{H}}(\zeta)\widehat{\mathbf{J}}(\zeta)^{-1} \tag{4.254}$$

is a meromorphic function throughout the ζ-plane, that is, the jump on the real axis has disappeared. The matrix $\widehat{\mathbf{K}}(\zeta)$ has only simple poles. Taking into account the definition of $h^+(\zeta)$, we find that the singularities of $\widehat{\mathbf{K}}(\zeta)$ are such that near the points $\zeta = i(1/2+n)$ for $n = 0, 1, 2, 3, \ldots$,

$$\widehat{\mathbf{K}}(\zeta)\begin{bmatrix} 1 & 0 \\ ie^{(u+2i)\zeta}\frac{\Gamma(1/2+i\zeta)}{\Gamma(1/2-i\zeta)} & 1 \end{bmatrix} = \text{holomorphic function of } \zeta, \tag{4.255}$$

which indicates that the simple pole is only in the first column. Similarly, near the points $\zeta = -i(1/2+n)$ for $n = 0, 1, 2, 3, \ldots$,

$$\widehat{\mathbf{K}}(\zeta)\begin{bmatrix} 1 & ie^{(u^*-2i)\zeta}\frac{\Gamma(1/2-i\zeta)}{\Gamma(1/2+i\zeta)} \\ 0 & 1 \end{bmatrix} = \text{holomorphic function of } \zeta, \tag{4.256}$$

and thus the simple pole is in the second column only. It follows that the product

$$\hat{\mathbf{L}}(\zeta) := \widehat{\mathbf{K}}(\zeta)\begin{bmatrix} \frac{1}{\Gamma(1/2+i\zeta)} & 0 \\ 0 & \frac{1}{\Gamma(1/2-i\zeta)} \end{bmatrix} \tag{4.257}$$

is an entire analytic function of ζ. Using (4.255) we see that near $\zeta = i(1/2+n)$ for $n = 0, 1, 2, 3, \ldots$,

$$\hat{\mathbf{L}}(\zeta)\begin{bmatrix} 1 & 0 \\ ie^{(u+2i)\zeta} & 1 \end{bmatrix} = \mathbf{Z}_1(\zeta), \tag{4.258}$$

where $\mathbf{Z}_1(\zeta)$ denotes a locally defined analytic matrix whose first column vanishes when $\zeta = i(1/2+n)$. This equation tells us that in some neighborhood of $\zeta = i(1/2+n)$, the second column of $\hat{\mathbf{L}}(\zeta)$ agrees exactly with the second column of the matrix $\mathbf{Z}_1(\zeta)$. Then, since the first column of $\mathbf{Z}_1(\zeta)$ vanishes at $\zeta = i(1/2+n)$, we find that

$$\hat{\mathbf{l}}_1(i(1/2+n)) = -ie^{(u+2i)\zeta}\big|_{\zeta=i(1/2+n)}\hat{\mathbf{l}}_2(i(1/2+n)) \quad \text{for } n = 0, 1, 2, 3, \ldots, \tag{4.259}$$

where $\hat{\mathbf{l}}_1(\zeta)$ and $\hat{\mathbf{l}}_2(\zeta)$ denote the columns of $\hat{\mathbf{L}}(\zeta)$. Similarly, using (4.256) we find that near $\zeta = -i(1/2+n)$,

$$\hat{\mathbf{L}}(\zeta)\begin{bmatrix} 1 & ie^{(u^*-2i)\zeta} \\ 0 & 1 \end{bmatrix} = \mathbf{Z}_2(\zeta), \quad \text{for} \quad n = 0, 1, 2, 3, \ldots, \tag{4.260}$$

where $\mathbf{Z}_2(\zeta)$ is a locally defined analytic matrix whose second column vanishes when $\zeta = -i(1/2+n)$. From this we see that the first column of $\hat{\mathbf{L}}(\zeta)$ agrees exactly with the first column of $\mathbf{Z}_2(\zeta)$ in a neighborhood of $\zeta = -i(1/2+n)$. Then, using the fact that the second column of $\mathbf{Z}_2(\zeta)$ vanishes for $\zeta = -i(1/2+n)$, we find that

$$\hat{\mathbf{l}}_2(-i(1/2+n)) = -ie^{(u^*-2i)\zeta}\big|_{\zeta=-i(1/2+n)}\hat{\mathbf{l}}_1(-i(1/2+n)), \quad \text{for } n = 0, 1, 2, 3, \ldots. \tag{4.261}$$

From the asymptotic behavior of $\widehat{\mathbf{H}}(\zeta)$ asserted in Riemann-Hilbert Problem 4.4.4 and using the explicit transformations from $\widehat{\mathbf{H}}(\zeta)$ to $\widehat{\mathbf{K}}(\zeta)$ and finally to $\hat{\mathbf{L}}(\zeta)$ as well as Stirling's formula to compute the asymptotics of the gamma functions introduced in going from $\widehat{\mathbf{K}}(\zeta)$ to $\hat{\mathbf{L}}(\zeta)$, we see that the entire function $\hat{\mathbf{L}}(\zeta)$ satisfies

$$\hat{\mathbf{L}}(\zeta)\begin{bmatrix} e^{(\pi/2-i)\zeta}\sqrt{2\pi} & 0 \\ 0 & e^{(\pi/2+i)\zeta}\sqrt{2\pi} \end{bmatrix} \to \mathbb{I} \quad \text{as } \zeta \to \infty \text{ in } \Omega_{\text{V}}. \tag{4.262}$$

Since the matrix involved in normalizing $\hat{\mathbf{L}}(\zeta)$ as $\zeta \to \infty$ is itself an entire analytic invertible matrix, we can introduce a final transformation

$$\widehat{\mathbf{M}}(\zeta) := \hat{\mathbf{L}}(\zeta)\begin{bmatrix} e^{(\pi/2-i)\zeta}\sqrt{2\pi} & 0 \\ 0 & e^{(\pi/2+i)\zeta}\sqrt{2\pi} \end{bmatrix} \tag{4.263}$$

so that the matrix $\widehat{\mathbf{M}}(\zeta)$ is again an entire analytic function of ζ that has identity asymptotics as $\zeta \to \infty$ in the sector Ω_{V}. The relations (4.259) and (4.261) then immediately imply that

$$\begin{aligned} \hat{\mathbf{m}}_1(i(1/2+n)) &= -ie^{u\zeta}|_{\zeta=i(1/2+n)}\hat{\mathbf{m}}_2(i(1/2+n)), \\ \hat{\mathbf{m}}_1(-i(1/2+n)) &= ie^{-u^*\zeta}|_{\zeta=-i(1/2+n)}\hat{\mathbf{m}}_2(-i(1/2+n)) \end{aligned} \tag{4.264}$$

for $n = 0, 1, 2, 3, \dots$, where $\hat{\mathbf{m}}_1(\zeta)$ and $\hat{\mathbf{m}}_2(\zeta)$ denote the columns of $\widehat{\mathbf{M}}(\zeta)$. Recalling from (4.249) that $u = \pi - i\beta$ with $\beta > 0$, it is not difficult to check that the matrix

$$\widehat{\mathbf{M}}(\zeta) = \begin{bmatrix} 1 & -ie^{(\pi+i\beta)\zeta} \\ -ie^{(\pi-i\beta)\zeta} & 1 \end{bmatrix} \tag{4.265}$$

satisfies all of the necessary conditions. Consequently,

$$\hat{\mathbf{L}}(\zeta) = \begin{bmatrix} \dfrac{e^{i\zeta}e^{-\pi\zeta/2}}{\sqrt{2\pi}} & -i\dfrac{e^{-i\zeta}e^{\pi\zeta/2}e^{i\beta\zeta}}{\sqrt{2\pi}} \\ -i\dfrac{e^{i\zeta}e^{\pi\zeta/2}e^{-i\beta\zeta}}{\sqrt{2\pi}} & \dfrac{e^{-i\zeta}e^{-\pi\zeta/2}}{\sqrt{2\pi}} \end{bmatrix}, \tag{4.266}$$

$$\widehat{\mathbf{K}}(\zeta) = \begin{bmatrix} \dfrac{e^{i\zeta}e^{-\pi\zeta/2}\Gamma(1/2+i\zeta)}{\sqrt{2\pi}} & -i\dfrac{e^{-i\zeta}e^{\pi\zeta/2}e^{i\beta\zeta}\Gamma(1/2-i\zeta)}{\sqrt{2\pi}} \\ -i\dfrac{e^{i\zeta}e^{\pi\zeta/2}e^{-i\beta\zeta}\Gamma(1/2+i\zeta)}{\sqrt{2\pi}} & \dfrac{e^{-i\zeta}e^{-\pi\zeta/2}\Gamma(1/2-i\zeta)}{\sqrt{2\pi}} \end{bmatrix}, \tag{4.267}$$

and therefore the solution of Riemann-Hilbert Problem 4.4.4 is given by

$$\widehat{\mathbf{H}}(\zeta) = \begin{bmatrix} \dfrac{\Gamma(1/2+i\zeta)}{\sqrt{2\pi}(-\zeta)^{i\zeta}e^{-i\zeta}e^{\pi\zeta/2}} & -ie^{\pi\zeta}e^{i\beta\zeta}\dfrac{\Gamma(1/2-i\zeta)}{\sqrt{2\pi}(-\zeta)^{-i\zeta}e^{i\zeta}e^{\pi\zeta/2}} \\ -ie^{\pi\zeta}e^{-i\beta\zeta}\dfrac{\Gamma(1/2+i\zeta)}{\sqrt{2\pi}(-\zeta)^{i\zeta}e^{-i\zeta}e^{\pi\zeta/2}} & \dfrac{\Gamma(1/2-i\zeta)}{\sqrt{2\pi}(-\zeta)^{-i\zeta}e^{i\zeta}e^{\pi\zeta/2}} \end{bmatrix}. \tag{4.268}$$

Setting $\widehat{\mathbf{F}}^+(\zeta) = \widehat{\mathbf{H}}(\zeta)$ for $\zeta \in \Omega_{\text{V}}$, using the jump conditions to obtain $\widehat{\mathbf{F}}^+(\zeta)$ in the remaining sectors, and rewriting the power functions appropriately, the following result is now easily verified.

THEOREM 4.4.1 *The explicit solution of Riemann-Hilbert Problem* 4.4.5 *for* $\sigma = +1$, *subject to the conditions* (4.249), *is given by*

$$\widehat{\mathbf{F}}^{+}(\zeta) := \begin{bmatrix} \dfrac{\Gamma(1/2-i\zeta)}{\sqrt{2\pi}(-i\zeta)^{-i\zeta}e^{i\zeta}} & ie^{-\pi\zeta}e^{i\beta\zeta}\dfrac{\Gamma(1/2+i\zeta)}{\sqrt{2\pi}(i\zeta)^{i\zeta}e^{-i\zeta}} \\ ie^{-\pi\zeta}e^{-i\beta\zeta}\dfrac{\Gamma(1/2-i\zeta)}{\sqrt{2\pi}(-i\zeta)^{-i\zeta}e^{i\zeta}} & \dfrac{\Gamma(1/2+i\zeta)}{\sqrt{2\pi}(i\zeta)^{i\zeta}e^{-i\zeta}} \end{bmatrix}, \quad \zeta \in \Omega_{\mathrm{I}}, \tag{4.269}$$

$$\widehat{\mathbf{F}}^{+}(\zeta) := \begin{bmatrix} \dfrac{\Gamma(1/2-i\zeta)}{\sqrt{2\pi}(-i\zeta)^{-i\zeta}e^{i\zeta}} + e^{-2\pi\zeta}\dfrac{\Gamma(1/2+i\zeta)}{\sqrt{2\pi}(i\zeta)^{i\zeta}e^{-i\zeta}} & ie^{-\pi\zeta}e^{i\beta\zeta}\dfrac{\Gamma(1/2+i\zeta)}{\sqrt{2\pi}(i\zeta)^{i\zeta}e^{-i\zeta}} \\ ie^{-\pi\zeta}e^{-i\beta\zeta}\left(\dfrac{\Gamma(1/2-i\zeta)}{\sqrt{2\pi}(-i\zeta)^{-i\zeta}e^{i\zeta}} - \dfrac{\Gamma(1/2+i\zeta)}{\sqrt{2\pi}(i\zeta)^{i\zeta}e^{-i\zeta}}\right) & \dfrac{\Gamma(1/2+i\zeta)}{\sqrt{2\pi}(i\zeta)^{i\zeta}e^{-i\zeta}} \end{bmatrix}, \quad \zeta \in \Omega_{\mathrm{II}}, \tag{4.270}$$

$$\widehat{\mathbf{F}}^{+}(\zeta) := \begin{bmatrix} \dfrac{\sqrt{2\pi}(-i\zeta)^{-i\zeta}e^{i\zeta}}{\Gamma(1/2-i\zeta)} & ie^{\pi\zeta}e^{i\beta\zeta}\left(\dfrac{\sqrt{2\pi}(-i\zeta)^{-i\zeta}e^{i\zeta}}{\Gamma(1/2-i\zeta)} - \dfrac{\Gamma(1/2-i\zeta)}{\sqrt{2\pi}(-i\zeta)^{-i\zeta}e^{i\zeta}}\right) \\ 0 & \dfrac{\Gamma(1/2-i\zeta)}{\sqrt{2\pi}(-i\zeta)^{-i\zeta}e^{i\zeta}} \end{bmatrix}, \quad \zeta \in \Omega_{\mathrm{III}}, \tag{4.271}$$

$$\widehat{\mathbf{F}}^{+}(\zeta) := \begin{bmatrix} \dfrac{\sqrt{2\pi}(-i\zeta)^{-i\zeta}e^{i\zeta}}{\Gamma(1/2-i\zeta)} & -ie^{\pi\zeta}e^{i\beta\zeta}\dfrac{\Gamma(1/2-i\zeta)}{\sqrt{2\pi}(-i\zeta)^{-i\zeta}e^{i\zeta}} \\ 0 & \dfrac{\Gamma(1/2-i\zeta)}{\sqrt{2\pi}(-i\zeta)^{-i\zeta}e^{i\zeta}} \end{bmatrix}, \quad \zeta \in \Omega_{\mathrm{IV}}, \tag{4.272}$$

$$\widehat{\mathbf{F}}^{+}(\zeta) := \begin{bmatrix} \dfrac{\Gamma(1/2+i\zeta)}{\sqrt{2\pi}(i\zeta)^{i\zeta}e^{-i\zeta}} & -ie^{\pi\zeta}e^{i\beta\zeta}\dfrac{\Gamma(1/2-i\zeta)}{\sqrt{2\pi}(-i\zeta)^{-i\zeta}e^{i\zeta}} \\ -ie^{\pi\zeta}e^{-i\beta\zeta}\dfrac{\Gamma(1/2+i\zeta)}{\sqrt{2\pi}(i\zeta)^{i\zeta}e^{-i\zeta}} & \dfrac{\Gamma(1/2-i\zeta)}{\sqrt{2\pi}(-i\zeta)^{-i\zeta}e^{i\zeta}} \end{bmatrix}, \quad \zeta \in \Omega_{\mathrm{V}}, \tag{4.273}$$

with $\widehat{\mathbf{F}}^{+}(\zeta)$ *being defined explicitly in the lower half-plane by the symmetry formula* $\widehat{\mathbf{F}}^{+}(\zeta) := \sigma_2\widehat{\mathbf{F}}^{+}(\zeta^*)^*\sigma_2$.

The asymptotic behavior of the solution may be obtained using the following Stirling-type formulae:

$$\Gamma(1/2 + i\zeta) = \sqrt{2\pi}(i\zeta)^{i\zeta}e^{-i\zeta}(1 + O(\zeta^{-1})) \tag{4.274}$$

as $\zeta \to \infty$, avoiding the positive imaginary axis, and

$$\Gamma(1/2 - i\zeta) = \sqrt{2\pi}(-i\zeta)^{-i\zeta}e^{i\zeta}(1 + O(\zeta^{-1})) \tag{4.275}$$

as $\zeta \to \infty$, avoiding the negative imaginary axis. Using the placement of the rays in the contour $\Sigma^{+}_{\widehat{\mathbf{F}}}$, one sees easily that uniformly for $\zeta \in \Omega_{\mathrm{I}}$,

$$\widehat{\mathbf{F}}^{+}(\zeta) = \mathbb{I} + O_{\mathrm{diag}}(\zeta^{-1}) + \text{exponentially small} \tag{4.276}$$

as $\zeta \to \infty$, where $O_{\mathrm{diag}}(\zeta^{-1})$ denotes a diagonal error matrix. Uniformly for $\zeta \in \Omega_{\mathrm{II}}$,

$$\widehat{\mathbf{F}}^{+}(\zeta) = \mathbb{I} + O_{\mathrm{lower}}(\zeta^{-1}) + \text{exponentially small} \tag{4.277}$$

as $\zeta \to \infty$, where $O_{\mathrm{lower}}(\zeta^{-1})$ refers to a lower-triangular matrix. In fact, away from the boundary $I_0^{+\prime}$ of Ω_{II}, this estimate becomes sharper, of the diagonally dominant form (4.276). Uniformly for $\zeta \in \Omega_{\mathrm{III}}$,

$$\widehat{\mathbf{F}}^{+}(\zeta) = \mathbb{I} + O_{\mathrm{upper}}(\zeta^{-1}) + \text{exponentially small} \tag{4.278}$$

as $\zeta \to \infty$, where $O_{\mathrm{upper}}(\zeta^{-1})$ refers to an upper-triangular matrix. Away from the boundary $I_0^{+\prime}$ of Ω_{III}, this estimate also becomes sharper, of the same diagonally dominant form as (4.276). In both Ω_{IV} and Ω_{V}, one has the uniform estimate (4.276) as $\zeta \to \infty$.

4.5 ESTIMATING THE ERROR

We have now completed our analysis of the Riemann-Hilbert problem for the matrix $\mathbf{N}^{\sigma}(\lambda)$ corresponding to arbitrary soliton ensembles connected with initial data $\psi_0(x) = A(x)$ via formal WKB theory. We are now in a position to *prove* that the various approximations we have constructed in different parts of the complex λ-plane are indeed valid. We begin by patching together the various approximations to obtain a global approximation that we will establish is indeed uniformly valid. Such a global approximation is called a *parametrix.*

4.5.1 Defining the Parametrix

We define the global parametrix, a model for $\mathbf{N}^{\sigma}(\lambda)$ in each part of the complex plane, as follows:

$$\widehat{\mathbf{N}}^{\sigma}(\lambda) := \begin{cases} \widehat{\mathbf{N}}^{\sigma}_{2k}(\lambda), & \lambda \in D_{2k},\ k = 0, \ldots, \\ & \qquad G/2 \text{ (cf. (4.157)–(4.160))}, \\ \sigma_2\widehat{\mathbf{N}}^{\sigma}_{2k}(\lambda^*)^*\sigma_2, & \lambda \in D^*_{2k},\ k = 0, \ldots, G/2, \\ \widehat{\mathbf{N}}^{\sigma}_{2k-1}(\lambda), & \lambda \in D_{2k-1},\ k = 1, \ldots, \\ & \qquad G/2+1 \text{ (cf. (4.186)–(4.189))}, \\ \sigma_2\widehat{\mathbf{N}}^{\sigma}_{2k-1}(\lambda^*)^*\sigma_2, & \lambda \in D^*_{2k-1},\ k = 1, \ldots, G/2+1, \\ \widehat{\mathbf{N}}^{\sigma}_{\mathrm{origin}}(\lambda), & \lambda \in U_{\hbar} \text{ (cf. (4.243))}, \\ \widehat{\mathbf{N}}^{\sigma}_{\mathrm{out}}(\lambda), & \text{otherwise (cf. (4.109))}. \end{cases} \tag{4.279}$$

The parametrix $\widehat{\mathbf{N}}^{\sigma}(\lambda)$ defined is analytic except on a complicated union of contours that we refer to simply as Σ_0. The contour Σ_0 is illustrated qualitatively in figure 4.18 for $\sigma = +1$ and $G = 2$. See the caption for details about the orientation. By contrast, we recall that the matrix $\mathbf{N}^{\sigma}(\lambda)$ is analytic for all $\lambda \in \mathbb{C}\backslash C \cup C^*$.

Along with the contour Σ_0, we consider a "completed" contour Σ defined as follows. As a set of points, Σ is the union of Σ_0 with the restrictions of the lens boundaries $C^{+}_{k\pm}$ and $C^{-}_{k\pm}$ to the interior of the disks D_k and D^*_k. The contour Σ now can easily be seen to divide the plane into two disjoint regions Ω^{+} and Ω^{-}, and consequently the orientation of each smooth arc of Σ may be chosen so that Σ forms the positively oriented boundary of Ω^{+} and therefore also the negatively oriented boundary of Ω^{-}. The orientation of corresponding arcs of Σ and Σ_0 will

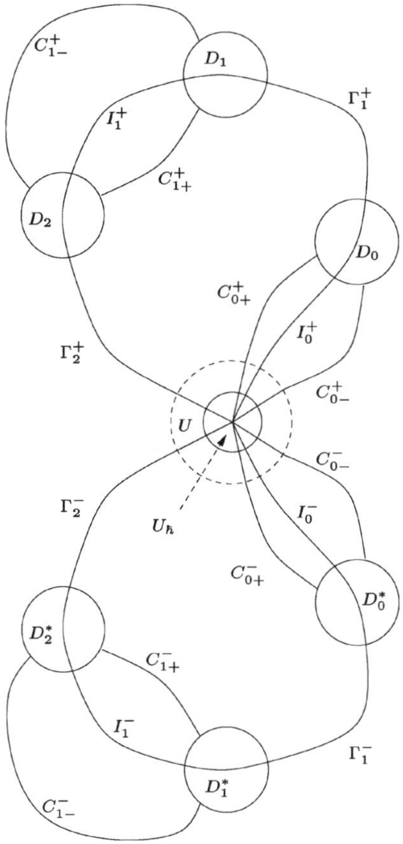

Figure 4.18 The domain of analyticity of the parametrix $\widehat{\mathbf{N}}^\sigma(\lambda)$ is $\mathbb{C}\setminus\Sigma_0$. The dashed circle is the boundary of the disk U and is not considered to be part of Σ_0. The contour Σ_0 is oriented as follows. The loop components C and C^* have orientation σ, the lens boundaries $C_{k\pm}^+$ (respectively, $C_{k\pm}^-$) are all oriented parallel to C_σ (respectively, $[C^*]_\sigma$), and all disk boundaries are oriented in the clockwise direction.

not necessarily agree. The parametrix $\widehat{\mathbf{N}}^\sigma(\lambda)$ may also be considered to be analytic for $\lambda \in \mathbb{C}\setminus\Sigma$, with a jump matrix on $\Sigma\setminus\Sigma_0$ equal to the identity matrix.

4.5.2 Asymptotic Validity of the Parametrix

The *error* in replacing $\mathbf{N}^\sigma(\lambda)$ by its parametrix is the matrix defined for $\lambda \in \mathbb{C}\setminus\Sigma$ by

$$\mathbf{E}(\lambda) := \mathbf{N}^\sigma(\lambda)\widehat{\mathbf{N}}^\sigma(\lambda)^{-1}. \tag{4.280}$$

This matrix depends on σ, as does the contour Σ, but we suppress this dependence in this section. Note that for each $\hbar_N$, $\mathbf{E}(\lambda)$ assumes boundary values on Σ that are uniformly continuous on the boundary of each connected component of $\mathbb{C}\setminus\Sigma$, since this is a property of both $\mathbf{N}^\sigma(\lambda)$ and of all approximations making up the sectionally holomorphic definition (4.279) of the parametrix. Therefore, we may define a jump matrix relative to the oriented contour Σ by setting

$$\mathbf{v}(\lambda) := \mathbf{E}_-(\lambda)^{-1}\mathbf{E}_+(\lambda), \quad \lambda \in \Sigma. \tag{4.281}$$

Under the conditions we assumed in the analysis of §4.3 and §4.4, we can easily establish the following.

Lemma 4.5.1 *The matrix $\mathbf{v}(\lambda)$ defined by (4.281) is uniformly bounded on Σ, and there exists an $M > 0$ such that for all sufficiently small $\hbar_N$ we have the estimate*

$$\sup_{\lambda\in\Sigma} \|\mathbf{v}(\lambda) - \mathbb{I}\| \leq M\hbar_N^{1/3}. \tag{4.282}$$

Proof. We begin with the smaller oriented contour Σ_0 and consider its smooth arc components one type at a time. For all $\lambda \in \Sigma_0$, let $\mathbf{v}_0(\lambda)$ denote the jump matrix for $\mathbf{E}(\lambda)$ relative to the orientation of Σ_0 rather than that of Σ. On each arc of Σ_0, we have either $\mathbf{v}(\lambda) = \mathbf{v}_0(\lambda)$ or $\mathbf{v}(\lambda) = \mathbf{v}_0(\lambda)^{-1}$.

First, consider the clockwise-oriented boundaries of the circular disks $U_\hbar$, $D_0, \dots,$
D_G, and $D_0^*, \dots, D_G^*$. On these boundaries, the jump matrix is simply

$$\mathbf{v}_0(\lambda) = \widehat{\mathbf{N}}^\sigma_-(\lambda)\mathbf{v}^\sigma_{\mathbf{N}}(\lambda)\mathbf{v}^\sigma_{\tilde{\mathbf{N}}}(\lambda)^{-1}\widehat{\mathbf{N}}^\sigma_-(\lambda)^{-1} = \widehat{\mathbf{N}}^\sigma_-(\lambda)\widehat{\mathbf{N}}^\sigma_+(\lambda)^{-1}, \tag{4.283}$$

since $\mathbf{N}^\sigma(\lambda)$ is analytic at these boundaries and thus the corresponding jump matrix is the identity. The "+" boundary value refers here to the exterior of the disk, and therefore by Lemma 4.4.5 and Lemma 4.4.10, we obtain

$$\|\mathbf{v}_0(\lambda) - \mathbb{I}\| \leq M\hbar_N^{1/3}, \quad \lambda \in \partial D_0 \cup \cdots \cup \partial D_G, \tag{4.284}$$

with similar estimates holding for the clockwise-oriented boundaries of $D_0^*, \dots, D_G^*$. Likewise, by Lemma 4.4.16,

$$\|\mathbf{v}_0(\lambda) - \mathbb{I}\| \leq M\hbar_N^{1-\delta}, \quad \lambda \in \partial U_\hbar. \tag{4.285}$$

Next, we consider the parts of Σ_0 in the interiors of the various circular disks. First consider one of the fixed disks D_j. Here we obtain

$$\begin{aligned}\|\mathbf{v}_0(\lambda) - \mathbb{I}\| &= \|\widehat{\mathbf{N}}^\sigma_-(\lambda)\mathbf{v}^\sigma_{\mathbf{N}}(\lambda)\mathbf{v}^\sigma_{\tilde{\mathbf{N}}}(\lambda)^{-1}\widehat{\mathbf{N}}^\sigma_-(\lambda)^{-1} - \mathbb{I}\| \\ &\leq \|\widehat{\mathbf{N}}^\sigma_-(\lambda)\| \cdot \|\widehat{\mathbf{N}}^\sigma_-(\lambda)^{-1}\| \cdot \|\mathbf{v}^\sigma_{\mathbf{N}}(\lambda)\mathbf{v}^\sigma_{\tilde{\mathbf{N}}}(\lambda)^{-1} - \mathbb{I}\| \\ &= \|\widehat{\mathbf{N}}^\sigma_-(\lambda)\| \cdot \|\widehat{\mathbf{N}}^\sigma_-(\lambda)^{-1}\| \cdot \|\mathbf{v}^\sigma_{\mathbf{N}}(\lambda)\mathbf{v}^\sigma_{\tilde{\mathbf{N}}}(\lambda)^{-1} - \mathbb{I}\| \\ &\leq M\hbar_N^{-2/3}\|\mathbf{v}^\sigma_{\mathbf{N}}(\lambda)\mathbf{v}^\sigma_{\tilde{\mathbf{N}}}(\lambda)^{-1} - \mathbb{I}\|,\end{aligned} \tag{4.286}$$

where we have used Lemma 4.4.3 and Lemma 4.4.4 (or their counterparts for odd-numbered endpoints Lemma 4.4.8 and Lemma 4.4.9). Now, since each disk is bounded away from the origin as $\hbar \downarrow 0$, it follows from Theorem 3.2.2 that

$$\|\mathbf{v}^\sigma_{\mathbf{N}}(\lambda)\mathbf{v}^\sigma_{\tilde{\mathbf{N}}}(\lambda)^{-1} - \mathbb{I}\| \leq M\hbar_N, \quad \lambda \in \Sigma_0 \cap D_j. \tag{4.287}$$

Consequently, we have

$$\|\mathbf{v}_0(\lambda) - \mathbb{I}\| \leq M\hbar_N^{1/3}, \quad \lambda \in \Sigma_0 \cap D_j, \tag{4.288}$$

with same estimate holding in D_j^*. Next, we examine the interior of the shrinking disk $U_\hbar$. For $\Sigma_0 \cap U_\hbar$, we find, applying first Lemma 4.4.14 and then Lemma 4.4.15, that

$$\begin{aligned}\|\mathbf{v}_0(\lambda) - \mathbb{I}\| &= \|\widehat{\mathbf{N}}_-^\sigma(\lambda)\mathbf{v}_{\mathbf{N}}^\sigma(\lambda)\mathbf{v}_{\widetilde{\mathbf{N}}}^\sigma(\lambda)^{-1}\widehat{\mathbf{N}}_-^\sigma(\lambda)^{-1} - \mathbb{I}\| \\ &\leq \|\widehat{\mathbf{N}}_-^\sigma(\lambda)\| \cdot \|\widehat{\mathbf{N}}_-^\sigma(\lambda)^{-1}\| \cdot \|\mathbf{v}_{\mathbf{N}}^\sigma(\lambda)\mathbf{v}_{\widetilde{\mathbf{N}}}^\sigma(\lambda)^{-1} - \mathbb{I}\| \\ &\leq M^2\|\mathbf{v}_{\mathbf{N}}^\sigma(\lambda)\mathbf{v}_{\widetilde{\mathbf{N}}}^\sigma(\lambda)^{-1} - \mathbb{I}\| \\ &\leq M'\hbar_N^{\min(1/3,2\delta-1)}, \quad \lambda \in \Sigma_0 \cap U_\hbar. \end{aligned} \tag{4.289}$$

It remains to examine the components of Σ_0 that lie strictly outside the closures of all disks. We call this exterior component Σ_0^{out}. In all of Σ_0^{out} we have a uniform bound on $\widehat{\mathbf{N}}^\sigma(\lambda) = \widehat{\mathbf{N}}_{\text{out}}^\sigma(\lambda)$ and in particular on the boundary values, that is independent of $\hbar$. Therefore, we have

$$\|\mathbf{v}_0(\lambda) - \mathbb{I}\| \leq M\|\mathbf{v}_{\mathbf{N}}^\sigma(\lambda)\mathbf{v}_{\widetilde{\mathbf{N}}}^\sigma(\lambda)^{-1} - \mathbb{I}\|, \quad \lambda \in \Sigma_0^{\text{out}}. \tag{4.290}$$

For λ in a band of $(C \cup C^*) \cap \Sigma_0^{\text{out}}$, we have from Lemma 4.3.1 that

$$\|\mathbf{v}_0(\lambda) - \mathbb{I}\| \leq M\|\mathbf{v}_{\mathbf{N}}^\sigma(\lambda)\mathbf{v}_{\widetilde{\mathbf{N}}}^\sigma(\lambda)^{-1} - \mathbb{I}\|, \quad \lambda \in I_k^\pm \cap \Sigma_0^{\text{out}}. \tag{4.291}$$

This quantity is $O(\hbar)$ for fixed λ but becomes larger near the outer boundary of the shrinking disk $U_\hbar$. From Theorem 3.2.2, we see that the error achieved at the cost of the shrinking radius of $U_\hbar$ is the estimate

$$\|\mathbf{v}_0(\lambda) - \mathbb{I}\| \leq M'\hbar_N^{1-\delta}, \quad \lambda \in I_k^\pm \cap \Sigma_0^{\text{out}}. \tag{4.292}$$

For λ in a gap of $(C \cup C^*) \cap \Sigma_0^{\text{out}}$, we have from Lemma 4.3.1 that

$$\|\mathbf{v}_0(\lambda) - \mathbb{I}\| \leq M\|\mathbf{v}_{\mathbf{N}}^\sigma(\lambda)\exp(-iJ\theta^\sigma(\lambda)\sigma_3/\hbar_N) - \mathbb{I}\|, \quad \lambda \in \Gamma_k^\pm \cap \Sigma_0^{\text{out}}. \tag{4.293}$$

The function $\theta^\sigma(\lambda)$ evaluates to a real constant in each gap. Using the definition of $\mathbf{v}_{\mathbf{N}}^\sigma(\lambda)$, we find that

$$\mathbf{v}_{\mathbf{N}}^\sigma(\lambda)\exp(-iJ\theta^\sigma(\lambda)\sigma_3/\hbar_N) - \mathbb{I} = e(\lambda) \cdot \sigma_1^{\frac{1-J}{2}} \begin{bmatrix} 0 & 0 \\ 1 & 0 \end{bmatrix} \sigma_1^{\frac{1-J}{2}}, \tag{4.294}$$

where the scalar $e(\lambda)$ is defined by

$$e(\lambda) := i\exp(-i\theta^\sigma(\lambda)/\hbar_N)\exp(\tilde{\phi}^\sigma(\lambda)/\hbar_N)\exp((\phi^\sigma(\lambda) - \tilde{\phi}^\sigma(\lambda))/\hbar_N). \tag{4.295}$$

Now $\exp(-i\theta^\sigma(\lambda)/\hbar_N)$ is uniformly bounded, and by Theorem 3.2.2, we find that for $\lambda \in \Sigma_0^{\text{out}}$ we have $\exp((\phi^\sigma(\lambda) - \tilde{\phi}^\sigma(\lambda))/\hbar_N) = 1 + O(\hbar_N^{1-\delta})$. The term $\exp(\tilde{\phi}^\sigma(\lambda)/\hbar_N)$ is, however, exponentially small with its maximum size $O(\exp(-M'\hbar_N^{\delta-1}))$ being attained on the outside boundary of $U_\hbar$. Consequently, we have the estimate

$$\|\mathbf{v}_0(\lambda) - \mathbb{I}\| \leq M\exp(-M'\hbar_N^{\delta-1}), \quad \lambda \in \Gamma_k^\pm \cap \Sigma_0^{\text{out}}, \tag{4.296}$$

for some $M' > 0$. Similarly, on the boundaries of the "lenses" in Σ_0^{out}, the matrix $\mathbf{N}^\sigma(\lambda)$ is analytic, so we have

$$\begin{aligned}\|\mathbf{v}_0(\lambda) - \mathbb{I}\| &\leq M\|\mathbf{v}_{\widetilde{\mathbf{N}}}^\sigma(\lambda)^{-1} - \mathbb{I}\| \\ &= M\|\mathbf{D}_+^\sigma(\lambda)\mathbf{D}_-^\sigma(\lambda)^{-1} - \mathbb{I}\|, \quad \lambda \in \{\text{lens boundaries}\} \cap \Sigma_0^{\text{out}}. \end{aligned} \tag{4.297}$$

Here, $\mathbf{D}^{\sigma}(\lambda)$ is the explicit "lens transformation" matrix, which differs from the identity matrix only inside the lenses, where it is defined by (4.111) and (4.112). These factors are exponentially small perturbations of the identity matrix away from the endpoints and the origin. It is again the proximity of the origin at the outside boundary of $U_{\hbar}$ that dominates this error; ultimately we obtain an estimate of the form

$$\|\mathbf{v}_0(\lambda) - \mathbb{I}\| \leq M \exp(-M'\hbar_N^{\delta-1}), \quad \lambda \in \{\text{lens boundaries}\} \cap \Sigma_0^{\text{out}}, \tag{4.298}$$

for some $M' > 0$.

We now combine the estimates (4.284), (4.285), (4.288), (4.289), (4.292), (4.296), and (4.298). The overall bound is optimized by taking $\delta = 2/3$, which determines the radius of the neighborhood $U_{\hbar}$ as $R = \hbar^{2/3}$. With this choice, all bounds except for the exponentially small contributions (4.296) and (4.298) are $O(\hbar_N^{1/3})$. Finally, we note that the required estimate for the jump matrix $\mathbf{v}(\lambda)$ for λ in the completed contour Σ follows from the corresponding result for Σ_0 along with the facts that $\det(\mathbf{v}_0(\lambda)) = 1$ for all $\lambda \in \Sigma_0$ and $\mathbf{v}(\lambda) \equiv \mathbb{I}$ for all $\lambda \in \Sigma \setminus \Sigma_0$. This completes the proof. □

Using the matrix $\mathbf{v}(\lambda)$ and the contour Σ, we can pose a final Riemann-Hilbert problem.

Riemann-Hilbert Problem 4.5.1 (Global error problem). *Find a matrix function* $\mathbf{R}(\lambda)$ *satisfying the following:*

1. **Analyticity:** $\mathbf{R}(\lambda)$ *is analytic for* $\lambda \in \mathbb{C}\setminus\Sigma$;
2. **Boundary behavior:** $\mathbf{R}(\lambda)$ *assumes continuous boundary values on* Σ *from each component of the complement, with continuity also at self-intersection points;*
3. **Jump condition:** *On the oriented contour* $\Sigma\setminus\{$*self-intersection points*$\}$, *the boundary values satisfy*
$$\mathbf{R}_+(\lambda) = \mathbf{R}_-(\lambda)\mathbf{v}(\lambda); \tag{4.299}$$
4. **Normalization:** $\mathbf{R}(\lambda)$ *is normalized at infinity:*
$$\mathbf{R}(\lambda) \to \mathbb{I} \quad \text{as } \lambda \to \infty. \tag{4.300}$$

On the one hand, we already "have" the solution to this Riemann-Hilbert problem.

Lemma 4.5.2 *The global error Riemann-Hilbert Problem* 4.5.2 *has a unique solution, namely,*

$$\mathbf{R}(\lambda) \equiv \mathbf{E}(\lambda), \tag{4.301}$$

where $\mathbf{E}(\lambda)$ *is defined by* (4.280).

Proof. The analyticity, boundary behavior, and normalization properties follow directly from the definition (4.280). The jump condition is equivalent to the definition (4.281) of the matrix $\mathbf{v}(\lambda)$. Uniqueness of the solution follows from the continuity of the boundary values and Liouville's theorem. □

But on the other hand, we can consider constructing the solution $\mathbf{R}(\lambda)$ of the global error Riemann-Hilbert Problem 4.5.2 directly, using *only* the available information about the jump matrix $\mathbf{v}(\lambda)$ contained in Lemma 4.5.1. Whatever we learn about $\mathbf{R}(\lambda)$ in this pursuit is then trivially a fact about the error matrix $\mathbf{E}(\lambda)$.

In order to construct $\mathbf{R}(\lambda)$ and obtain the desired estimates, we need to establish a uniformity result concerning the $\hbar$-dependence Cauchy integral operators on the contour Σ entering through the shrinking circle $\partial U_\hbar$ of radius $\hbar^\delta$. The main result we need is Lemma 4.5.4, but in order to prove this we first need the following technical lemma.

Lemma 4.5.3 *Let $I(s)$, $0 \le s \le s_{\max} < \infty$, be a C^2-curve in the complex plane parametrized by arc length, and suppose that $I(0) = 0$ and $I(s_{\max}) \neq 0$. Let C_ϵ denote the clockwise-oriented circle centered at the origin with radius ϵ. See figure 4.19. Then, for $\mathbf{f} \in L^2(C_\epsilon, |dz|)$ and $w \in I$, the Cauchy integral*

$$(\mathcal{C}^{C_\epsilon}\mathbf{f})(w) := \frac{1}{2\pi i}\int_{C_\epsilon}(z-w)^{-1}\mathbf{f}(z)\,dz \tag{4.302}$$

defines a bounded operator from $L^2(C_\epsilon, |dz|)$ to $L^2(I, ds)$ with a norm that is uniformly bounded above by a constant for all sufficiently small ϵ. Similarly, for $\mathbf{f} \in L^2(I, ds)$ and $w \in C_\epsilon$, the Cauchy integral

$$(\mathcal{C}^I\mathbf{f})(w) := \frac{1}{2\pi i}\int_0^{s_{\max}}(I(s)-w)^{-1}\mathbf{f}(I(s))\frac{dI}{ds}(s)\,ds \tag{4.303}$$

is a bounded operator from $L^2(I, ds)$ to $L^2(C_\epsilon, |dz|)$ with a norm that is uniformly bounded above by a constant for all ϵ sufficiently small.

Proof. For ϵ sufficiently small there is a unique $s_\epsilon > 0$ such that $|I(s_\epsilon)| = \epsilon$. This value satisfies $s_\epsilon/\epsilon \to 1$ as $\epsilon \downarrow 0$. Let $\phi := \arg(I'(0))$. We divide each contour

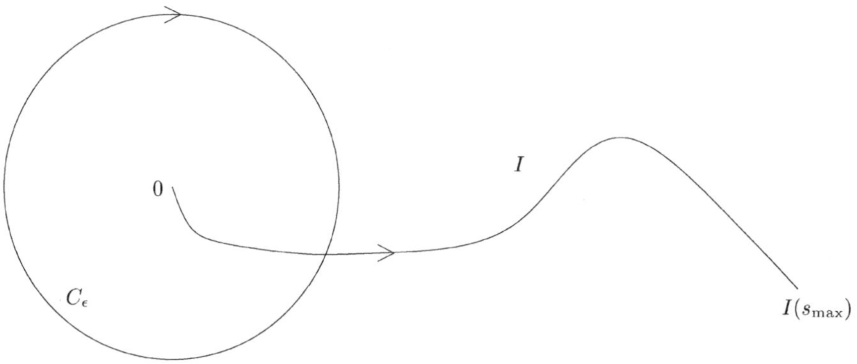

Figure 4.19 The contours of Lemma 4.5.3.

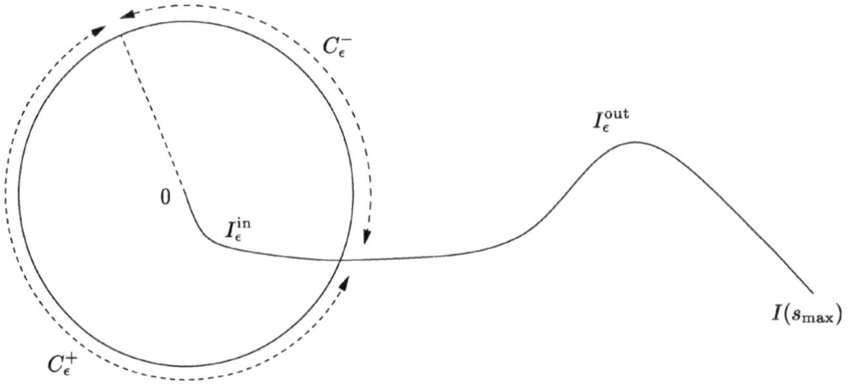

Figure 4.20 The subdivided contours.

into two pieces as follows. Let I_ϵ^{in} denote the contour parametrized by $I(s)$ for $s \in [0, s_\epsilon]$ and I_ϵ^{out} denote the contour parametrized by $I(s)$ for $s \in [s_\epsilon, s_{\max}]$. Then let C_ϵ^+ denote the part of C_ϵ with $\phi - \pi \leq \arg(z) \leq \arg(s_\epsilon)$ and C_ϵ^- the part of C_ϵ with $\arg(s_\epsilon) \leq \arg(z) \leq \phi + \pi$. Note that $\arg(s_\epsilon)$ tends to ϕ as ϵ tends to zero, so for small ϵ we are nearly dividing the circle in half along the tangent line to I at the origin. These subdivisions of the contours are illustrated in figure 4.20.

Let $\chi[\Sigma](z)$ denote the characteristic function of a contour segment Σ; then for $\mathbf{f} \in L^2(C_\epsilon)$ and $w \in I$, the Cauchy operator $\mathcal{C}^{C_\epsilon}$ is decomposed as

$$\begin{aligned}(\mathcal{C}^{C_\epsilon}\mathbf{f})(w) = {} & \chi[I_\epsilon^{\text{out}}](w) \cdot (\mathcal{C}^{C_\epsilon}(\chi[C_\epsilon^+](\cdot)\mathbf{f}))(w) \\ & + \chi[I_\epsilon^{\text{out}}](w) \cdot (\mathcal{C}^{C_\epsilon}(\chi[C_\epsilon^-](\cdot)\mathbf{f}))(w) \\ & + \chi[I_\epsilon^{\text{in}}](w) \cdot (\mathcal{C}^{C_\epsilon}(\chi[C_\epsilon^+](\cdot)\mathbf{f}))(w) \\ & + \chi[I_\epsilon^{\text{in}}](w) \cdot (\mathcal{C}^{C_\epsilon}(\chi[C_\epsilon^-](\cdot)\mathbf{f}))(w). \end{aligned} \tag{4.304}$$

Likewise, for $\mathbf{f} \in L^2(I, ds)$ and $w \in C_\epsilon$, we have

$$\begin{aligned}(\mathcal{C}^{I}\mathbf{f})(w) = {} & \chi[C_\epsilon^+](w) \cdot (\mathcal{C}^{I}(\chi[I_\epsilon^{\text{out}}](\cdot)\mathbf{f}))(w) \\ & + \chi[C_\epsilon^+](w) \cdot (\mathcal{C}^{I}(\chi[I_\epsilon^{\text{in}}](\cdot)\mathbf{f}))(w) \\ & + \chi[C_\epsilon^-](w) \cdot (\mathcal{C}^{I}(\chi[I_\epsilon^{\text{out}}](\cdot)\mathbf{f}))(w) \\ & + \chi[C_\epsilon^-](w) \cdot (\mathcal{C}^{I}(\chi[I_\epsilon^{\text{in}}](\cdot)\mathbf{f}))(w). \end{aligned} \tag{4.305}$$

It is therefore sufficient to prove the uniform boundedness of each operator on the right-hand side of each expression. We give details only for the terms involving C_ϵ^+, since the reader will see that the bounds involving C_ϵ^- are obtained in exactly the same way.

We now introduce arc-length parametrizations of C_ϵ^+, I_ϵ^{in}, and I_ϵ^{out} and at the same time use the invariance of the Cauchy integral under translations and rotations to bring the intersection point $I(s_\epsilon)$ to the origin and to make C_ϵ^+ tangent to the positive real axis there. Therefore, using tildes to denote the translation and rotation (see figure 4.21), for $\widetilde{C}_\epsilon^+$ we have the parametrization

$$\widetilde{C}_\epsilon^+(y) := i\epsilon(\exp(-iy/\epsilon) - 1), \quad 0 \leq y \leq \epsilon\xi_\epsilon^+, \tag{4.306}$$

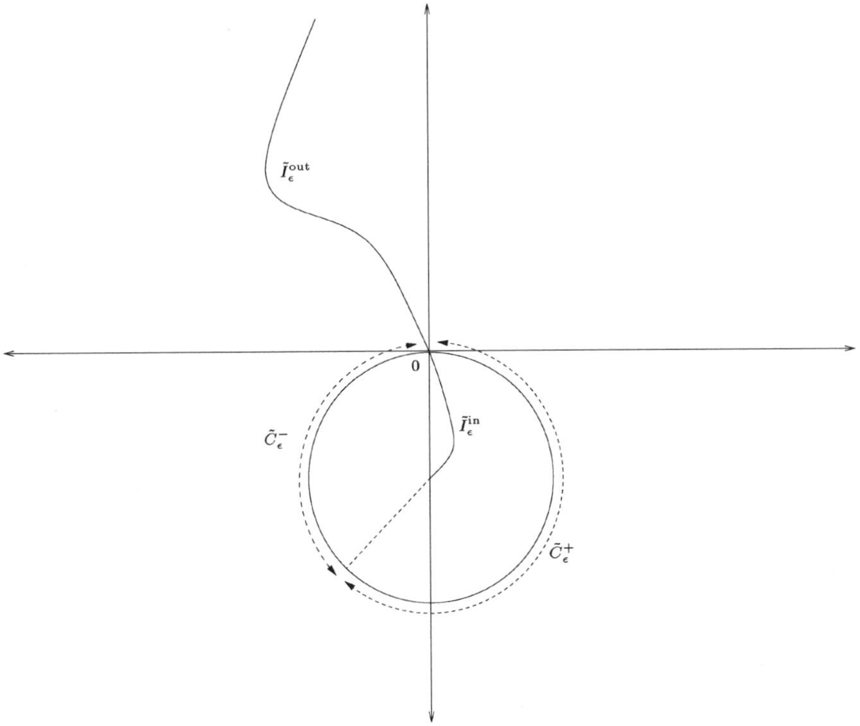

Figure 4.21 After rotation and translation.

where $\xi_\epsilon^+ := \pi + \arg(I(s_\epsilon)) - \phi$ is an angle converging to π as $\epsilon \downarrow 0$. For $\widetilde{I}_\epsilon^{\text{in}}$ we have

$$\widetilde{I}_\epsilon^{\text{in}}(x) := (I(s_\epsilon - x) - I(s_\epsilon))i\exp(-i\arg(I(s_\epsilon))), \quad 0 \le x \le s_\epsilon, \tag{4.307}$$

and for $\widetilde{I}_\epsilon^{\text{out}}$ we have

$$\widetilde{I}_\epsilon^{\text{out}}(x) := (I(s_\epsilon + x) - I(s_\epsilon))i\exp(-i\arg(I(s_\epsilon))), \quad 0 \le x \le S_\epsilon, \tag{4.308}$$

where $S_\epsilon := s_{\max} - s_\epsilon$ converges to $s_{\max} > 0$ as $\epsilon \downarrow 0$. Note that at the origin, $\widetilde{I}_\epsilon^{\text{in}}$ is tangent to a ray making an angle of $\zeta_\epsilon^{\text{in}} := -\pi/2 + \arg(I'(s_\epsilon)) - \arg(I(s_\epsilon))$ with the positive real axis and $\widetilde{I}_\epsilon^{\text{out}}$ is tangent to a ray making an angle of $\zeta_\epsilon^{\text{out}} := \pi/2 + \arg(I'(s_\epsilon)) - \arg(I(s_\epsilon))$ with the positive real axis. These angles converge to $-\pi/2$ and $\pi/2$, respectively, as $\epsilon \downarrow 0$. The reader should note that the scales have not been changed by these transformations. The circle $\widetilde{C}_\epsilon^+ \cup \widetilde{C}_\epsilon^-$ still has radius ϵ, and the subsequent estimates will be uniform for ϵ sufficiently small.

For $\mathbf{f} \in L^2(\widetilde{C}_\epsilon^+, dy)$ and $0 \le x \le s_\epsilon$, we define a kernel $K^{\text{in}}(x, y)$ by writing

$$\begin{aligned}\frac{1}{2\pi i}\int_0^{\epsilon\xi_\epsilon^+} &(\widetilde{C}_\epsilon^+(y) - \widetilde{I}_\epsilon^{\text{in}}(x))^{-1}\mathbf{f}(\widetilde{C}_\epsilon^+(y))\widetilde{C}_\epsilon^{+\prime}(y)\,dy \\ &= \frac{1}{2\pi i}\int_0^{\epsilon\xi_\epsilon^+}(y - x\exp(i\zeta_\epsilon^{\text{in}}))^{-1}(\mathcal{U}_1\mathbf{f})(y)\,dy \\ &\quad + \frac{1}{2\pi i}\int_0^{\epsilon\xi_\epsilon^+} K^{\text{in}}(x, y)(\mathcal{U}_1\mathbf{f})(y)\,dy,\end{aligned} \tag{4.309}$$

where the map defined by $(\mathcal{U}_1\mathbf{f})(y) := \mathbf{f}(\widetilde{C}_\epsilon^+(y))\widetilde{C}_\epsilon^{+\prime}(y)$ is a unitary isomorphism between $L^2(\widetilde{C}_\epsilon^+, dy)$ and $L^2([0, \epsilon\xi_\epsilon^+], dy)$. The reader will observe that the left-hand side of (4.309) is the third term on the right-hand side of (4.304). Therefore, we have written this term as a sum of an explicit Cauchy integral in the new coordinate system, plus an error term involving the kernel $K^{\text{in}}(x, y)$. Then, for $\mathbf{f} \in L^2(\widetilde{I}_\epsilon^{\text{in}}, dx)$ and $0 \le y \le \epsilon\xi_\epsilon^+$, we have an expression in terms of the same kernel of the "reciprocal" Cauchy integral:

$$\begin{aligned}\frac{1}{2\pi i}\int_0^{S_\epsilon} & (\widetilde{I}_\epsilon^{\text{in}}(x) - \widetilde{C}_\epsilon^+(y))^{-1}\mathbf{f}(\widetilde{I}_\epsilon^{\text{in}}(x))\widetilde{I}_\epsilon^{\text{in}\prime}(x)\,dx \\ &= \frac{1}{2\pi i}\int_0^{S_\epsilon}(x\exp(i\zeta_\epsilon^{\text{in}}) - y)^{-1}(\mathcal{U}_2\mathbf{f})(x)\,dx \\ &\quad - \frac{1}{2\pi i}\int_0^{S_\epsilon} K^{\text{in}}(x, y)(\mathcal{U}_2\mathbf{f})(x)\,dx, \end{aligned} \tag{4.310}$$

where the map $(\mathcal{U}_2\mathbf{f})(x) := \mathbf{f}(\widetilde{I}_\epsilon^{\text{in}}(x))\widetilde{I}_\epsilon^{\text{in}\prime}(x)$ is a unitary isomorphism between $L^2(\widetilde{I}_\epsilon^{\text{in}}, dx)$ and $L^2([0, s_\epsilon], dx)$. The left-hand side of (4.310) is just the second term on the right-hand side of (4.305) written in the new coordinates, which has similarly been split into a Cauchy integral and a remainder term involving $K^{\text{in}}(x, y)$.

Similarly, we define a kernel $K^{\text{out}}(x, y)$ by writing for $\mathbf{f} \in L^2(\widetilde{C}_\epsilon^+, dy)$ and $0 \le x \le S_\epsilon$,

$$\begin{aligned}\frac{1}{2\pi i}\int_0^{\epsilon\xi_\epsilon^+} & (\widetilde{C}_\epsilon^+(y) - \widetilde{I}_\epsilon^{\text{out}}(x))^{-1}\mathbf{f}(\widetilde{C}_\epsilon^+(y))\widetilde{C}_\epsilon^{+\prime}(y)\,dy \\ &= \frac{1}{2\pi i}\int_0^{\epsilon\xi_\epsilon^+}(y - x\exp(i\zeta_\epsilon^{\text{out}}))^{-1}(\mathcal{U}_1\mathbf{f})(y)\,dy \\ &\quad + \frac{1}{2\pi i}\int_0^{\epsilon\xi_\epsilon^+} K^{\text{out}}(x, y)(\mathcal{U}_1\mathbf{f})(y)\,dy, \end{aligned} \tag{4.311}$$

from which we then obtain for the reciprocal Cauchy integral, for $\mathbf{f} \in L^2(\widetilde{I}_\epsilon^{\text{out}}, dx)$ and $0 \le y \le \epsilon\xi_\epsilon^+$,

$$\begin{aligned}\frac{1}{2\pi i}\int_0^{S_\epsilon} & (\widetilde{I}_\epsilon^{\text{out}}(x) - \widetilde{C}_\epsilon^+(y))^{-1}\mathbf{f}(\widetilde{I}_\epsilon^{\text{out}}(x))\widetilde{I}_\epsilon^{\text{out}\prime}(x)\,dx \\ &= \frac{1}{2\pi i}\int_0^{S_\epsilon}(x\exp(i\zeta_\epsilon^{\text{out}}) - y)^{-1}(\mathcal{U}_3\mathbf{f})(x)\,dx \\ &\quad - \frac{1}{2\pi i}\int_0^{S_\epsilon} K^{\text{out}}(x, y)(\mathcal{U}_3\mathbf{f})(x)\,dx, \end{aligned} \tag{4.312}$$

where $\mathcal{U}_3$ denotes the unitary isomorphism from $L^2(\widetilde{I}_\epsilon^{\text{out}}, dx)$ to $L^2([0, S_\epsilon], dx)$ defined by $(\mathcal{U}_3\mathbf{f})(x) := \mathbf{f}(\widetilde{I}_\epsilon^{\text{out}}(x))\widetilde{I}_\epsilon^{\text{out}\prime}(x)$. The decomposition (4.311) is a representation of the first term on the right-hand side of (4.304). Likewise, (4.312) is a representation of the first term on the right-hand side of (4.305).

Having represented each term involving C_ϵ^+ on the right-hand sides of (4.304) and (4.305) as a sum of a Cauchy integral operator in the new coordinate system and a remainder-type integral operator, we now need to prove that these are in fact all bounded operators. The Cauchy operators are handled by an argument of Beals,

Deift, and Tomei [BDT88] that uses the theory of Mellin transforms in L^2 spaces on straight rays. Their methods show that regardless of the value of ϵ,

$$\begin{aligned}
&\int_0^{s_\epsilon}\left\|\frac{1}{2\pi i}\int_0^{\epsilon\xi_\epsilon^+}\frac{dy}{y-x\exp(i\zeta_\epsilon^{\mathrm{in}})}(\mathcal{U}_1\mathbf{f})(y)\right\|^2 dx \\
&\quad\le\int_0^{\epsilon\xi_\epsilon^+}\|\mathbf{f}(\widetilde{C}_\epsilon^+(y))\|^2\,dy,\quad\forall\mathbf{f}\in L^2(\widetilde{C}_\epsilon^+,dy),\\
&\int_0^{S_\epsilon}\left\|\frac{1}{2\pi i}\int_0^{\epsilon\xi_\epsilon^+}\frac{dy}{y-x\exp(i\zeta_\epsilon^{\mathrm{out}})}(\mathcal{U}_1\mathbf{f})(y)\right\|^2 dx \\
&\quad\le\int_0^{\epsilon\xi_\epsilon^+}\|\mathbf{f}(\widetilde{C}_\epsilon^+(y))\|^2\,dy,\quad\forall\mathbf{f}\in L^2(\widetilde{C}_\epsilon^+,dy),\\
&\int_0^{\epsilon\xi_\epsilon^+}\left\|\frac{1}{2\pi i}\int_0^{s_\epsilon}\frac{dx}{x\exp(i\zeta_\epsilon^{\mathrm{in}})-y}(\mathcal{U}_2\mathbf{f})(x)\right\|^2 dy \\
&\quad\le\int_0^{s_\epsilon}\|\mathbf{f}(\widetilde{I}_\epsilon^{\mathrm{in}}(x))\|^2\,dx,\quad\forall\mathbf{f}\in L^2(\widetilde{I}_\epsilon^{\mathrm{in}},dx),\\
&\int_0^{\epsilon\xi_\epsilon^+}\left\|\frac{1}{2\pi i}\int_0^{S_\epsilon}\frac{dx}{x\exp(i\zeta_\epsilon^{\mathrm{out}})-y}(\mathcal{U}_3\mathbf{f})(x)\right\|^2 dy \\
&\quad\le\int_0^{S_\epsilon}\|\mathbf{f}(\widetilde{I}_\epsilon^{\mathrm{out}}(x))\|^2\,dx,\quad\forall\mathbf{f}\in L^2(\widetilde{I}_\epsilon^{\mathrm{out}},dx),
\end{aligned}\tag{4.313}$$

as long as $\zeta_\epsilon^{\mathrm{in}}$ and $\zeta_\epsilon^{\mathrm{out}}$ are both nonzero. Since these angles converge to $-\pi/2$ and $\pi/2$, respectively, as $\epsilon\downarrow 0$, this is the case for all sufficiently small ϵ. Thus, the Cauchy integral operators appearing as the first terms on the right-hand sides of (4.309), (4.310), (4.311), and (4.312) are bounded, with bounds that are independent of ϵ.

We now turn to the estimation of the "remainder" operators with kernels $K^{\mathrm{in}}(x,y)$ and $K^{\mathrm{out}}(x,y)$. In this connection, we first note that these kernels, which are explicitly written as

$$\begin{aligned}
K^{\mathrm{in}}(x,y)&:=\frac{1}{\widetilde{C}_\epsilon^+(y)-\widetilde{I}_\epsilon^{\mathrm{in}}(x)}-\frac{1}{y-x\exp(i\zeta_\epsilon^{\mathrm{in}})},\\
K^{\mathrm{out}}(x,y)&:=\frac{1}{\widetilde{C}_\epsilon^+(y)-\widetilde{I}_\epsilon^{\mathrm{out}}(x)}-\frac{1}{y-x\exp(i\zeta_\epsilon^{\mathrm{out}})},
\end{aligned}\tag{4.314}$$

are bounded functions on their respective domains of definition and

$$\limsup_{x,y\to 0}|K^{\mathrm{in}}(x,y)|<\infty,\qquad\limsup_{x,y\to 0}|K^{\mathrm{out}}(x,y)|<\infty.\tag{4.315}$$

Therefore, for all $\mathbf{f}\in L^2(\widetilde{C}_\epsilon^+,dy)$,

$$\begin{aligned}
&\int_0^{s_\epsilon}\left\|\int_0^{\epsilon\xi_\epsilon^+}K^{\mathrm{in}}(x,y)(\mathcal{U}_1\mathbf{f})(y)\,dy\right\|^2 dx\\
&\quad\le\left(\int_0^{s_\epsilon}dx\int_0^{\epsilon\xi_\epsilon^+}dy\,|K^{\mathrm{in}}(x,y)|^2\right)\int_0^{\epsilon\xi^+}\|\mathbf{f}(\widetilde{C}_\epsilon^+(y))\|^2\,dy
\end{aligned}\tag{4.316}$$

and

$$\int_0^{S_\epsilon}\left\|\int_0^{\epsilon\xi_\epsilon^+} K^{\rm out}(x,y)(\mathcal{U}_1\mathbf{f})(y)\,dy\right\|^2 dx$$
$$\le \left(\int_0^{S_\epsilon} dx \int_0^{\epsilon\xi_\epsilon^+} dy\,|K^{\rm out}(x,y)|^2\right)\int_0^{\epsilon\xi^+} \|\mathbf{f}(\widetilde{C}_\epsilon^+(y))\|^2\,dy, \quad (4.317)$$

while for $\mathbf{f}\in L^2(\widetilde{I}_\epsilon^{\rm in}, dx)$,

$$\int_0^{\epsilon\xi_\epsilon^+}\left\|\int_0^{S_\epsilon} K^{\rm in}(x,y)(\mathcal{U}_2\mathbf{f})(x)\,dx\right\|^2 dy$$
$$\le \left(\int_0^{S_\epsilon} dx \int_0^{\epsilon\xi_\epsilon^+} dy\,|K^{\rm in}(x,y)|^2\right)\int_0^{S_\epsilon} \|\mathbf{f}(\widetilde{I}_\epsilon^{\rm in}(x))\|^2\,dx, \quad (4.318)$$

and for $\mathbf{f}\in L^2(\widetilde{I}_\epsilon^{\rm out}, dx)$,

$$\int_0^{\epsilon\xi_\epsilon^+}\left\|\int_0^{S_\epsilon} K^{\rm out}(x,y)(\mathcal{U}_3\mathbf{f})(x)\,dx\right\|^2 dy$$
$$\le \left(\int_0^{S_\epsilon} dx \int_0^{\epsilon\xi_\epsilon^+} dy\,|K^{\rm out}(x,y)|^2\right)\int_0^{S_\epsilon} \|\mathbf{f}(\widetilde{I}_\epsilon^{\rm out}(x))\|^2\,dx. \quad (4.319)$$

These norm estimates are finite for all $\epsilon > 0$, and we must control their dependence on ϵ as $\epsilon \downarrow 0$.

We now claim that

$$\lim_{\epsilon\downarrow 0}\int_0^{S_\epsilon} dx \int_0^{\epsilon\xi_\epsilon^+} dy|K^{\rm in}(x,y)|^2 = \int_0^{\pi} dz \int_0^1 dw\left|\frac{1}{i(\exp(-iz)-1)+iw} - \frac{1}{z+iw}\right|^2 < \infty,$$
$$\lim_{\epsilon\downarrow 0}\int_0^{S_\epsilon} dx \int_0^{\epsilon\xi_\epsilon^+} dy|K^{\rm out}(x,y)|^2 = \int_0^{\pi} dz \int_0^\infty dw\left|\frac{1}{i(\exp(-iz)-1)-iw} - \frac{1}{z-iw}\right|^2 < \infty. \quad (4.320)$$

These limits are finite because the integrands are bounded near $z = w = 0$ (in particular they are both less than 1 for all $z < 1$), decay like w^{-4} for large w, and are uniformly bounded elsewhere. To prove the claim, first rescale in both integrals by setting $z = y/\epsilon$ and $w = x/\epsilon$. This modifies the integrand through the Jacobian by multiplication by ϵ^2. For the $K^{\rm in}(x, y)$ integral, the region of integration tends to the fixed rectangle $[0, 1] \times [0, \pi]$, while for the $K^{\rm out}(x, y)$ integral, the region of integration tends to the semi-infinite strip $[0, \infty] \times [0, \pi]$. Using the fact that the curve $I(s)$ is twice differentiable, one sees that the integrands $\epsilon^2|K^{\rm in}(\epsilon w, \epsilon z)|^2$ and $\epsilon^2|K^{\rm out}(\epsilon w, \epsilon z)|^2$ converge pointwise as $\epsilon \downarrow 0$ to the integrands on the right-hand side of (4.320). The convergence is in fact uniform for the $K^{\rm in}(x, y)$ integral, and therefore this part of the claim follows immediately. For the $K^{\rm out}(x, y)$ integral, the claim follows from a dominated convergence argument.

Since the limits (4.320) exist, the operators having kernels $K^{\rm in}(x, y)$, $-K^{\rm in}(y, x)$, $K^{\rm out}(x, y)$, and finally $-K^{\rm out}(y, x)$ are bounded in the appropriate L^2 spaces uniformly as $\epsilon \downarrow 0$. Combining these estimates with the Beals-Deift-Tomei estimates of the Cauchy kernels and the results of a parallel analysis involving the circular arc C_ϵ^- completes the proof of the lemma. □

For each fixed value of $\hbar$, we can define operators $\mathcal{C}_\pm^\Sigma$ on $L^2(\Sigma)$ taking a function $\mathbf{f}(z)$ to the boundary values taken on each oriented segment of Σ from the "+" and "−" sides, respectively, of the Cauchy contour integral

$$(\mathcal{C}^\Sigma\mathbf{f})(w) := \frac{1}{2\pi i}\int_\Sigma (z-w)^{-1}\mathbf{f}(z)\,dz. \tag{4.321}$$

With a suitable interpretation of convergence to the boundary values, for each fixed $\hbar$, these operators are bounded on $L^2(\Sigma)$.

Lemma 4.5.4 *There exists an $M > 0$ such that for all sufficiently small $\hbar$,*

$$\|\mathcal{C}_+^\Sigma\|_{L^2(\Sigma)} < M \quad \textit{and} \quad \|\mathcal{C}_-^\Sigma\|_{L^2(\Sigma)} < M. \tag{4.322}$$

Proof. We first note that, modulo self-intersection points, the contour Σ can be written as a union of an $\hbar$-independent part $\Sigma\backslash\partial U_\hbar$ and several arcs making up the shrinking circle $\partial U_\hbar$. Let $\mathbf{f} \in L^2(\Sigma)$, and decompose it into a sum $\mathbf{g}+\mathbf{h}$, where the support of $\mathbf{g}$ is contained in $\Sigma\backslash\partial U_\hbar$ and that of $\mathbf{h}$ is contained in $\partial U_\hbar$. Then, for almost every $z \in \Sigma\backslash\partial U_\hbar$,

$$(\mathcal{C}_\pm^\Sigma\mathbf{f})(z) = (\mathcal{C}_\pm^{\Sigma\backslash\partial U_\hbar}\mathbf{g})(z) + (\mathcal{C}^{\partial U_\hbar}\mathbf{h})(z), \tag{4.323}$$

and for almost every $z \in \partial U_\hbar$,

$$(\mathcal{C}_\pm^\Sigma\mathbf{f})(z) = (\mathcal{C}_\pm^{\partial U_\hbar}\mathbf{h})(z) + (\mathcal{C}^{\Sigma\backslash\partial U_\hbar}\mathbf{g})(z). \tag{4.324}$$

Integrating to compute the norm, we first estimate

$$\int_\Sigma \|(\mathcal{C}_\pm^\Sigma\mathbf{f})(z)\|^2\,|dz| \le I_{gg} + I_{gh} + I_{hg} + I_{hh}, \tag{4.325}$$

where

$$\begin{aligned} I_{gg} &:= \int_{\Sigma\backslash\partial U_\hbar} \|(\mathcal{C}_\pm^{\Sigma\backslash\partial U_\hbar}\mathbf{g})(z)\|^2\,|dz|, \quad I_{gh} := \int_{\Sigma\backslash\partial U_\hbar} \|(\mathcal{C}^{\partial U_\hbar}\mathbf{h})(z)\|^2\,|dz|,\\ I_{hg} &:= \int_{\partial U_\hbar} \|(\mathcal{C}^{\Sigma\backslash\partial U_\hbar}\mathbf{g})(z)\|^2\,|dz|, \quad I_{hh} := \int_{\partial U_\hbar} \|(\mathcal{C}_\pm^{\partial U_\hbar}\mathbf{h})(z)\|^2\,|dz|. \end{aligned} \tag{4.326}$$

First, we estimate the "diagonal" terms I_{gg} and I_{hh}. Because the contour $\Sigma\backslash\partial U_\hbar$ is independent of $\hbar$, there is some $\hbar$-independent constant $C_{gg} > 0$ such that

$$I_{gg} \le C_{gg}\int_{\Sigma\backslash\partial U_\hbar} \|\mathbf{g}(z)\|^2\,|dz| \le C_{gg}\int_\Sigma \|\mathbf{f}(z)\|^2\,|dz|. \tag{4.327}$$

Now for the integral I_{hh} the contour depends on $\hbar$, but in a simple way that can be scaled out. Thus, rescaling,

$$I_{hh} = \hbar^\delta \int_{\partial U_1} \left\| \frac{1}{2\pi i}\int_{\partial U_1} (t-w_\pm)^{-1}\mathbf{h}(\hbar^\delta t)\,dt \right\|^2 |dw|. \tag{4.328}$$

With the contour rescaled to a radius independent of $\hbar$, we see that there exists an $\hbar$-independent constant $C_{hh} > 0$ such that

$$I_{hh} \leq \hbar^\delta C_{hh} \int_{\partial U_1} \|\mathbf{h}(\hbar^\delta t)\|^2 \, |dt| = C_{hh} \int_{\partial U_\hbar} \|\mathbf{h}(z)\|^2 \, |dz| \leq C_{hh} \int_\Sigma \|\mathbf{f}(z)\|^2 \, |dz|. \tag{4.329}$$

Next, we turn to the estimation of the "cross terms" I_{gh} and I_{hg}. For this purpose, we again decompose the $\hbar$-independent contour $\Sigma \backslash \partial U_\hbar$ into two $\hbar$-independent parts by cutting it at the boundary of the fixed but small disk U (this disk is illustrated in figure 4.18). Thus, let $\Gamma^{\rm in}$ (respectively, $\Gamma^{\rm out}$) denote the part of $\Sigma \backslash \partial U_\hbar$ inside (respectively, outside) U. With this decomposition, we have by Cauchy-Schwarz,

$$\begin{aligned} I_{gh} &= \int_{\Gamma^{\rm in}} \|(C^{\partial U_\hbar}\mathbf{h})(z)\|^2 \, |dz| + \int_{\Gamma^{\rm out}} \|(C^{\partial U_\hbar}\mathbf{h})(z)\|^2 \, |dz| \\ &\leq \int_{\Gamma^{\rm in}} \|(C^{\partial U_\hbar}\mathbf{h})(z)\|^2 \, |dz| \\ &\quad + \frac{\hbar^\delta |\Gamma^{\rm out}|}{2\pi} \sup_{s \in \Gamma^{\rm out}, z \in \partial U_\hbar} |s - z|^{-2} \int_{\partial U_\hbar} \|\mathbf{h}(s)\|^2 \, |ds|, \end{aligned} \tag{4.330}$$

where $|\Gamma^{\rm out}|$ is the $\hbar$-independent total arc length of $\Gamma^{\rm out}$. The supremum in the last line is bounded as $\hbar$ tends to zero because distance between $\Gamma^{\rm out}$ and $\partial U_\hbar$ increases as $\hbar$ decreases. Therefore, there is an $\hbar$-independent constant $C^{\rm out}_{\rm cross} > 0$ such that

$$I_{gh} \leq \int_{\Gamma^{\rm in}} \|(C^{\partial U_\hbar}\mathbf{h})(z)\|^2 \, |dz| + \hbar^\delta C^{\rm out}_{\rm cross} \int_\Sigma \|\mathbf{f}(z)\|^2 \, |dz| \tag{4.331}$$

for all sufficiently small $\hbar$. We momentarily delay the estimation of the term involving $\Gamma^{\rm in}$. Applying the same decomposition to I_{hg} and using Cauchy-Schwarz, we find with the same constant $C^{\rm out}_{\rm cross}$

$$\begin{aligned} I_{hg} &\leq \int_{\partial U_\hbar} \|(C^{\Gamma^{\rm in}}\mathbf{g})(z)\|^2 \, |dz| + \int_{\partial U_\hbar} \|(C^{\Gamma^{\rm out}}\mathbf{g})(z)\|^2 \, |dz| \\ &\leq \int_{\partial U_\hbar} \|(C^{\Gamma^{\rm in}}\mathbf{g})(z)\|^2 \, |dz| + \hbar^\delta C^{\rm out}_{\rm cross} \int_\Sigma \|\mathbf{f}(z)\|^2 \, |dz|. \end{aligned} \tag{4.332}$$

Finally, to estimate the remaining terms in I_{gh} and I_{hg}, we appeal to Lemma 4.5.3. Note that $\Gamma^{\rm in}$ is a union of eight smooth curve segments (in fact all but two of the segments are exactly straight ray segments) meeting at the origin. Therefore, the lemma applies to the interaction of each curve segment with the circle $\partial U_\hbar$ of radius $\epsilon = \hbar^\delta$. Summing the $\hbar$-independent estimates guaranteed by Lemma 4.5.3 finally gives constants $C^{\rm in}_{gh} > 0$ and $C^{\rm in}_{hg} > 0$ such that

$$\int_{\Gamma^{\rm in}} \|(C^{\partial U_\hbar}\mathbf{h})(z)\|^2 \, |dz| \leq C^{\rm in}_{gh} \int_{\partial U_\hbar} \|\mathbf{h}(z)\|^2 \, |dz| \leq C^{\rm in}_{gh} \int_\Sigma \|\mathbf{f}(z)\|^2 \, |dz| \tag{4.333}$$

and

$$\int_{\partial U_\hbar} \|(C^{\Gamma^{\rm in}}\mathbf{g})(z)\|^2 \, |dz| \leq C^{\rm in}_{hg} \int_{\Gamma^{\rm in}} \|\mathbf{g}(z)\|^2 \, |dz| \leq C^{\rm in}_{hg} \int_\Sigma \|\mathbf{f}(z)\|^2 \, |dz|. \tag{4.334}$$

Assembling the estimates of the diagonal terms and the cross terms finally completes the proof. □

With these results in hand, we can estimate the error $\mathbf{E}(\lambda)$ under the condition that we have found a complex phase function $g^{\sigma}(\lambda)$. We do this by constructing the solution $\mathbf{R}(\lambda)$ of the global error Riemann-Hilbert Problem 4.5.2 directly from its jump matrix $\mathbf{v}(\lambda)$, for which we have a uniform estimate from Lemma 4.5.1.

Theorem 4.5.1 Conditional error bound). *Given the existence of a complex phase function $g^{\sigma}(\lambda)$, there exists some $M > 0$ such that for all sufficiently small $\hbar_N$, the error satisfies the estimate*

$$\|\mathbf{E}(\lambda) - \mathbb{I}\| \leq M\hbar_N^{1/3} \sup_{s\in\Sigma} |s-\lambda|^{-1}. \tag{4.335}$$

Proof. For $\lambda \in \Sigma$, set $\mathbf{w}^{+}(\lambda) := \mathbf{v}(\lambda) - \mathbb{I}$ and $\mathbf{w}^{-}(\lambda) := \mathbf{0}$. Then from Lemma 4.5.1 and Lemma 4.5.4, we observe that the conditions of Lemma B.0.2 in appendix B are met. This guarantees the existence of a matrix $\mathbf{R}(\lambda)$ satisfying the global error Riemann-Hilbert Problem 4.5.2 in the $L^2(\Sigma)$ sense. According to Lemma B.0.2, this solution $\mathbf{R}(\lambda)$ satisfies the error estimate (B.11), which by virtue of the bound on $\|\mathbf{v}(\lambda) - \mathbb{I}\|$ afforded by Lemma 4.5.1 and the fact that the total length of the contour Σ remains uniformly bounded as $\hbar_N$ tends to zero implies (4.335) via the equivalence $\mathbf{E}(\lambda) \equiv \mathbf{R}(\lambda)$. □

Remark. In principle, the representation $\mathbf{E}(\lambda) \equiv \mathbf{R}(\lambda)$ and the asymptotic control on $\mathbf{w}^{\pm}(\lambda)$ imply not only the estimate given here but also an explicit series representation of $\mathbf{R}(\lambda)$ obtained via the solution of the associated singular integral equation (cf. Lemma B.0.1) by Neumann series. Making effective the calculation of correction terms in the expansion $\mathbf{R}(\lambda) = \mathbb{I} + \cdots$ requires explicit knowledge of the matrices $\mathbf{w}^{\pm}(\lambda)$. Because the matrix $\mathbf{v}(\lambda)$ is constructed in terms of boundary values of the parametrix $\widehat{\mathbf{N}}^{\sigma}(\lambda)$ on Σ, the practical computation of higher-order corrections for $\mathbf{R}(\lambda)$ amounts to representing the various approximations of $\mathbf{N}^{\sigma}(\lambda)$ in terms of known functions. Such a representation is of course available for the "outer" approximation $\widehat{\mathbf{N}}^{\sigma}_{\text{out}}(\lambda)$. Also, as remarked in §4.4.1 and §4.4.2, it is possible to obtain explicit formulae for the approximations $\widehat{\mathbf{N}}^{\sigma}_{2k}(\lambda)$ and $\widehat{\mathbf{N}}^{\sigma}_{2k-1}(\lambda)$ in terms of Airy functions (although we did not pursue this particular path). Therefore, the obstruction to calculating explicit higher-order corrections to $\mathbf{R}(\lambda)$ is really an *explicit* special function representation of the matrix $\widehat{\mathbf{F}}^{\sigma}(\lambda)$ used to build the local approximation $\widehat{\mathbf{N}}^{\sigma}_{\text{origin}}(\lambda)$ near the origin. (Note added: this obstruction has been removed. See §4.4.4.)

Recall the definition (2.28) of the exact solution ψ of the nonlinear Schrödinger equation corresponding to the semiclassical soliton ensemble that is the nonlinear superposition of the individual solitons indexed by the WKB eigenvalues $\lambda^{\text{WKB}}_{\hbar_N,n}$, as well as the definitions (4.96) and (4.108) of the explicit approximation $\tilde{\psi}$ obtained in §4.3. Using the relation between $\mathbf{N}^{\sigma}(\lambda)$ and $\mathbf{M}(\lambda)$ given in (4.1), we obtain the following theorem.

THEOREM 4.5.2 (Conditional strong asymptotics for semiclassical soliton ensembles). *Given the existence of a complex phase function $g^\sigma(\lambda)$, there exists a positive constant M such that for all sufficiently small $\hbar_N$,*

$$|\psi - \tilde{\psi}| \le M\hbar_N^{1/3}. \tag{4.336}$$

The asymptotics are strong because x and t are held fixed.

Proof. Start with the exact representation of $\mathbf{M}(\lambda)$:

$$\mathbf{M}(\lambda) = \mathbf{E}(\lambda)\widehat{\mathbf{N}}^\sigma(\lambda)\exp(g^\sigma(\lambda)\sigma_3/\hbar_N), \tag{4.337}$$

which implies for the $(1,2)$ entry

$$M_{12}(\lambda) = \hat{N}^\sigma_{12}(\lambda)\exp(-g^\sigma(\lambda)/\hbar_N) + \left[(\mathbf{E}(\lambda) - \mathbb{I})\cdot\widehat{\mathbf{N}}^\sigma(\lambda)\right]_{12}\exp(-g^\sigma(\lambda)/\hbar_N). \tag{4.338}$$

Multiplying by $2i\lambda$ and passing to the limit $\lambda \to \infty$ for fixed $\hbar_N$, using the fact that for large λ, $\widehat{\mathbf{N}}^\sigma(\lambda) \equiv \widehat{\mathbf{N}}^\sigma_{\text{out}}(\lambda)$ and the fact that $g^\sigma(\lambda) = O(|\lambda|^{-1})$, yields

$$\psi = \tilde{\psi} + \lim_{\lambda\to\infty}\left(2i\lambda\exp(-g^\sigma(\lambda)/\hbar_N)\left[(\mathbf{E}(\lambda) - \mathbb{I})\cdot\widehat{\mathbf{N}}^\sigma(\lambda)\right]_{12}\right). \tag{4.339}$$

The theorem is thus established upon using the estimate stated in Theorem 4.5.1. □

Remark. Theorem 4.5.2 presumes the existence for fixed x and t of an appropriate complex phase function $g^\sigma(\lambda)$ characterized by an admissible density function $\rho^\sigma(\eta)$ as in Definition 4.2.5 for which:

1. The contour C is smooth except at the origin, where it has well-defined tangents on both sides, neither vertical nor horizontal;
2. The density $\rho^\sigma(\eta)$ vanishes like a square root at the endpoints $\lambda_0, \ldots, \lambda_G$.

Under these conditions, Theorem 4.5.2 shows that the function $\tilde{\psi}$ provides explicit strong asymptotics for ψ at the particular x and t values under consideration. Uniformity with respect to x and t in certain compact sets can be obtained from continuity properties of $g^\sigma(\lambda)$, which are established in chapter 5.

The utility of these results in the semiclassical analysis of semiclassical soliton ensembles rests upon finding the complex phase function $g^\sigma(\lambda)$, a task we now pursue.

Chapter Five

Direct Construction of the Complex Phase

In this chapter, we turn to the question of reducing Riemann-Hilbert Problem 2.0.2 for the matrix $\mathbf{M}(\lambda)$ ultimately to the simple form of the outer model Riemann-Hilbert Problem 4.2.3 for the matrix $\widetilde{\mathbf{O}}(\lambda)$, which was solved exactly in §4.3. Achieving the reduction required finding an appropriate complex phase function $g^{\sigma}(\lambda)$ on an appropriate contour $C \cup C^*$. We describe here a construction of good "candidate" complex phase functions. It is then often possible to prove directly that an appropriate candidate actually satisfies all of the criteria (cf. Definition 4.2.5) required for reducing the Riemann-Hilbert problem to an analytically tractable form, as in chapter 4.

5.1 POSTPONING THE INEQUALITIES. GENERAL CONSIDERATIONS

We try to construct a complex phase function $g^{\sigma}(\lambda)$ on the basis of two ideas:

1. Suppose the number of bands that will ultimately lie on the yet-to-be-determined contour C is given as $G/2 + 1$ for some even integer $G \geq 0$. Suppose further that the initial part of C coming out of $\lambda = 0$ is part of a band and that the final part of C going into $\lambda = 0$ is part of a gap. Here, "initial" and "final" are defined by the index σ.
2. Ignore, for the moment, *all inequalities* that go into the specification of $g^{\sigma}(\lambda)$. They are checked later.

So, for each fixed x and t we can try to use the remaining conditions on $g^{\sigma}(\lambda)$ to construct a "candidate" phase function for each nonnegative even integer G. It will soon be clear that this construction of the candidates is completely systematic. In chapter 6, we will need to determine which of the candidates, if any, satisfy the necessary inequalities that we are about to throw out. Motivated by the exact solution of the outer model problem presented in §4.3, which involved hyperelliptic Riemann surfaces of genus G, we refer to a candidate complex phase function corresponding to some even integer G as a *genus-G ansatz*.

5.1.1 Collapsing the Loop Contour C

The first step in constructing a genus-G ansatz for the complex phase is to temporarily assume that the contour C passes through the point $\lambda = iA$, which is the top of the support of the asymptotic eigenvalue measure $\rho^0(\eta)\,d\eta$ on the imaginary

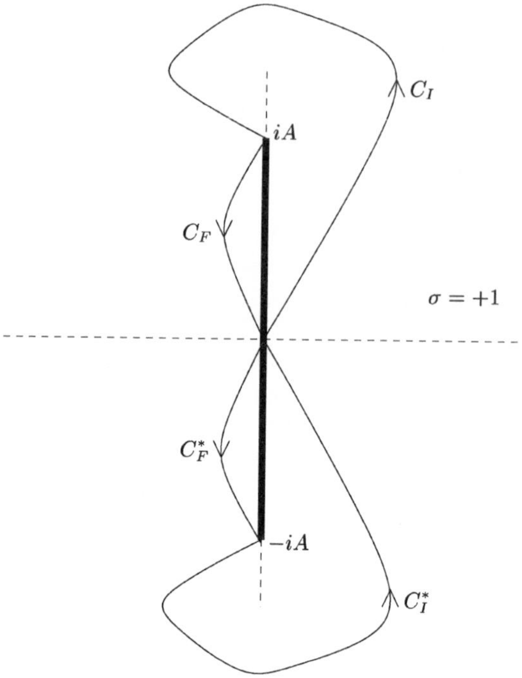

Figure 5.1 The contour C is temporarily assumed to pass through $\lambda = iA$.

axis. This effectively divides C into two halves: an "initial" half C_I and a "final" half C_F. See figure 5.1. We suppose for the time being that the support of the function $\rho^\sigma(\eta)$ on C is contained in C_I. This means that the entire contour $C_F \cup C_F^*$ is being assumed to be part of a gap of $C \cup C^*$. The gap condition $\rho^\sigma(\eta) \equiv 0$ in $C_F \cup C_F^*$ is satisfied by assumption, and the remaining gap condition $\Re(\tilde{\phi}^\sigma) < 0$ is an inequality that we are temporarily putting aside. Therefore, until we restore the inequalities, we concern ourselves strictly with the contour $C_I \cup C_I^*$. *Note that C_I is always considered to be oriented from* 0 *to* iA *and that* C_I^* *is oriented from* $-iA$ *to* 0.

The function that is our main concern is defined for $\lambda \in C_I$ as

$$\tilde{\phi}^\sigma(\lambda) = \int_0^{iA} L_\eta^0(\lambda)\rho^0(\eta)\,d\eta + \int_{-iA}^0 L_\eta^0(\lambda)\rho^0(\eta^*)^*\,d\eta$$
$$+ J\left(2i\lambda x + 2i\lambda^2 t - (2K+1)i\pi\int_\lambda^{iA}\rho^0(\eta)\,d\eta - g_+^\sigma(\lambda) - g_-^\sigma(\lambda)\right). \quad (5.1)$$

If we define for $\lambda \in C_I$

$$L_{\eta\pm}^{C,\sigma}(\lambda) = \lim_{\mu\to\lambda_\pm} L_\eta^{C,\sigma}(\mu), \quad (5.2)$$

indicating the nontangential boundary values from the left (+) and right (−) sides of C_σ and set

$$\overline{L_\eta^{C,\sigma}}(\lambda) = \frac{1}{2}(L_{\eta+}^{C,\sigma}(\lambda) + L_{\eta-}^{C,\sigma}(\lambda)) \quad (5.3)$$

to denote the average, then for $\lambda \in C_I$ we can use the presumed complex-conjugation symmetry of $\rho^\sigma(\eta)$ to write

$$\begin{aligned}\tilde{\phi}^\sigma(\lambda) = &\int_0^{iA} L_\eta^0(\lambda)\rho^0(\eta)\,d\eta - \int_{C_I} \overline{L_\eta^{C,\sigma}}(\lambda)\rho^\sigma(\eta)\,d\eta \\ &+ \int_{-iA}^0 L_\eta^0(\lambda)\rho^0(\eta^*)^*\,d\eta - \int_{C_I^*} \overline{L_\eta^{C,\sigma}}(\lambda)\rho^\sigma(\eta^*)^*\,d\eta \\ &+ J\left(2i\lambda x + 2i\lambda^2 t - (2K+1)i\pi \int_\lambda^{iA} \rho^0(\eta)\,d\eta\right). \end{aligned} \tag{5.4}$$

Now, we need to suppose that *the asymptotic eigenvalue density* $\rho^0(\eta)$ *is analytic in the region bounded by the imaginary interval* $[0, iA]$ *and the curve* C_I. From the formula (3.1), this amounts to the condition that the function $A(x)$ characterizing the initial data for (1.1) is real-analytic (in addition to being bell-shaped and even and having nonzero curvature at the peak and sufficient decay in the tails). Singularities of $\rho^0(\eta)$ off of the imaginary interval $(0, iA)$ represent obstructions to the free positioning of the contour C. However, we choose to neglect the possibility that these singularities could constrain our analysis at this point; of course there are nontrivial cases (such as the case of the Satsuma-Yajima soliton ensemble, for which $\rho^0(\eta) \equiv i$) where $\rho^0(\eta)$ is *entire*. With the assumption that $\rho^0(\eta)$ is analytic, we may rewrite the integrals over the imaginary axis for $\lambda \in C_I$ as

$$\int_0^{iA} L_\eta^0(\lambda)\rho^0(\eta)\,d\eta = \begin{cases} \displaystyle\int_{C_I} L_{\eta-}^{C,\sigma}(\lambda)\rho^0(\eta)\,d\eta, & \sigma = +1, \\ \displaystyle\int_{C_I} L_{\eta+}^{C,\sigma}(\lambda)\rho^0(\eta)\,d\eta, & \sigma = -1, \end{cases} \tag{5.5}$$

and similarly, by symmetry,

$$\int_{-iA}^0 L_\eta^0(\lambda)\rho^0(\eta^*)^*\,d\eta = \begin{cases} \displaystyle\int_{C_I^*} L_{\eta-}^{C,\sigma}(\lambda)\rho^0(\eta^*)^*\,d\eta, & \sigma = +1, \\ \displaystyle\int_{C_I^*} L_{\eta+}^{C,\sigma}(\lambda)\rho^0(\eta^*)^*\,d\eta, & \sigma = -1. \end{cases} \tag{5.6}$$

Next, note that $L_{\eta+}^{C,\sigma}(\lambda) = L_{\eta-}^{C,\sigma}(\lambda)$ for all $\eta \in C_I \cup C_I^*$ "below" $\lambda \in C_I$ (that is, all $\eta \in C_I^*$ and all η in the oriented portion of C_I from 0 to λ), and at the same time $L_{\eta+}^{C,\sigma}(\lambda) = 2\pi i + L_{\eta-}^{C,\sigma}(\lambda)$ for $\eta \in C_I$ "above" λ (that is, in the oriented portion of C_I from λ to iA). This means that for $\lambda \in C_I$,

$$\begin{aligned}&\int_{C_I} L_{\eta\pm}^{C,\sigma}(\lambda)\rho^0(\eta)\,d\eta + \int_{C_I^*} L_{\eta\pm}^{C,\sigma}(\lambda)\rho^0(\eta^*)^*\,d\eta \\ &\quad = \int_{C_I} \overline{L_\eta^{C,\sigma}}(\lambda)\rho^0(\eta)\,d\eta + \int_{C_I^*} \overline{L_\eta^{C,\sigma}}(\lambda)\rho^0(\eta^*)^*\,d\eta \pm \pi i \int_\lambda^{iA} \rho^0(\eta)\,d\eta, \end{aligned} \tag{5.7}$$

with the final integral being taken along C_I. Assembling these results gives the expression

$$\tilde{\phi}^\sigma(\lambda) = \int_{C_I} \overline{L_\eta^{C,\sigma}}(\lambda)\bar{\rho}^\sigma(\eta)\,d\eta + \int_{C_I^*} \overline{L_\eta^{C,\sigma}}(\lambda)\bar{\rho}^\sigma(\eta^*)^*\,d\eta$$
$$+ J(2i\lambda x + 2i\lambda^2 t) - (J(2K+1)+\sigma)i\pi \int_\lambda^{iA} \rho^0(\eta)\,d\eta, \quad (5.8)$$

valid for $\lambda \in C_I$, where we have introduced the *complementary density* for $\eta \in C_I$,

$$\bar{\rho}^\sigma(\eta) := \rho^0(\eta) - \rho^\sigma(\eta). \quad (5.9)$$

As a matter of future convenience, we now determine the arbitrary integer K that indexes the interpolants of the proportionality constants by choosing

$$K := -\frac{1}{2}(J\sigma + 1). \quad (5.10)$$

Since $J = \pm 1$ and $\sigma = \pm 1$, one is taking either $K = 0$ or $K = -1$. Consequently the function $\tilde{\phi}$ becomes simply

$$\tilde{\phi}^\sigma(\lambda) = \int_{C_I} \overline{L_\eta^{C,\sigma}}(\lambda)\bar{\rho}^\sigma(\eta)\,d\eta + \int_{C_I^*} \overline{L_\eta^{C,\sigma}}(\lambda)\bar{\rho}^\sigma(\eta^*)^*\,d\eta + J(2i\lambda x + 2i\lambda^2 t). \quad (5.11)$$

Remark. Note that the relation (5.10) implies that for all four combinations of values for J and σ, we have

$$i^J(-1)^K\sigma = -i, \quad (5.12)$$

which we assumed in stating the outer model Riemann-Hilbert Problem 4.2.3.

Until we restore our interest in the inequalities, our task is to find a system of $G/2+1$ bands and $G/2+1$ gaps on C_I such that the following are true:

1. The initial part of C_I is contained in a band, and the final part of C_I is contained in a gap;
2. In each band of C_I, $\tilde{\phi}^\sigma(\lambda)$ is pure imaginary and independent of λ, while $\rho^\sigma(\eta)\,d\eta$ is a real differential;
3. In each gap of C_I, $\rho^\sigma(\eta) \equiv 0$ or, equivalently, $\bar{\rho}^\sigma(\eta) \equiv \rho^0(\eta)$.

5.1.2 The Scalar Boundary Value Problem for Genus G. Moment Conditions

Let us begin by imposing just two of the conditions:

$$\frac{\partial\tilde{\phi}^\sigma}{\partial\lambda}(\lambda) \equiv 0, \quad \lambda \text{ in a band of } C_I,$$
$$\rho^\sigma(\lambda) \equiv 0, \quad \lambda \text{ in a gap of } C_I. \quad (5.13)$$

In the first condition, by the derivative with respect to λ we mean the derivative along the contour C_I. We want to think of these as equations for the unknown function $\bar{\rho}^\sigma(\eta)$ for $\eta \in C_I$. What auxiliary properties do we demand of any solution $\bar{\rho}^\sigma(\eta)$?

We recall that the analysis in chapter 4 required the following of $\rho^\sigma(\eta)$:

1. $\rho^\sigma(\eta)$ should admit analytic continuation to the left and right of any band or gap (this is of course trivial in the gaps where $\rho^\sigma(\eta) \equiv 0$);
2. $\rho^\sigma(\eta)$ should take a finite value in the limit $\eta \to 0$ for $\eta \in C_I$;
3. $\rho^\sigma(\eta)$ should vanish exactly like a square root at each band endpoint λ_k and λ_k^*.

We now recall that our assumption that the function $A(x)$ decays sufficiently rapidly for large $|x|$ guarantees via the definition (3.1) *that the function $\rho^0(\eta)$ is bounded as $\eta \to 0$ for $\eta \in C_I$.* Then, the three just stated conditions on $\rho^\sigma(\eta)$ imply in particular that

$$\bar{\rho}^\sigma(\eta) \text{ is uniformly Hölder-continuous on } C_I \text{ with exponent } 1/2. \tag{5.14}$$

We further impose the condition that the limit as η tends to zero along C_I of $\bar{\rho}^\sigma(\eta)$ is real:

$$\bar{\rho}^\sigma(0) \in \mathbb{R}. \tag{5.15}$$

In this case, $\bar{\rho}^\sigma(\eta)$ extends by the definition $\bar{\rho}^\sigma(\eta^*)^*$ for $\eta \in C_I^*$ to a function that satisfies the Hölder condition with exponent $1/2$ on the whole contour $C_I \cup C_I^*$.

Suppose that a function $\bar{\rho}^\sigma(\eta)$ is given for $\eta \in C_I$ satisfying (5.14) and (5.15) and for which the corresponding function $\tilde{\phi}^\sigma(\lambda)$ defined by (5.11) satisfies (5.13). Let $F(\lambda)$ be the Cauchy integral:

$$F(\lambda) := \int_{C_I} \frac{\bar{\rho}^\sigma(\eta)\, d\eta}{\lambda - \eta} + \int_{C_I^*} \frac{\bar{\rho}^\sigma(\eta^*)^*\, d\eta}{\lambda - \eta}. \tag{5.16}$$

Then, from the Plemelj-Sokhotski formula the function $\bar{\rho}^\sigma(\eta)$ is recovered as

$$\bar{\rho}^\sigma(\eta) = -\frac{1}{2\pi i}(F_+(\eta) - F_-(\eta)), \tag{5.17}$$

and $F(\lambda)$ has the following properties:

1. $F(\lambda)$ is analytic for $\lambda \in \mathbb{C} \setminus (C_I \cup C_I^*)$;
2. $F(\lambda)$ satisfies the decay condition
$$F(\lambda) = O(1/\lambda), \qquad \lambda \to \infty; \tag{5.18}$$
3. For all λ in the domain of analyticity, $F(\lambda)$ satisfies the symmetry property
$$F(\lambda^*) = -F(\lambda)^*; \tag{5.19}$$
4. The boundary values taken by $F(\lambda)$ on both sides of $C_I \cup C_I^*$ are Hölder-continuous with exponent $1/2$ and for $\lambda \in C_I$ satisfy
$$\begin{aligned} F_+(\lambda) + F_-(\lambda) &= -4iJ(x + 2\lambda t), \quad \lambda \text{ in a band}, \\ F_+(\lambda) - F_-(\lambda) &= -2\pi i \rho^0(\lambda), \quad \lambda \text{ in a gap}. \end{aligned} \tag{5.20}$$

These properties follow from well-known properties of Cauchy integrals for Hölder-continuous functions [M53] and from the Plemelj-Sokhotski formula.

Conversely, we define a boundary value problem as follows. Seek a function $F(\lambda)$ that satisfies conditions 1, 2, and 4 above and additionally takes Hölder 1/2 boundary values for $\lambda \in C_I^*$ that satisfy the conjugate boundary conditions:

$$\begin{aligned} F_+(\lambda) + F_-(\lambda) &= -(F_+(\lambda^*)^* + F_-(\lambda^*)^*), \quad \lambda^* \text{ in a band,} \\ F_+(\lambda) - F_-(\lambda) &= -(F_+(\lambda^*)^* - F_-(\lambda^*)^*), \quad \lambda^* \text{ in a gap.} \end{aligned} \tag{5.21}$$

We call this the *scalar boundary value problem for genus G.*

Remark. Although the complex phase function $g^\sigma(\lambda)$ depends on the index σ, this parameter enters into the properties of $F(\lambda)$ only in that for $\sigma = +1$ (respectively, $\sigma = -1$) the contour C_I lies to the right (respectively, left) of the imaginary interval $[0, iA]$. Thus, to simplify notation we do not reproduce the superscript σ on the function $F(\lambda)$.

Lemma 5.1.1 *There exists at most one solution of the scalar boundary value problem for genus G. If it exists, the solution satisfies the symmetry property* (5.19), *and the function $\bar{\rho}^\sigma(\eta)$ defined for $\lambda \in C_I$ by* (5.17) *gives rise to a function $\tilde{\phi}^\sigma(\lambda)$ via* (5.11) *that satisfies* (5.13) *and $\bar{\rho}^\sigma(0) \in \mathbb{R}$.*

Proof. The uniqueness and the symmetry condition (5.19) are proved in exactly the same way. Consider the related boundary value problem of seeking a function $Z(\lambda)$ that is analytic in $C_I \cup C_I^*$, satisfies (5.18), and takes Hölder-continuous boundary values on both sides of $C_I \cup C_I^*$ with exponent 1/2 that satisfy $Z_+(\lambda) + Z_-(\lambda) \equiv 0$ for λ or λ^* in a band of C_I and $Z_+(\lambda) - Z_-(\lambda) \equiv 0$ for λ or λ^* in a gap of C_I. Recall the function $R(\lambda)$, first used in §4.3, that satisfies

$$R(\lambda)^2 = \prod_{k=0}^{G} (\lambda - \lambda_k)(\lambda - \lambda_k^*), \tag{5.22}$$

is analytic except at the bands of $C_I \cup C_I^*$ (where the branch cuts are placed) and satisfies $R(\lambda) \sim -\lambda^{G+1}$ as $\lambda \to \infty$. With this choice, the boundary value $R_+(0)$ relative to the oriented contour $C_I \cup C_I^*$ is the positive value

$$R_+(0) = \prod_{k=0}^{G} |\lambda_k|^2. \tag{5.23}$$

The function defined from any solution of this boundary value problem by the formula $W(\lambda) := Z(\lambda)/R(\lambda)$ is again analytic in $\mathbb{C} \setminus (C_I \cup C_I^*)$ with at worst inverse square-root singularities at the endpoints of the bands and gaps on $C_I \cup C_I^*$. The boundary values on $C_I \cup C_I^*$, continuous except possibly at the isolated band and gap endpoints, satisfy $W_+(\lambda) = W_-(\lambda)$. The function $W(\lambda)$ decays like $O(\lambda^{-(G+2)})$ as $\lambda \to \infty$. The relatively mild nature of the singularities, together with the agreement of the boundary values where they are continuous, implies that $W(\lambda)$ is actually entire, and then by Liouville's theorem the decay condition implies that $W(\lambda) \equiv 0$. It then follows that $Z(\lambda) \equiv 0$, so that all solutions are trivial. To use this result to prove uniqueness for $F(\lambda)$, one considers $Z(\lambda)$ to be the difference of two solutions and finds that this difference satisfies the above problem and is therefore zero. Similarly, to prove the symmetry property (5.19), one considers the function $Z(\lambda) = F(\lambda) + F(\lambda^*)^*$ and again finds that $Z \equiv 0$.

Now, the function defined for $\eta \in C_I \cup C_I^*$ by $u(\eta) := -(F_+(\eta) - F_-(\eta))/(2\pi i)$ is by construction Hölder-continuous on $C_I \cup C_I^*$ with exponent $1/2$ and satisfies $u(\eta^*)^* = u(\eta)$. It follows that $u(0) \in \mathbb{R}$. The proof is complete upon observing that by continuity of boundary values and decay at infinity $F(\lambda)$ necessarily agrees with the Cauchy integral of the function $u(\eta)$. That is, the relation

$$F(\lambda) = \int_{C_I \cup C_I^*} \frac{u(\eta)\, d\eta}{\lambda - \eta} \tag{5.24}$$

is a consequence of Liouville's theorem. It follows that the function $\bar{\rho}^\sigma(\eta) := u(\eta)$ for $\eta \in C_I$ leads to a solution of (5.13) as required. □

So, it is sufficient to seek a solution to the scalar boundary value problem for genus G, and we know that the solution will be unique *if it exists*. This will be the case only if the endpoints of the bands and gaps satisfy certain explicit conditions.

Lemma 5.1.2 *Fix $J = \pm 1$, $\sigma = \pm 1$, and an even integer $G \geq 0$. Let a contour C_I be given in the upper half-plane that connects 0 to iA and whose interior points all lie to the right (respectively, left) of the vertically oriented segment $[0, iA]$ for $\sigma = +1$ (respectively, $\sigma = -1$), and let points $\lambda_0, \dots, \lambda_G$ be given in order on C_I. Let the bands and gaps on $C_I \cup C_I^*$ separated by these points be denoted according to the scheme illustrated in figure 4.4, and let Γ_I denote $(\bigcup_k \Gamma_k^\pm) \cap (C_I \cup C_I^*)$. Then, there exists a unique solution to the scalar boundary value problem for genus G if and only if for $p = 0, \dots, G$ the endpoints satisfy the real equations*

$$M_p := J \int_{\cup_k I_k^\pm} \frac{2ix + 4i\eta t}{R_+(\eta)} \eta^p \, d\eta + 2\Re\left(\int_{\Gamma_I \cap C_I} \frac{\pi i \rho^0(\eta)}{R(\eta)} \eta^p \, d\eta \right) = 0. \tag{5.25}$$

Moreover, the function $\bar{\rho}^\sigma(\eta)$ defined from the solution $F(\lambda)$ by (5.17) is analytic in the interior of each band $I_k^\pm$ and each component of Γ_I.

Proof. The proof is by direct construction. In the scalar boundary value problem for genus G, use $R(\lambda)$ to make the change of variables

$$H(\lambda) := \frac{F(\lambda)}{R(\lambda)}. \tag{5.26}$$

Note that $R(\lambda^*) = R(\lambda)^*$, and since $R(\lambda)$ is analytic in the gaps while in the bands satisfies $R_+(\lambda) + R_-(\lambda) = 0$, the conditions satisfied by the boundary values of $H(\lambda)$ on the oriented contour C_I are then

$$H_+(\lambda) - H_-(\lambda) = \begin{cases} \dfrac{-4iJ(x + 2\lambda t)}{R_+(\lambda)}, & \lambda \text{ in a band,} \\[2ex] -\dfrac{2\pi i \rho^0(\lambda)}{R(\lambda)}, & \lambda \text{ in a gap,} \end{cases} \tag{5.27}$$

while for $\lambda \in C_I^*$, the conjugate boundary conditions hold,

$$H_+(\lambda) - H_-(\lambda) = -(H_+(\lambda^*)^* - H_-(\lambda^*)^*). \tag{5.28}$$

Since the quotient $H(\lambda)$ defined by (5.26) necessarily decays at infinity if $F(\lambda)$ exists and has at worst inverse square-root singularities at the isolated endpoints λ_k

and λ_k^*, by the same kind of reasoning as in the proof of Lemma 5.1.1 it follows that $H(\lambda)$ must agree with the Cauchy integral of the difference of its boundary values (5.27) and (5.28). That is, $H(\lambda)$ must be given by

$$H(\lambda) = \frac{1}{\pi i}\int_{\cup_k I_k^{\pm}} \frac{2iJ(x+2\eta t)}{(\lambda-\eta)R_+(\eta)}\,d\eta + \frac{1}{\pi i}\int_{\Gamma_I\cap C_I} \frac{\pi i\rho^0(\eta)}{(\lambda-\eta)R(\eta)}\,d\eta$$
$$+\frac{1}{\pi i}\int_{\Gamma_I\cap C_I^*} \frac{\pi i\rho^0(\eta^*)^*}{(\lambda-\eta)R(\eta)}\,d\eta. \tag{5.29}$$

Now, the function $F(\lambda)$ defined from such a solution $H(\lambda)$ by (5.26) will only satisfy the decay condition $F(\lambda) = O(1/\lambda)$ as $\lambda \to \infty$ if $H(\lambda) = O(\lambda^{-(G+2)})$ in the same limit. Expanding the explicit formula (5.29) for large λ gives a Laurent series whose leading term is $O(1/\lambda)$. Therefore, for existence it is necessary that the first $G+1$ coefficients in this Laurent series vanish identically. These coefficients are computed as *moments* of the densities. Expanding $(\lambda-\eta)^{-1}$ inside the integrals in geometric series, one sees that the Laurent series of $H(\lambda)$, convergent for $|\lambda|$ sufficiently large, is

$$H(\lambda) = \frac{1}{\pi i}\sum_{p=0}^{\infty}\frac{M_p}{\lambda^{p+1}}, \tag{5.30}$$

where the quantities M_p, easily seen to be real-valued by using the complex-conjugation symmetry of the contours, are defined in (5.25). Thus, the conditions for existence are exactly the *moment conditions* recorded in (5.25).

When the moment conditions are satisfied, the function $F(\lambda) := H(\lambda)R(\lambda)$ is the solution (unique, by Lemma 5.1.1) of the scalar boundary value problem for genus G. The claimed analyticity of $\bar{\rho}^\sigma(\eta)$ defined by (5.17) then follows from the explicit formula (5.29) for $H(\lambda)$ and the analyticity of the boundary values of $R(\lambda)$. □

Remark. Given the endpoints $\lambda_0, \ldots, \lambda_G$, the moments M_p have the same value for any contour C_I that connects 0 to iA in the domain of analyticity of $\rho^0(\eta)$ and interpolates these points in order. This follows because the integrands have analytic continuations to either side of the contour $C_I \cup C_I^*$. Therefore, the moments M_p are functions of the *ordered sequence* of endpoints $\lambda_0, \ldots, \lambda_G$ alone and thus *the moment conditions are only constraints on the endpoints* and not on the interpolating contour. This statement is strengthened in §5.4, when we show that the ordering is irrelevant; the moments are in fact seen to be symmetric functions of the endpoints.

Remark. The moment conditions amount to $G+1$ real equations for $G+1$ complex unknowns. Intuitively, one expects the solution space to be $G+1$ real-dimensional. On the other hand, proving the existence of solutions is a nontrivial analytical issue. Indeed, in the simplest of cases, solutions can fail to exist for given x and t except for certain values of the parameters J and σ.

Remark. Even when the endpoints do not satisfy the moment conditions (5.25), we still consider a function $F(\lambda)$ to be defined by the relation $F(\lambda) := H(\lambda)R(\lambda)$ with $H(\lambda)$ given by the explicit formula (5.29). Exactly the same arguments as in the proof of Lemma 5.1.1 then show that the function so defined for arbitrary endpoints is the unique solution of the corresponding boundary value problem in which the decay condition (5.18) is replaced with the growth condition

$$F(\lambda) = O(\lambda^G), \qquad \lambda \to \infty, \tag{5.31}$$

with all other conditions on $F(\lambda)$ exactly as before. The moment conditions (5.25) are then interpreted as the conditions that must be satisfied in order for $F(\lambda)$ to have the form of a Cauchy integral (5.16) corresponding to the complementary density $\bar{\rho}^\sigma(\eta)$.

The solution $\bar{\rho}^\sigma(\lambda)$ given by (5.17) can be calculated explicitly. One can check that, as specified, the formula gives $\bar{\rho}^\sigma(\lambda) \equiv \rho^0(\lambda)$ for all λ in $\Gamma_I \cap C_I$. For λ in any band I_k^+ of C_I, the same formula (5.17) gives

$$\begin{aligned}\bar{\rho}^\sigma(\lambda) = \frac{4Jt}{\pi} R_+(\lambda)\delta_{G,0} - \frac{R_+(\lambda)}{\pi i}\int_{\Gamma_I \cap C_I} \frac{\rho^0(\eta)\, d\eta}{(\lambda-\eta)R(\eta)} \\ - \frac{R_+(\lambda)}{\pi i}\int_{\Gamma_I \cap C_I^*} \frac{\rho^0(\eta^*)^*\, d\eta}{(\lambda-\eta)R(\eta)}.\end{aligned} \tag{5.32}$$

Using the symmetry relation (5.23), it follows from this formula that the condition (5.15) holds.

Remark. This latter calculation takes advantage of the useful fact that

$$\int_{\cup_k I_k^\pm} \frac{f(\eta)\, d\eta}{R_+(\eta)} = -\frac{1}{2}\oint \frac{f(\eta)\, d\eta}{R(\eta)} \tag{5.33}$$

for all meromorphic $f(\eta)$, where the integral on the right-hand side is over a closed counterclockwise-oriented contour surrounding all the bands. This allows such integrals to be evaluated in many cases exactly by residues. Such a procedure can be applied to the first term in M_p, for example.

From these explicit formulae for the complementary density $\bar{\rho}^\sigma(\lambda)$, we can of course write down the corresponding formulae for the density $\rho^\sigma(\lambda)$. In the gaps of C_I, one verifies that $\rho^\sigma(\lambda) \equiv 0$, while in the bands of C_I,

$$\begin{aligned}\rho^\sigma(\lambda) = \rho^0(\lambda) - \frac{4Jt}{\pi} R_+(\lambda)\delta_{G,0} + \frac{R_+(\lambda)}{\pi i}\int_{\Gamma_I \cap C_I} \frac{\rho^0(\eta)\, d\eta}{(\lambda-\eta)R(\eta)} \\ + \frac{R_+(\lambda)}{\pi i}\int_{\Gamma_I \cap C_I^*} \frac{\rho^0(\eta^*)^*\, d\eta}{(\lambda-\eta)R(\eta)}.\end{aligned} \tag{5.34}$$

For λ on the conjugate contour C_I^*, the function $\rho^\sigma(\lambda)$ is defined by conjugation: $\rho^\sigma(\lambda) := \rho^\sigma(\lambda^*)^*$.

This formula has the following useful property.

Lemma 5.1.3 *The formula* (5.34) *can be written in the form*

$$\rho^\sigma(\lambda) = R_+(\lambda)Y(\lambda), \tag{5.35}$$

where $Y(\lambda)$ is analytic in a simply connected region of the upper half-plane containing all bands I_k^+ in C_I, including the endpoints $\lambda_0, \dots, \lambda_G$, provided $\lambda_G \neq iA$.

Proof. It suffices to show that the sum of the first and third terms in (5.34) have this property. Consider the integral

$$I(\lambda) := \frac{1}{\pi i}\int_{\Gamma_I \cap C_I} \frac{\rho^0(\eta)\,d\eta}{(\lambda-\eta)R(\eta)}. \tag{5.36}$$

This integral defines an analytic function of $\lambda \in \mathbb{C}\backslash(\Gamma_I \cap C_I)$. Therefore, for λ in a band I_k^+, we have in particular

$$I(\lambda) = \lim_{\mu\to\lambda} I(\mu), \tag{5.37}$$

where for concreteness we suppose μ to lie to the left of the band I_k^+. For such μ, we can augment the contour of integration by writing

$$\begin{aligned} I(\mu) = &\frac{1}{\pi i}\int_{\Gamma_I\cap C_I} \frac{\rho^0(\eta)\,d\eta}{(\mu-\eta)R(\eta)} + \frac{1}{2\pi i}\int_{\cup_k I_k^+} \frac{\rho^0(\eta)\,d\eta}{(\mu-\eta)R_+(\eta)} \\ &+ \frac{1}{2\pi i}\int_{\cup_k I_k^+} \frac{\rho^0(\eta)\,d\eta}{(\mu-\eta)R_-(\eta)}, \end{aligned} \tag{5.38}$$

since μ is not contained in any band I_k^+ and since $R_+(\eta)+R_-(\eta)=0$. By standard analyticity deformations, this expression can be written as

$$I(\mu) = \frac{1}{2\pi i}\int_{C_{I+}\cup C_{I-}} \frac{\rho^0(\eta)\,d\eta}{(\mu-\eta)R(\eta)}, \tag{5.39}$$

where $C_{I\pm}$ are contours lying just to the left and right of C_I (see figure 5.2). Now, passing to the limit (5.37), there is a residue contribution as μ crosses C_{I+}, and one obtains the formula

$$I(\lambda) = -\frac{\rho^0(\lambda)}{R_+(\lambda)} + \frac{1}{2\pi i}\int_{C_{I+}\cup C_{I-}} \frac{\rho^0(\eta)\,d\eta}{(\lambda-\eta)R(\eta)}. \tag{5.40}$$

Using this expression for $I(\lambda)$ in the formula (5.34) for $\rho^\sigma(\lambda)$, we finally obtain the desired representation with

$$Y(\lambda) := -\frac{4Jt}{\pi}\delta_{G,0} + \frac{1}{2\pi i}\int_{C_{I+}\cup C_{I-}} \frac{\rho^0(\eta)\,d\eta}{(\lambda-\eta)R(\eta)} + \frac{1}{\pi i}\int_{\Gamma_I\cap C_I^*} \frac{\rho^0(\eta^*)^*\,d\eta}{(\lambda-\eta)R(\eta)}. \tag{5.41}$$

This function is clearly analytic for all λ in between C_{I+} and C_{I-}, which completes the proof. Note that the final term in this formula for $Y(\lambda)$ can be rewritten as

$$\frac{1}{\pi i}\int_{\Gamma_I\cap C_I^*} \frac{\rho^0(\eta^*)^*\,d\eta}{(\lambda-\eta)R(\eta)} = \frac{1}{2\pi i}\int_{C_{I+}^*\cup C_{I-}^*} \frac{\rho^0(\eta^*)^*\,d\eta}{(\lambda-\eta)R(\eta)}, \tag{5.42}$$

where the conjugate contours are presumed to be oriented from $-iA$ toward the origin. With this substitution, the function $Y(\lambda)$ is also defined and analytic for λ in the lower half-plane between C_{I+} and C_{I-}, where it satisfies $Y(\lambda^*) = Y(\lambda)^*$. □

Remark. Since it turns out that when $x=t=0$ the $G=0$ ansatz satisfies $\lambda_0 = iA$, this representation of the density $\rho^\sigma(\lambda)$ does not hold at this point in the (x,t)-plane.

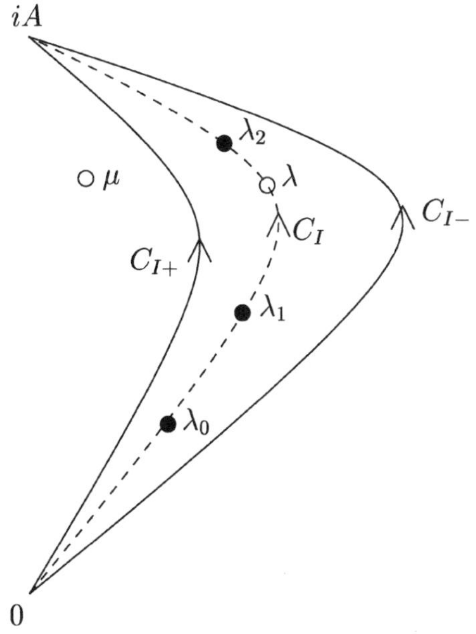

Figure 5.2 The contours $C_{I\pm}$. The point μ is outside the enclosed region and $\lambda \in I_k^+$ is inside. All three contours C_I, C_{I+}, and C_{I-} are oriented from 0 to iA.

From the formula (5.35), it is clear that $\rho^\sigma(\lambda)$ vanishes at least like a square root at the band endpoints. Higher-order vanishing of $\rho^\sigma(\lambda)$ at an endpoint λ_k corresponds to a zero of the analytic function $Y(\lambda)$ at $\lambda = \lambda_k$. Since $\rho^\sigma(\lambda)$ is identically zero in the gaps, the candidate density $\rho^\sigma(\lambda)$ is continuous at the nonzero endpoints $\lambda_0, \ldots, \lambda_G$. These formulae therefore desingularize the expression (5.34) for the candidate density.

Finally, recall that $\bar{\rho}^\sigma(\lambda)$ satisfies (5.15) and therefore extends by the definition $\bar{\rho}^\sigma(\lambda^*) = \bar{\rho}^\sigma(\lambda)^*$ to a continuous function on $C_I \cup C_I^*$. Similarly, being a density function for a real measure on the imaginary axis, the function $\rho^0(\lambda)$ satisfies $\rho^0(0) \in i\mathbb{R}$ and therefore does not generally extend in this way to all of $C_I \cup C_I^*$. Consequently, the symmetric extension of $\rho^\sigma(\lambda)$ to the contour $C_I \cup C_I^*$ is generally discontinuous at $\lambda = 0$. The origin is its *only* point of discontinuity.

Remark. An important observation that goes back to the papers of Lax and Levermore [LL83] is that the partial derivatives of the function $F(\lambda)$ with respect to the parameters x and t satisfy very simple boundary value problems. As an application we refer to in chapter 6, let us obtain simple formulae for the functions $\partial F/\partial x$ and $\partial F/\partial t$ valid for the assumption of $G = 0$.

First consider $\partial F/\partial x$. By differentiating the jump relations (5.20), one observes that $\partial F/\partial x$ is analytic in the whole λ-plane except for $\lambda \in I_0$, where we have

$$\frac{\partial F_+}{\partial x}(\lambda) + \frac{\partial F_-}{\partial x}(\lambda) \equiv -4iJ. \tag{5.43}$$

Also, $\partial F/\partial x$ has at worst inverse square-root singularities at λ_0 and λ_0^* and decays like $1/\lambda$ for large $|\lambda|$. This simple problem is solved by defining a new unknown $X(\lambda)$ according to

$$\frac{\partial F}{\partial x}(\lambda) = \frac{X(\lambda)}{R(\lambda)}, \tag{5.44}$$

where $R(\lambda)$ is the square root function first defined in §4.3. Then, $X(\lambda)$ is analytic except for $\lambda \in I_0$, where it satisfies the jump relation

$$X_+(\lambda) - X_-(\lambda) = -4iJR_+(\lambda). \tag{5.45}$$

This problem is solved by a Cauchy integral that can be evaluated explicitly by residues:

$$X(\lambda) = \frac{1}{2\pi i}\int_{I_0} \frac{4iJR_+(\eta)\,d\eta}{\lambda - \eta} = -2iJR(\lambda) + iJ(\lambda_0 + \lambda_0^* - 2\lambda). \tag{5.46}$$

Here, the term proportional to $R(\lambda)$ comes from a residue at $\eta = \lambda$, and the remaining terms come from a residue at $\eta = \infty$ tailored for the special case of genus $G = 0$. Therefore, we find the formula

$$\frac{\partial F}{\partial x}(\lambda) = -2iJ + \frac{2iJ}{R(\lambda)}\left(\frac{\lambda_0 + \lambda_0^*}{2} - \lambda\right). \tag{5.47}$$

It is easy to verify that this simple formula satisfies the jump relations exactly. Now, using the explicit formula for $R(\lambda)$ valid for genus $G = 0$, one sees that in fact

$$\frac{\partial F}{\partial x}(\lambda) = -2iJ\left(1 + \frac{\partial R}{\partial \lambda}(\lambda)\right). \tag{5.48}$$

Next, consider $\partial F/\partial t$ for $G = 0$, which is analytic except for $\lambda \in I_0$, where

$$\frac{\partial F_+}{\partial t}(\lambda) + \frac{\partial F_-}{\partial t}(\lambda) \equiv -8iJ\lambda. \tag{5.49}$$

Again, $\partial F/\partial t$ can have at worst inverse square-root singularities at λ_0 and λ_0^* and must decay like $1/\lambda$. One solves this problem in a similar way, introducing a new unknown $T(\lambda)$ by

$$\frac{\partial F}{\partial t}(\lambda) = \frac{T(\lambda)}{R(\lambda)}, \tag{5.50}$$

and then expressing $T(\lambda)$ in terms of a Cauchy integral over I_0 that one evaluates by residues. The final result one obtains for $G = 0$ is the formula

$$\frac{\partial F}{\partial t}(\lambda) = -iJ\left(4\lambda + \frac{\partial}{\partial \lambda}\left[(2\lambda + \lambda_0 + \lambda_0^*)R(\lambda)\right]\right). \tag{5.51}$$

These formulae show that the partial derivatives of F with respect to x and t actually decay faster at infinity than originally supposed; they are both $O(1/\lambda^2)$. By following similar reasoning, explicit formulas may be obtained easily for derivatives of F with respect to x and t for larger values of G.

5.1.3 Ensuring $\Re(\tilde{\phi}^\sigma) = 0$ in the Bands. Vanishing Conditions

We now turn to the question of determining what additional constraints are required to ensure that the constant value of $\tilde{\phi}^\sigma(\lambda)$ in each band $I_k^\pm$ is in fact purely imaginary. Since $\bar{\rho}^\sigma(\eta)$ satisfies the condition (5.15), it follows from (5.11) that $\tilde{\phi}^\sigma(\lambda)$ has a finite limit as λ tends to zero in C_I. It then follows directly from (5.11) that this limiting value is purely imaginary. Consequently, throughout the band $I_0 = I_0^+ \cup I_0^-$, $\tilde{\phi}^\sigma(\lambda)$ is automatically a purely imaginary constant. If one is constructing a genus-zero ansatz, then there are no more bands to consider. However, generally one must enforce the condition $\Re(\tilde{\phi}^\sigma) = 0$ in the other bands. This amounts to further conditions on the endpoints $\lambda_0, \ldots, \lambda_G$, conditions we refer to as *vanishing conditions*.

LEMMA 5.1.4 *Assume all the conditions of Lemma* 5.1.2 *and suppose also that the endpoints $\lambda_0, \ldots, \lambda_G$ satisfy the additional constraints*

$$V_k := \Re\left(\int_{\lambda_{2k}}^{\lambda_{2k+1}} \left[2iJx+4iJ\lambda t+\frac{1}{2}(F_+(\lambda)+F_-(\lambda))\right] d\lambda\right) = 0, \quad k = 0, \ldots, G/2-1. \tag{5.52}$$

Then the complementary density function $\bar{\rho}^\sigma(\eta)$ characterized by Lemma 5.1.2 *has the property that the associated function $\tilde{\phi}^\sigma(\lambda)$ defined by* (5.11) *agrees with a purely imaginary constant in each band of C_I.*

Proof. Starting from a terminal endpoint λ_{2k} of a band I_k^+ in which the condition is satisfied, we can ensure that the condition is satisfied as well in the next band along C_I by integrating $d\tilde{\phi}^\sigma$ along the intermediate gap Γ_{k+1}^+ and insisting that the real part vanish. It is easy to see that since $\tilde{\phi}^\sigma(\lambda)$ is defined in terms of the average of the boundary values of a logarithmic integral, its derivative with respect to λ is the average of boundary values of a Cauchy integral, which explains the integrand in the expression (5.52). Although the integral is taken over the gap Γ_{k+1}^+ of the contour C_I between the endpoints λ_{2k} and λ_{2k+1}, the conditions (5.52) depend only on the ordered sequence of endpoints $\lambda_0, \ldots, \lambda_G$ and not on the particular contour gaps Γ_{k+1}^+. This is because the integrand has an analytic continuation from each gap of C_I to either side; using the jump condition satisfied by the boundary values of $F(\lambda)$ in the gaps, one finds that

$$\frac{1}{2}(F_+(\lambda) + F_-(\lambda)) = F_\pm(\lambda) \pm \pi i \rho^0(\lambda). \tag{5.53}$$

Therefore, the integrand continues to the left as $2iJx + 4iJ\lambda t + F(\lambda) + \pi i \rho^0(\lambda)$ and to the right as $2iJx + 4iJ\lambda t + F(\lambda) - \pi i \rho^0(\lambda)$. □

Remark. Taken together, the moment conditions and the vanishing conditions place $3G/2+1$ real constraints on the $G+1$ complex endpoints $\lambda_0, \ldots, \lambda_G$. Intuitively, one expects the set of admissible endpoints to be $G/2+1$ real-dimensional. Existence of solutions remains to be shown, but certainly at this point the solution is not expected to be unique without imposing further conditions.

5.1.4 Determination of the Contour Bands. Measure Reality Conditions

Given G and the choices of $J = \pm 1$ and $\sigma = \pm 1$, there still remains some freedom in specifying the endpoints, and the contour bands and gaps that connect the endpoints remain completely unspecified. Now, we turn to the question of identifying further constraints sufficient to ensure that the differential measure $\rho^\sigma(\eta)\,d\eta$ is real-valued in the bands, where it does not vanish identically. As remarked in the discussion of the conditions on the complex phase function in §4.2, the reality of this differential is as much a condition on the contour bands (making up its support) through the differential $d\eta$ as on the function $\rho^\sigma(\eta)$ itself. We therefore expect that we may have to choose the bands carefully in order to achieve the required reality. In fact, the reality condition further constrains the endpoints as well.

The function $\rho^\sigma(\lambda)$ is defined in the bands of C_I. But in each separate oriented band I_k^+, it has an analytic continuation to the left and right sides. To extend to the left, write

$$\begin{aligned}\rho^\sigma(\lambda) &= \rho^0(\lambda) + \frac{1}{2\pi i}(F_+(\lambda) - F_-(\lambda))\\ &= \rho^0(\lambda) + \frac{1}{\pi i}F_+(\lambda) - \frac{1}{2\pi i}(F_+(\lambda) + F_-(\lambda))\\ &= \rho^0(\lambda) + \frac{1}{\pi i}F_+(\lambda) + \frac{2Jx}{\pi} + \frac{4J\lambda t}{\pi},\end{aligned} \tag{5.54}$$

while to extend to the right,

$$\begin{aligned}\rho^\sigma(\lambda) &= \rho^0(\lambda) + \frac{1}{2\pi i}(F_+(\lambda) - F_-(\lambda))\\ &= \rho^0(\lambda) - \frac{1}{\pi i}F_-(\lambda) + \frac{1}{2\pi i}(F_+(\lambda) + F_-(\lambda))\\ &= \rho^0(\lambda) - \frac{1}{\pi i}F_-(\lambda) - \frac{2Jx}{\pi} - \frac{4J\lambda t}{\pi}.\end{aligned} \tag{5.55}$$

These calculations use the fact that $F_+(\lambda) + F_-(\lambda)$ is given in (5.20) as an explicit analytic function in the bands. Since $\rho^\sigma(\lambda)$ is the same analytic function no matter what precise contour is taken to be the band I_k^+, the reality condition

$$\Im(\rho^\sigma(\eta)\,d\eta) = \Im(\rho^\sigma(u + iv)(du + i\,dv)) = 0 \tag{5.56}$$

may be viewed as a differential equation for the band contour I_k^+ *in the real* (u, v)-*plane*. Given the endpoints, the differential equation (5.56) is explicit, with $\rho^\sigma(\eta)$ being given by (5.34).

The obstruction to using this differential equation to define the bands I_k^+ (and thus the bands I_k^- by complex conjugation) is simply that the two endpoints of the band I_k^+ might not lie on the same integral curve of (5.56). However, we can try to exploit the remaining degrees of freedom in the choice of the endpoints λ_k to solve a kind of *connection problem* for the vector field (5.56) in the (u, v)-plane. Thus, we want to further constrain the set of endpoints $\lambda_0, \ldots, \lambda_G$ precisely so that for each band there exists an integral curve of (5.56) that joins the two endpoints of the band. Each nonzero endpoint of a band is a fixed point of the vector field (5.56). Assuming the generic case of exact square-root vanishing of the candidate density

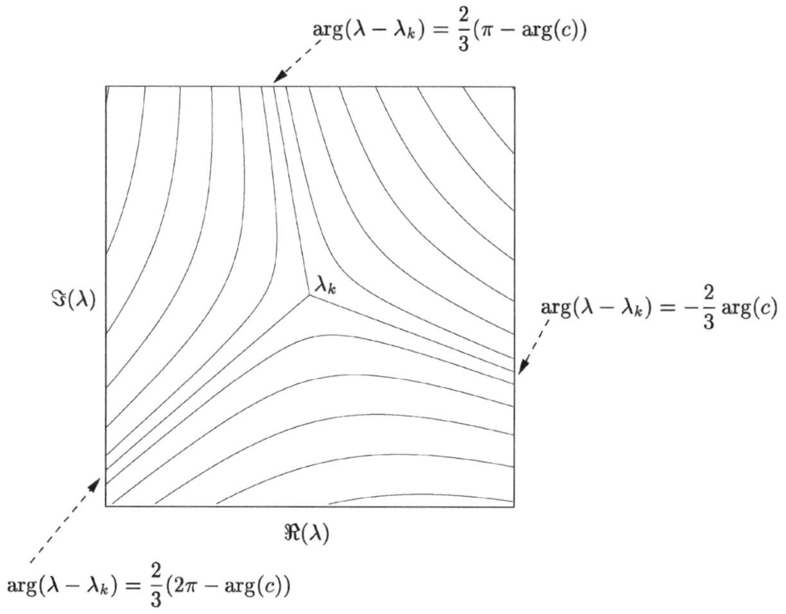

Figure 5.3 The local orbit structure of the differential equation (5.56) for the bands near a nonzero band endpoint λ_k. Locally, we have $\rho^\sigma(\lambda) = c\sqrt{\lambda - \lambda_k}(1 + O(\lambda - \lambda_k))$ for some nonzero constant c.

$\rho^\sigma(\lambda)$ given by (5.34) at each endpoint, there are locally three orbits of (5.56) that meet the fixed point at 120-degree angles. See figure 5.3. If the endpoints are chosen correctly, each band in the upper half-plane except for I_0^+ is then a heteroclinic orbit of the system (5.56) in the (u, v)–phase plane. The band I_0^+ is exceptional because one of the endpoints is $\lambda = 0$, which is not a fixed point of (5.56). Thus, this band can be found in principle by the process of *shooting*. That is, one solves (5.56) as a well-defined initial-value problem starting from $u = v = 0$ with the correct limiting value of $\rho^\sigma(\eta)$ taken as $\eta \to 0$ within I_0^+ and insists that the orbit terminate (in infinite "time") at λ_0.

Remark. Note that this problem is also known in geometric function theory as the characterization of *trajectories of quadratic differentials*.

Remark. Note that if a *single orbit* of the vector field (5.56) connects two endpoints, then not only is the candidate measure $\rho^\sigma(\eta)\,d\eta$ real on the orbit, but *it is also strictly of one sign on the whole interior of the orbit*. This follows from the fact that internal zeros of the candidate measure necessarily correspond to fixed points of the vector field (5.56). Therefore, it is possible to replace the analytical constraint that the inequality $\rho^\sigma(\eta)\,d\eta < 0$ holds strictly throughout the interior of each band with a kind of topological constraint on the orbits of (5.56), together with the verification of the inequality at an isolated interior point. In this sense, for steepest-descent analysis of Riemann-Hilbert problems where the correct contour must be determined as part of the solution of the problem, the correct generalization of the notion of strictness of inequality is the notion of *connectivity*. We will revisit this theme shortly.

How does one find the endpoint configurations for which the integral curves of (5.56) form the bands in practice? By analyticity of $\rho^\sigma(\lambda)$, for each band $I_k^+ \subset C_I$, we can define an analytic function by the formulae

$$B_0(\lambda) = \int_0^\lambda \rho^\sigma(\eta)\, d\eta, \qquad B_k(\lambda) = \int_{\lambda_{2k-1}}^\lambda \rho^\sigma(\eta)\, d\eta, \quad k = 1, \ldots, G/2, \quad (5.57)$$

where $\rho^\sigma(\eta)$ is given by (5.34). The main point is that *the contours along which the measure $\rho^\sigma(\eta)\, d\eta$ is real are exactly the zero-level sets of the real-analytic functions $\Im(B_k(\lambda))$*. By construction, the "lower" endpoint of each band lies on the corresponding zero level. A necessary condition for there to exist a single branch of the zero-level curve that connects this endpoint of each band to its partner is that the *measure reality conditions*

$$R_k := \Im(B_k(\lambda_{2k})) = 0, \qquad k = 0, \ldots, G/2, \qquad (5.58)$$

hold. Once again, it is clear by analyticity that the integrals are path-independent, and therefore the R_k are manifestly real-valued functions of the endpoints $\lambda_0, \ldots, \lambda_G$ alone.

We therefore have the following lemma.

Lemma 5.1.5 *Suppose the conditions of Lemma* 5.1.2 *and Lemma* 5.1.4 *are satisfied. Then a necessary condition for the candidate measure $\rho^\sigma(\eta)\, d\eta$ to be real in its support is that the endpoints $\lambda_0, \ldots, \lambda_G$ satisfy the reality conditions $R_k(\lambda_0, \ldots, \lambda_G) = 0$ for $k = 0, \ldots, G/2$.*

Remark. With the addition of the reality conditions to the moment conditions and the vanishing conditions, we at last have $2G + 2$ real equations in $2G + 2$ real unknowns, the real and imaginary parts of $\lambda_0, \ldots, \lambda_G$. Intuitively, one expects the set of solutions to be discrete. Given values of J and σ, the equations for the endpoints involve x and t as real parameters. If for some x and t a solution exists and the Jacobian matrix of derivatives of the conditions with respect to the endpoints and their complex conjugates is nonsingular, then by the implicit function theorem the endpoints will locally be continuous functions of x and t. This property is then inherited by the function $\rho^\sigma(\eta)$.

Note that it is by no means clear that the reality condition $R_k = 0$ guarantees the existence of a real level connecting the endpoints of the band I_k^+. To establish the existence we would need to have some discrete topological information in addition, like the connectedness of the real-level set of the analytic function $B_k(\lambda)$. Furthermore, if $R_k = 0$, then there may be more than one contour connecting the two endpoints of the interval, since locally there are three possible real paths emerging from each nonzero endpoint. These are the central difficulties in the characterization of trajectories of quadratic differentials in geometric function theory. We do not pursue these questions further in the general context, since it is clear from examples to follow how the procedure works in practice. We are in fact able to find the precise contours I_k^+ connecting the band endpoints, as long as $J = \pm 1$ and $\sigma = \pm 1$ are chosen correctly. Note that if the measure $\rho^\sigma(\eta)\,d\eta$ is real in I_k^+, then the conjugate measure $\rho^\sigma(\eta^*)^*\,d\eta$ is automatically real and of the opposite sign in $I_k^- = I_k^{+*}$ with the orientation of C_I^*.

5.1.5 Restoring the Loop Contour C

Let us suppose that this construction has been successful, so that we have found a set of admissible endpoints $\lambda_0, \ldots, \lambda_G$ that satisfy the $2G+2$ real conditions we have imposed, and we have shown that for each band I_k^+, the two endpoints of the band (including $\lambda = 0$ for the band I_0^+) are contained in the *same connected component* of the level set $\Im(B_k(\lambda)) = 0$. We have then constructed a genus-G ansatz. At this point, the bands of the contour $C_I \cup C_I^*$ are completely specified as contours in the complex plane. But the gaps have not been constrained at all by this construction. In particular, as long as $\lambda_G \neq iA$, the final portion of C_I is part of a gap that may be chosen freely. We now show that under this generic condition the temporary assumption we made at the beginning of this chapter—that the contour C passes through $\lambda = iA$—can be removed.

Without changing the value of $\tilde{\phi}^\sigma(\lambda)$ on the contour C_I, we may rewrite it in its original form:

$$\begin{aligned}\tilde{\phi}^\sigma(\lambda) = &\int_0^{iA} L_\eta^0(\lambda)\rho^0(\eta)\,d\eta + \int_{-iA}^0 L_\eta^0(\lambda)\rho^0(\eta^*)^*\,d\eta \\ &- \int_{\cup_k I_k^+} \overline{L_\eta^{C,\sigma}}(\lambda)\rho^\sigma(\eta)\,d\eta - \int_{\cup_k I_k^-} \overline{L_\eta^{C,\sigma}}(\lambda)\rho^\sigma(\eta^*)^*\,d\eta \\ &+ J(2i\lambda x + 2i\lambda^2 t) + i\pi\sigma\int_\lambda^{iA} \rho^0(\eta)\,d\eta. \end{aligned} \tag{5.59}$$

But now it is clear that the jump matrix $\mathbf{v}_{\tilde{\mathbf{N}}}^\sigma(\lambda)$ is analytic in λ from the final interval endpoint $\lambda_G \in C_I$, along C_I into $\lambda = iA$, and then out again from $\lambda = iA$ down along C_F to $\lambda = 0$. This means that in a distorted triangular region Δ (see figure 5.4) with corners at $\lambda = \lambda_G$, $\lambda = iA$, and $\lambda = 0$, it is possible to redefine the matrix $\tilde{\mathbf{N}}^\sigma(\lambda)$ by the analytic transformation

$$\tilde{\mathbf{N}}^\sigma(\lambda) \to \tilde{\mathbf{N}}^\sigma(\lambda)\mathbf{v}_{\tilde{\mathbf{N}}}^\sigma(\lambda), \qquad \lambda \in \Delta. \tag{5.60}$$

In the region Δ^*, the transformation that preserves the complex-conjugation symmetry of $\tilde{\mathbf{N}}^\sigma(\lambda)$ is used; for $\lambda \in \Delta^*$ we set $\tilde{\mathbf{N}}^\sigma(\lambda) := \sigma_2\tilde{\mathbf{N}}^\sigma(\lambda^*)^*\sigma_2$. As a consequence, the jump matrix is restored to the identity on the final gap portions of C_I and C_I^* and also along all of $C_F \cup C_F^*$. On the third boundary curve of Δ, there is now a jump, which is given by exactly the same formula for $\mathbf{v}_{\tilde{\mathbf{N}}}^\sigma(\lambda)$. This third boundary curve, which is shown dashed in figure 5.4, is now the final gap $\Gamma_{G/2+1}^+$ on the contour C that genuinely encircles the imaginary interval $[0, iA]$. The corresponding conjugated contour in the lower half-plane is $\Gamma_{G/2+1}^-$.

Thus, all reference to a contour C that is required to pass through the point $\lambda = iA$ disappears. It is important to correctly interpret this fact in the context of the various

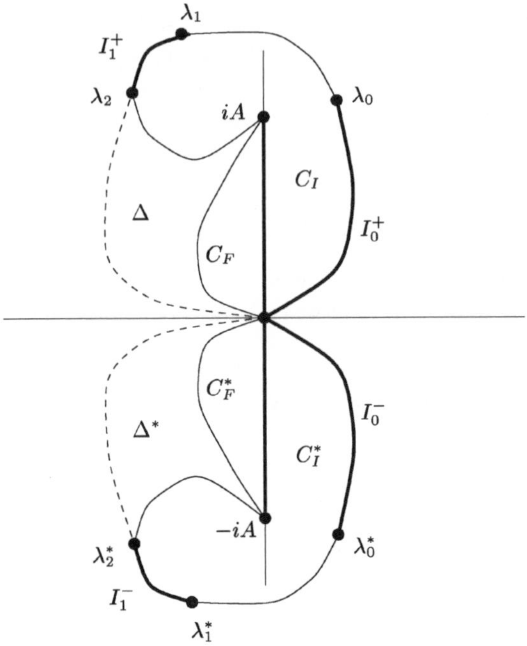

Figure 5.4 A sketch of how the matrix $\widetilde{\mathbf{N}}^\sigma(\lambda)$ is redefined as $\widetilde{\mathbf{N}}^\sigma(\lambda)\mathbf{v}^\sigma_{\widetilde{\mathbf{N}}}(\lambda)$ in the region Δ. This sketch is for a genus-$G = 2$ ansatz with orientation $\sigma = +1$. In the region Δ^*, the matrix is redefined to preserve the original reflection symmetry of $\widetilde{\mathbf{N}}^\sigma(\lambda)$.

integrals that have been introduced to characterize the conditions on the endpoints $\lambda_0, \ldots, \lambda_G$ and to provide a formula for the candidate density $\rho^\sigma(\lambda)$. In each case that we have integrated over the set $\Gamma_I \cap C_I$, we simply have a sum of integrals

$$\lambda_0 \to \lambda_1, \lambda_2 \to \lambda_3, \ldots, \lambda_G \to iA, \tag{5.61}$$

and when we have integrated over $\Gamma_I \cap C_I^*$, we have a sum of integrals

$$-iA \to \lambda_G^*, \lambda_{G-2}^* \to \lambda_{G-3}^*, \ldots, \lambda_1^* \to \lambda_0. \tag{5.62}$$

These integrals may be taken over any paths in $\mathbb{C} \setminus [-iA, iA]$ that when combined with the precisely specified band contours $I_k^\pm$ make up a non-self-intersecting contour C_I.

5.2 IMPOSING THE INEQUALITIES. LOCAL AND GLOBAL CONTINUATION THEORY

In principle, the algorithm described in §5.1 can be carried out for any even G, and it results in a discrete number of real candidate measures $\rho^\sigma(\eta)\,d\eta$ supported on a system of well-defined bands in the complex plane with endpoints $(0, \lambda_0), (\lambda_2, \lambda_3), \ldots, (\lambda_{G-1}, \lambda_G)$ and their complex conjugates. Within each band,

$\tilde{\phi}^\sigma(\lambda)$ is an imaginary constant. If one of these candidate measures is to generate a complex phase function $g^\sigma(\lambda)$ that asymptotically simplifies the Riemann-Hilbert problem for $\widetilde{\mathbf{N}}^\sigma(\lambda)$, then two additional conditions need to be satisfied by the candidate density:

1. The candidate measure $\rho^\sigma(\eta)\,d\eta$ must be strictly negative in the interior of each band $I_0^+,\ldots,I_{G/2}^+$ of the loop contour C;
2. It must be possible to choose the gaps $\Gamma_1^\pm,\ldots,\Gamma_{G/2+1}^\pm$ so that in the interior of each the real part of the function $\tilde{\phi}^\sigma(\lambda)$ constructed from the candidate measure $\rho^\sigma(\eta)\,d\eta$ is strictly negative.

If these additional conditions can be satisfied for some genus-G ansatz and for some values of J and σ, then $\rho^\sigma(\eta)\,d\eta$ is promoted from a candidate measure to a bona fide measure that generates a true complex phase function $g^\sigma(\lambda)$, and the rigorous asymptotic analysis described in chapter 4 is valid. This in turn yields a rigorous pointwise-asymptotic description, in terms of genus-G Riemann theta functions, of the sequence of solutions $\psi(x,t)$ of the nonlinear Schrödinger equation that for each value $\hbar_N$ of $\hbar$ is the soliton ensemble connected with the WKB approximations $\lambda_{\hbar_N,n}^{\rm WKB}$ of the true discrete eigenvalues for the initial condition $\psi(x,0)=A(x)$. One expects that there will be only one value of G (possibly depending on the parameters x and t) for which a candidate satisfies the inequalities.

In this section, we show that under certain conditions the existence of a genus-G ansatz that satisfies all inequalities for some x and t implies that a successful genus G-ansatz in fact exists for nearby values of x and t. Let $\mathbb{H}$ denote the open upper half-plane minus the imaginary interval $[0,iA]$. First, we show that, essentially, the existence of band contours connecting pairs of endpoints is an open condition in the (x,t)-plane.

Lemma 5.2.1 *Fix σ and J. Let x_0 and t_0 be given in the (x,t)-plane such that the following conditions hold.*

1. *For each (x,t) in some disk E centered at (x_0,t_0), there is a solution $\Lambda(x,t):=\{\lambda_0(x,t),\ldots,\lambda_G(x,t)\}$ of the moment conditions* (5.25), *the vanishing conditions* (5.52), *and the measure reality conditions* (5.58), *for which*

   - *each $\lambda_k(x,t)$ lies in $\mathbb{H}$ for all $(x,t)\in E$,*
   - *each $\lambda_k(x,t)$ is continuously differentiable in E,*
   - *the $\lambda_k(x,t)$ are distinct for all $(x,t)\in E$.*

   Let the candidate density constructed from the endpoints $\Lambda(x,t)$ for $(x,t)\in E$ be denoted by $\rho^\sigma(\eta;x,t)$.
2. *The function $\rho^\sigma(\eta;x_0,t_0)$ admits for all $k=1,\ldots,G/2$ a smooth orbit $I_k^+(x_0,t_0)$ of the differential equation* (5.56) *connecting the pair of consecutive endpoints $\lambda_{2k-1}(x_0,t_0)$ and $\lambda_{2k}(x_0,t_0)$ and lying entirely in $\mathbb{H}$, as well as a smooth orbit $I_0^+(x_0,t_0)$ of* (5.56) *connecting the origin to $\lambda_0(x_0,t_0)$ and lying in $\mathbb{H}\cup\{0\}$. Furthermore, the function $\rho^\sigma(\eta;x_0,t_0)$ is nonzero in the interior of each band $I_k^+(x_0,t_0)$ with $\rho^\sigma(\eta;x_0,t_0)\,d\eta$ being a negative (real) differential*

and with

$$\inf_{\eta \in I_k^+(x_0,t_0)} \left| \frac{\rho^\sigma(\eta; x_0, t_0)}{R_+(\eta; x_0, t_0)} \right| > 0, \tag{5.63}$$

where $R(\eta; x_0, t_0)$ denotes the square root function defined relative to the endpoints $\Lambda(x_0, t_0)$ and the given bands $I_k^\pm(x_0, t_0)$. That is, $\rho^\sigma(\eta; x_0, t_0)$ vanishes exactly like a square root at the endpoints and not to higher order.

Then, there exists a disk $D \subset E$ centered at (x_0, t_0) such that for all $(x, t) \in D$ and corresponding to the candidate density function $\rho^\sigma(\eta; x, t)$, there is for each $k = 1, \ldots, G/2$ a smooth orbit $I_k^+(x, t)$ of (5.56) connecting $\lambda_{2k-1}(x, t)$ to $\lambda_{2k}(x, t)$ and lying in $\mathbb{H}$, as well as a smooth orbit $I_0^+(x, t)$ of (5.56) connecting the origin to $\lambda_0(x, t)$ and lying in $\mathbb{H} \cup \{0\}$. Moreover, the differential $\rho^\sigma(\eta; x, t)\, d\eta$ is negative in each band $I_k^+(x, t)$ for all $(x, t) \in D$.

Proof. First observe that from the formula (5.35) for $\rho^\sigma(\eta; x, t)$ in terms of the function Y (cf. Lemma 5.1.3), the continuity of the endpoint functions $\lambda_k(x, t)$ in E and the continuous dependence of the analytic function Y on the endpoints, we immediately find that for (x, t) in some sufficiently small disk $D_{\rm sr} \subset E$, the statement that $\rho^\sigma(\eta; x_0, t_0)$ vanishes exactly like a square root at all endpoints $\lambda_0(x_0, t_0), \ldots, \lambda_G(x_0, t_0)$ carries over to $\rho^\sigma(\eta; x, t)$ as well.

Consider the deformation of a band $I_k^+(x_0, t_0)$ for $k = 1, \ldots, G/2$. We seek a map $\tau_k(\eta) : I_k^+(x_0, t_0) \to I_k^+(x, t)$ that satisfies the implicit relation

$$\int_{\lambda_{2k}(x,t)}^{\tau_k(\eta)} \rho^\sigma(\zeta; x, t)\, d\zeta = \alpha_k(x_0, t_0, x, t) \int_{\lambda_{2k}(x_0,t_0)}^{\eta} \rho^\sigma(\zeta; x_0, t_0)\, d\zeta, \tag{5.64}$$

where $\alpha_k(x_0, t_0, x, t)$ is a real constant chosen so that $\tau_k(\lambda_{2k-1}(x_0, t_0)) = \lambda_{2k-1} \times (x, t)$, that is,

$$\alpha_k(x_0, t_0, x, t) := \frac{\int_{\lambda_{2k}(x,t)}^{\lambda_{2k-1}(x,t)} \rho^\sigma(\zeta; x, t)\, d\zeta}{\int_{\lambda_{2k}(x_0,t_0)}^{\lambda_{2k-1}(x_0,t_0)} \rho^\sigma(\zeta; x_0, t_0)\, d\zeta}. \tag{5.65}$$

Also, we restrict attention to maps for which $\tau_k(\lambda_{2k}(x_0, t_0)) = \lambda_{2k}(x, t)$.

First, we show that for $(x - x_0)^2 + (t - t_0)^2$ sufficiently small,

$$|\alpha_k(x_0, t_0, x, t) - 1| \le C_{\alpha,k} \sqrt{(x - x_0)^2 + (t - t_0)^2} \tag{5.66}$$

for some $C_{\alpha,k} > 0$. Since the denominator of α_k is real and strictly nonzero by assumption, this will follow if we can argue that the numerator of α_k is differentiable at $(x, t) = (x_0, t_0)$. Differentiating the numerator with respect to x or t, we may take the derivative operator inside the integral, since the integrand vanishes at the endpoints for $x = x_0$ and $t = t_0$. The derivatives of the integrand with respect to x and t have contributions from explicit x- and t-dependence and from x- and t-dependence through the endpoints $\Lambda(x, t)$. From the explicit formula (5.34), it is easy to see that the partial derivatives with respect to x and t are both integrable in $I_k^+(x_0, t_0)$ for $x = x_0$ and $t = t_0$. Then, the chain rule terms are integrable in $I_k^+(x_0, t_0)$ because the endpoints are continuously differentiable by assumption and because the derivatives of $\rho^\sigma(\zeta; x, t)$ with respect to the endpoints are integrable in

$I_k^+(x_0, t_0)$, although they blow up like inverse square roots at the two endpoints of the contour of integration.

Next, introduce the change of variables

$$\mu_k(\eta) := \frac{\tau_k(\eta) - B_k(x_0, t_0, x, t)}{A_k(x_0, t_0, x, t)}, \tag{5.67}$$

where

$$\begin{aligned} A_k(x_0, t_0, x, t) &:= \frac{\lambda_{2k}(x, t) - \lambda_{2k-1}(x, t)}{\lambda_{2k}(x_0, t_0) - \lambda_{2k-1}(x_0, t_0)}, \\ B_k(x_0, t_0, x, t) &:= \frac{\lambda_{2k}(x_0, t_0)\lambda_{2k-1}(x, t) - \lambda_{2k-1}(x_0, t_0)\lambda_{2k}(x, t)}{\lambda_{2k}(x_0, t_0) - \lambda_{2k-1}(x_0, t_0)}. \end{aligned} \tag{5.68}$$

Note that from the differentiability and distinctness properties of the endpoints in the neighborhood E, we have the estimates

$$\begin{aligned} |A_k(x_0, t_0, x, t) - 1| &\leq C_{A,k}\sqrt{(x - x_0)^2 + (t - t_0)^2}, \\ |B_k(x_0, t_0, x, t)| &\leq C_{B,k}\sqrt{(x - x_0)^2 + (t - t_0)^2}, \end{aligned} \tag{5.69}$$

for some positive constants $C_{A,k}$ and $C_{B,k}$ and all sufficiently small $(x - x_0)^2 + (t - t_0)^2$. The implicit relation (5.64) therefore becomes

$$\int_{\lambda_{2k}(x_0, t_0)}^{\mu_k(\eta)} \rho^\sigma(A_k\zeta + B_k; x, t)\, d\zeta = \frac{\alpha_k}{A_k}\int_{\lambda_{2k}(x_0, t_0)}^{\eta} \rho^\sigma(\zeta; x_0, t_0)\, d\zeta. \tag{5.70}$$

Now, consider the function $h_k(\eta)$ defined by the integral

$$h_k(\eta) := \int_{\lambda_{2k}(x_0, t_0)}^{\eta} \rho^\sigma(\zeta; x_0, t_0)\, d\zeta, \tag{5.71}$$

for η in a lens-shaped neighborhood of $I_k^+(x_0, t_0)$. If the neighborhood is sufficiently thin, then the map $h_k(\eta)$ is one-to-one, since by assumption $\rho^\sigma(\eta; x_0, t_0)$ is strictly nonzero in the interior of $I_k^+(x_0, t_0)$. By the reality condition satisfied by $I_k^+(x_0, t_0)$, the image of the lens-shaped neighborhood of $I_k^+(x_0, t_0)$ is a lens-shaped neighborhood of the open real interval

$$h_k(I_k^+(x_0, t_0)) = \left(0, -\int_{I_k^+(x_0, t_0)} \rho^\sigma(\eta; x_0, t_0)\, d\eta\right) := (0, h_k^{\max}). \tag{5.72}$$

The inverse function $h_k^{-1}(\cdot)$ is defined and analytic in the open real interval $(0, h_k^{\max})$. Near the endpoints, we have $h_k^{-1}(\phi) \sim C_1\phi^{2/3}$ near $\phi = 0$ and $h_k^{-1}(\phi) \sim C_2 + C_3(\phi - h_k^{\max})^{2/3}$ near $\phi = h_k^{\max}$ for some constants C_1, C_2, and C_3. Letting $M := h_k(\mu_k(\eta))$ and $H := h_k(\eta)$; the relation (5.70) can be rewritten as

$$M = H + \int_0^M U_k(\phi)\, d\phi := T_k(M), \tag{5.73}$$

where

$$U_k(\phi) := \left[\rho^\sigma(h_k^{-1}(\phi); x_0, t_0) - \frac{A_k}{\alpha_k}\rho^\sigma(A_k h_k^{-1}(\phi) + B_k; x, t)\right]\frac{dh_k^{-1}(\phi)}{d\phi}. \tag{5.74}$$

We want to consider solving this equation for $M = M(H)$ by fixed-point iteration, that is by choosing some M_0 and constructing the sequence $\{M_n\}$ by the recursion $M_n := T_k(M_{n-1})$. If this sequence converges, then we have a solution.

For $\epsilon_k > 0$ consider the rectangular region R_k with corner points $-2\epsilon_k \pm 2i\epsilon_k$ and $h_k^{\max} + 2\epsilon_k \pm 2i\epsilon_k$. The interval $h_k(I_k^+(x_0, t_0))$ is contained in R_k. We claim that for $\epsilon_k > 0$ sufficiently small the function $U_k(\phi)$ in the integrand of (5.73) has an analytic extension as a function of ϕ to some open set containing R_k for which

$$\lim_{x\to x_0,\, t\to t_0} \left[\sup_{\phi\in R_k} |U_k(\phi)| \right] = 0. \tag{5.75}$$

To show the analyticity of $U_k(\phi)$, it suffices to examine the endpoints $\phi = 0$ and $\phi = h_k^{\max}$. On the one hand, $dh_k^{-1}/d\phi$ in (5.74) blows up exactly like a negative one-third power at each endpoint. But on the other hand, the inverse map $h_k^{-1}(\cdot)$ vanishes like a two-thirds power at each endpoint, and since we are working in D_{sr}, the function $\rho^\sigma(\eta; x, t)$ vanishes like a square root at each endpoint; thus, the function of ϕ in square brackets in (5.74) vanishes exactly like a one-third power at each endpoint. Analyticity at the endpoints thus follows for the product $U_k(\phi)$. Next, to establish (5.75), we note that by analyticity in R_k, there exists a uniform bound and the only question is its behavior as $(x, t) \to (x_0, t_0)$. Clearly, $U_k(\phi)$ converges pointwise to zero in this limit for all $\phi \in R_k$ except possibly at the endpoints $\phi = 0$ and $\phi = h_k^{\max}$. But by analyticity at the endpoints and compactness of R_k, the convergence to zero is in fact uniform for $\phi \in R_k$, and the result follows.

Note that for all $H \in h_k(I_k^+(x_0, t_0))$ the disk $|\phi - H| < \epsilon_k$ is contained in R_k. We claim that for sufficiently small $(x - x_0)^2 + (t - t_0)^2$, the transformation $T_k(\phi)$ maps this disk into itself. Indeed

$$|T_k(\phi) - H| = \left| \int_0^{H+(\phi-H)} U_k(\phi')\, d\phi' \right| \le (H + \epsilon_k) \sup_{\phi\in R_k} |U_k(\phi)|, \tag{5.76}$$

which can be made arbitrarily small and in particular less than ϵ_k for x and t close enough to x_0 and t_0, respectively, in view of (5.75).

Let $\widetilde{D}_k \subset E$ denote the disk in the (x, t)-plane centered at (x_0, t_0) in which the last line is bounded above by ϵ_k. Then, for ϕ_1 and ϕ_2 both in the disk $|\phi - H| < \epsilon_k$, we also have for $(x, t) \in \widetilde{D}_k$,

$$\begin{aligned} |T_k(\phi_2) - T_k(\phi_1)| &= \left| \int_{\phi_1}^{\phi_2} U_k(\phi')\, d\phi' \right| \le |\phi_2 - \phi_1| \sup_{\phi\in R_k} |U_k(\phi)| \\ &< \frac{\epsilon_k}{H + \epsilon_k} |\phi_2 - \phi_1| < |\phi_2 - \phi_1|. \end{aligned} \tag{5.77}$$

From (5.76) and (5.77), the contraction-mapping theorem guarantees that the iteration $M_n := T_k(M_{n-1})$ will converge when $(x, t) \in \widetilde{D}_k$ whenever M_0 is taken in the disk $|M_0 - H| < \epsilon_k$. Moreover, the limit $M = \lim_{n\to\infty} M_n$ is the unique solution of the equation (5.73) in this disk. We therefore have a function $M = M(H)$ defined for all H in the closure of the open interval $h_k(I_k^+(x_0, t_0))$. This function is continuously differentiable in the closed interval of its definition since for all $M(H)$ defined above and for $(x, t) \in \widetilde{D}_k$, we have

$$|U_k(M)| \le \frac{\epsilon_k}{H + \epsilon_k} < 1 \tag{5.78}$$

and consequently

$$\frac{\partial}{\partial M}[M - T_k(M)] = 1 - U_k(M) \neq 0, \tag{5.79}$$

holding even at the endpoints. At these endpoints, we know that the unique solution in the disk is given simply by $M(0) = 0$ and $M(h_k^{\max}) = h_k^{\max}$. The curve $M(H)$ is therefore homotopic to the closed real interval $[0, h_k^{\max}]$.

The function $h_k^{-1}(\cdot)$ is defined on the closed real interval $[0, h_k^{\max}]$ and has a unique analytic continuation to the curve $M(H)$. The inverse function so defined on $M(H)$ is continuous, and we then obtain

$$\mu_k(\eta) = h_k^{-1}(M(h_k(\eta))). \tag{5.80}$$

For each x and t in $\widetilde{D}_k$, we therefore obtain a curve with the same endpoints, $\lambda_{2k-1}(x_0, t_0)$ and $\lambda_{2k}(x_0, t_0)$. By our estimates, the curves contract uniformly to $I_k^+(x_0, t_0)$ as $(x, t) \to (x_0, t_0)$. Finally, set

$$\tau_k(\eta) := A_k(x_0, t_0, x, t) \cdot h_k^{-1}(M(h_k(\eta)) + B_k(x_0, t_0, x, t). \tag{5.81}$$

This is a continuous function of $\eta \in I_k^+(x_0, t_0)$. Each point in the image satisfies (5.64), and consequently the image is a smooth curve $I_k^+(x, t)$ connecting $\lambda_{2k-1}(x, t)$ to $\lambda_{2k}(x, t)$. Moreover, by continuity, $I_k^+(x, t)$ will lie in the set $\mathbb{H}$ for $(x - x_0)^2 + (t - t_0)^2$ sufficiently small, and to achieve this, we restrict x and t to some slightly smaller disk $D_k \subset \widetilde{D}_k$. Finally, to see that for all $(x, t) \in D_k$ the differential $\rho^\sigma(\eta; x, t)\, d\eta$ is nonvanishing in $I_k^+(x, t)$ and of the same sign as $\rho^\sigma(\eta; x_0, t_0)\, d\eta$ in $I_k^+(x_0, t_0)$, simply differentiate (5.64) to obtain

$$\rho^\sigma(\tau_k; x, t)\, d\tau_k = \rho^\sigma(\tau_k(\eta); x, t)\tau_k'(\eta)\, d\eta = \alpha_k(x_0, t_0, x, t)\rho^\sigma(\eta; x_0, t_0)\, d\eta, \tag{5.82}$$

from which the required result follows from the estimate (5.66).

To verify the continuity of the exceptional band I_0^+, one repeats the above arguments, substituting zero everywhere for λ_{2k-1}. Thus, one uses

$$\alpha_0(x_0, t_0, x, t) := \frac{\int_{\lambda_0(x,t)}^0 \rho^\sigma(\zeta; x, t)\, d\zeta}{\int_{\lambda_0(x_0,t_0)}^0 \rho^\sigma(\zeta; x_0, t_0)\, d\zeta} \tag{5.83}$$

and obtains an estimate analogous to (5.66). Also, one takes $B_0(x_0, t_0, x, t) := 0$ and

$$A_0(x_0, t_0, x, t) := \left[\frac{\lambda_0(x_0, t_0)}{\lambda_0(x, t)} - 1\right]^{-1} \tag{5.84}$$

and obtains estimates analogous to (5.69). By similar arguments based on contraction mapping, one verifies the continuity of $I_0^+(x, t)$ for x and t in some sufficiently small disk neighborhood D_0 of (x_0, t_0). Finally, we restrict (x, t) to lie in D, where

$$D = D_{\text{sr}} \cap \left[\bigcap_{k=0}^{G/2} D_k\right], \tag{5.85}$$

which is nonempty for finite G. This completes the proof. □

The existence of contour segments in which the gap inequalities may be satisfied is also an open condition in the (x, t)-plane.

Lemma 5.2.2 *Assume all the conditions of Lemma* 5.2.1, *and let* $\tilde{\phi}^\sigma(\lambda; x, t)$ *denote the function corresponding to the candidate density* $\rho^\sigma(\eta; x, t)$ *for* $(x, t) \in E$ *via* (5.1) *with* K *chosen according to* (5.10). *Furthermore, suppose that the bands* $I_k^\pm(x_0, t_0)$ *are complemented by a system of gap contours* $\Gamma_k^\pm(x_0, t_0)$ *making up a loop contour* $C(x_0, t_0) \subset \mathbb{H} \cup \{0\}$ *such that* $\Re(\tilde{\phi}^\sigma(\lambda; x_0, t_0)) < 0$ *strictly in the interior of all gaps* $\Gamma_k^+(x_0, t_0)$. *Then, there exists a disk* $D' \subset D \subset E$ *centered at* (x_0, t_0) *in the* (x, t)*-plane such that for all* $(x, t) \in D'$, *smooth gap contours may be chosen in* $\mathbb{H} \cup \{0\}$ *for which the relevant inequality persists. That is, there exist smooth paths* $\Gamma_k^+(x, t)$ *in* $\mathbb{H}$ *connecting* $\lambda_{2k-2}(x, t)$ *to* $\lambda_{2k-1}(x, t)$ *for* $k = 1, \ldots, G/2$ *and a path* $\Gamma_{G/2+1}^+(x, t)$ *in* $\mathbb{H} \cup \{0\}$ *connecting* $\lambda_G(x, t)$ *to the origin such that for all* λ *in the interior of a path* $\Gamma_k^+(x, t)$, *the strict inequality* $\Re(\tilde{\phi}^\sigma(\lambda; x, t)) < 0$ *holds.*

Proof. Using the general relations (4.32) and (4.33), we see that the function $\tilde{\phi}^\sigma(\lambda; x, t)$ may be expressed in terms of an integral of the corresponding candidate density $\rho^\sigma(\eta; x, t)$. In this connection, the desingularized representation (5.35) of the density is useful. Let $R(\lambda; x, t)$ and $Y(\lambda; x, t)$ denote the square-root function R (defined in §4.3) and the analytic function Y (cf. (5.41)) defined from the endpoints $\Lambda(x, t)$ for $(x, t) \in E$. Consider an "internal" gap $\Gamma_k^+(x, t)$ intended to connect the endpoints $\lambda_{2k-2}(x, t)$ and $\lambda_{2k-1}(x, t)$ for $k = 1, \ldots, G/2$. Since $D' \subset D$, the results of Lemma 5.2.1 hold, and throughout D' the band contours $I_k^+(x, t)$ exist as smooth curves. Therefore, in $\Gamma_k^+(x, t)$, the function $\tilde{\phi}^\sigma(\lambda; x, t)$ may be written as

$$\begin{aligned}\tilde{\phi}^\sigma(\lambda; x, t) &= \tilde{\phi}^\sigma(\lambda; x, t)|_{\lambda \in I_{k-1}^+(x,t)} + i\pi \int_{\lambda_{2k-2}(x,t)}^{\lambda} R(\eta; x, t) Y(\eta; x, t)\, d\eta \\ &= \tilde{\phi}^\sigma(\lambda; x, t)|_{\lambda \in I_k^+(x,t)} + i\pi \int_{\lambda_{2k-1}(x,t)}^{\lambda} R(\eta; x, t) Y(\eta; x, t)\, d\eta, \end{aligned} \tag{5.86}$$

where we recall that by construction $\tilde{\phi}^\sigma(\lambda; x, t)$ is an imaginary constant when restricted to each band $I_k^+(x, t)$.

Let $\lambda = w(s)$ for $0 \le s \le 1$ be a parametrization of the given gap path $\Gamma_k^+(x_0, t_0)$. Therefore, $w(0) = \lambda_{2k-2}(x_0, t_0)$ and $w(1) = \lambda_{2k-1}(x_0, t_0)$. We show that for all (x, t) in the sufficiently small disk D', the path parametrized by

$$\lambda = w(s; x, t) := A_k(x_0, t_0, x, t) w(s) + B_k(x_0, t_0, x, t), \tag{5.87}$$

where

$$\begin{aligned} A_k(x_0, t_0, x, t) &:= \frac{\lambda_{2k-1}(x, t) - \lambda_{2k-2}(x, t)}{\lambda_{2k-1}(x_0, t_0) - \lambda_{2k-2}(x_0, t_0)}, \\ B_k(x_0, t_0, x, t) &:= \frac{\lambda_{2k-1}(x_0, t_0)\lambda_{2k-2}(x, t) - \lambda_{2k-2}(x_0, t_0)\lambda_{2k-1}(x, t)}{\lambda_{2k-1}(x_0, t_0) - \lambda_{2k-2}(x_0, t_0)}, \end{aligned} \tag{5.88}$$

admits the relevant strict inequality for all $s \in (0, 1)$. By continuity of the endpoints in x and t, this is a near-identity linear transformation. We are given that $\Re(\tilde{\phi}^\sigma(w(s); x_0, t_0)) < 0$ and must show that $\Re(\tilde{\phi}^\sigma(A_k w(s) + B_k); x, t) < 0$ for D' sufficiently small.

First, we consider a neighborhood of the endpoint $s = 0$. Since by assumption we are working in the neighborhood D of Lemma 5.2.1, and therefore in the bigger neighborhood D_{sr} (cf. the proof of Lemma 5.2.1), the integrand $R(\lambda; x, t)Y(\lambda; x, t)$

vanishes at $\lambda_{2k-2}(x,t)$ exactly like $(\lambda-\lambda_{2k-2}(x,t))^{1/2}$ for all $(x,t)\in D'$. This implies that in a sufficiently small (independent of x and t) neighborhood V_0 in the complex plane that contains $\lambda=\lambda_{2k-2}(x,t)$ for all (x,t) close enough to (x_0,t_0), the region where $\Re(\tilde{\phi}^\sigma(\lambda;x,t))<0$ holds is a generalized sector whose boundary curves have tangents at $\lambda_{2k-1}(x,t)$ that meet at an angle of $2\pi/3$. The band contour $I_{k-1}^+(x,t)$ has a tangent at its upper endpoint $\lambda_{2k-2}(x,t)$ that bisects this angle. Without loss of generality, we now suppose that the given gap contour $\Gamma_k^+(x_0,t_0)$ has a tangent at $\lambda_{2k-2}(x_0,t_0)$ whose angle lies *strictly* between the tangents to the boundary curves. Indeed, if this is not true of the given gap contour, it may be achieved sacrificing neither smoothness nor the inequality $\Re(\tilde{\phi}^\sigma(w(s);x_0,t_0))<0$ by a small deformation near $s=0$. Now, the boundary curves in V_0 satisfy

$$\Im\left(\int_{\lambda_{2k-2}(x,t)}^{\lambda} R(\eta;x,t)Y(\eta;x,t)\,d\eta\right)=0, \tag{5.89}$$

and from the fixed-point theory used in the proof of Lemma 5.2.1, it follows that these boundary curves deform continuously in (x,t) near (x_0,t_0). Since the same is true of the path $\lambda=w(s;x,t)$ by construction, it is clear that the disk D' can be taken small enough that the inequality is satisfied on $\Gamma_k^+(x,t)\cap V_0$ for all $(x,t)\in D'$.

To handle the other endpoint, $s=1$, choose an analogous fixed neighborhood V_1 containing $\lambda_{2k-1}(x,t)$ for all (x,t) sufficiently close to (x_0,t_0). Then a similar argument can be used to show that, possibly by replacing D' with a smaller disk, the inequality is satisfied on $\Gamma_k^+(x,t)\cap V_1$ for all $(x,t)\in D'$.

Let $s_0(x,t)$ be defined so that the interval $(0,s_0(x,t))$ parametrizes the curve $\Gamma_k^+(x,t)\cap V_0$ by the function $w(s;x,t)$. Similarly, let $s_1(x,t)$ be defined so that $(s_1(x,t),1)$ parametrizes $\Gamma_k^+(x,t)\cap V_1$. Let

$$s_0:=\inf_{(x,t)\in D'} s_0(x,t)>0, \qquad s_1:=\sup_{(x,t)\in D'} s_1(x,t)<1. \tag{5.90}$$

It remains to verify the inequality (again, possibly by replacing D' with a smaller disk) for $s\in[s_0,s_1]$. Now, because we are avoiding the endpoints, there exists some $\epsilon<0$ depending only on s_0, s_1, x_0, and t_0 such that in this closed interval we have $\Re(\tilde{\phi}^\sigma(w(s);x_0,t_0))\le\epsilon$. Consequently, it is sufficient to show that $|\Re(\tilde{\phi}^\sigma(w(s;x,t);x,t))-\Re(\tilde{\phi}^\sigma(w(s);x_0,t_0))|<\epsilon$ for (x,t) close enough to (x_0,t_0). We have

$$\begin{aligned}
&|\Re(\tilde{\phi}^\sigma(w(s;x,t);x,t))-\Re(\tilde{\phi}^\sigma(w(s);x_0,t_0))|\\
&\quad\le\pi\left|\int_{\lambda_{2k-2}(x_0,t_0)}^{w(s)}[A_kR(A_k\eta+B_k;x,t)Y(A_k\eta+B_k;x,t)\right.\\
&\qquad\qquad\left.-R(\eta;x_0,t_0)Y(\eta;x_0,t_0)]\,d\eta\right|\\
&\quad\le\pi\sup_{s\in[s_0,s_1]}|w(s)-\lambda_{2k-2}(x_0,t_0)|\\
&\qquad\times\sup_{s\in[s_0,s_1]}|A_kR(w(s;x,t);x,t)Y(w(s;x,t);x,t)\\
&\qquad\qquad-R(w(s);x_0,t_0)Y(w(s);x_0,t_0)|.
\end{aligned} \tag{5.91}$$

The first factor is uniformly bounded, and by simple continuity arguments using the fact that the map $w(s)\to w(s;x,t)$ is a near-identity transformation, the second factor can be made arbitrarily small for (x,t) near (x_0,t_0) and in particular the product can be made less than ϵ. This completes the proof of existence of the "internal" gap $\Gamma_k^+(x,t)$.

Having established the persistence of the gaps connecting pairs of endpoints in Λ, we must now show that the "final" gap $\Gamma_{G/2+1}^+(x,t)$, which must connect $\lambda_G(x,t)$ to the origin, also persists for (x,t) near (x_0,t_0). In this case, the near-identity transformation of the path $\Gamma_{G/2+1}^+(x_0,t_0)$ parametrized by $w(s)$ is given simply by

$$w(s;x,t) := \frac{\lambda_G(x,t)}{\lambda_G(x_0,t_0)} w(s). \tag{5.92}$$

The local analysis near $s=0$ corresponding to the endpoint $\lambda=\lambda_G(x,t)$ goes through exactly as before.

For the local analysis near $s=1$ corresponding to $\lambda=0$ for all (x,t), we first consider the definition (5.1) of the function $\tilde{\phi}^\sigma(\lambda;x,t)$. For λ in the interior of the gap $\Gamma_{G/2+1}^+(x,t)$, we can use the analyticity of the given eigenvalue density $\rho^0(\eta)$ to rewrite the formula (5.1) for $\tilde{\phi}^\sigma(\lambda;x,t)$ in the form

$$\begin{aligned}\tilde{\phi}^\sigma(\lambda;x,t) = &\int_{C_I} L_\eta^{C,\sigma}(\lambda)\bar{\rho}^\sigma(\eta;x,t)\,d\eta + \int_{C_I^*} L_\eta^{C,\sigma}(\lambda)\bar{\rho}^\sigma(\eta^*;x,t)^*\,d\eta\\ &+ 2iJ(\lambda x+\lambda^2 t) + i\pi\sigma\int_\lambda^{iA}\rho^0(\eta)\,d\eta,\end{aligned} \tag{5.93}$$

where $\bar{\rho}^\sigma(\eta;x,t)$ is the complementary density function corresponding to $\rho^\sigma(\eta;x,t)$ via (5.9). Now it follows from the boundary value problem (5.20) that the function $\bar{\rho}^\sigma(\eta;x,t)$ extended by complex conjugation $\bar{\rho}^\sigma(\eta^*;x,t)^*$ to $C_I\cup C_I^*$ is analytic at $\eta=0$. Therefore, the first two integrals on the right-hand side of (5.93) can be combined and the path of integration may be deformed slightly either to the right (for $\sigma=+1$) or left (for $\sigma=-1$) in a small neighborhood of the origin. Thus, we deduce that $\tilde{\phi}^\sigma(\lambda;x,t)$ extends analytically to a neighborhood of the final endpoint $\lambda=0$ of the gap $\Gamma_{G/2+1}^+(x,t)$.

Now, for λ real, it follows from reality of the logarithm that

$$\Re(\tilde{\phi}^\sigma(\lambda;x,t)) = \Re\left(i\pi\sigma\int_\lambda^{iA}\rho^0(\eta)d\eta\right), \qquad \lambda\in\mathbb{R}, \tag{5.94}$$

which (a Cauchy-Riemann argument shows) has the same sign as $\sigma\lambda$ for λ near the origin. Since for $\sigma=+1$ (respectively, $\sigma=-1$) the portion of $\Gamma_{G/2+1}^+(x_0,t_0)$ near the origin necessarily lies in the second (respectively, first) quadrant, this together with the analyticity of $\tilde{\phi}^\sigma(\lambda;x,t)$ at $\lambda=0$ shows that in some neighborhood V_1 of the origin, the given gap contour $\Gamma_{G/2+1}^+(x_0,t_0)$ lies in some generalized sector bounded by the real axis and some boundary curve that makes a nonzero angle with the real axis at $x=x_0$ and $t=t_0$ (recall that $\Re(\tilde{\phi}^\sigma(0;x,t))=0$). Without loss of generality, we may assume that $\Gamma_{G/2+1}^+(x_0,t_0)$ has a tangent line at the origin making a nonzero angle with both the real axis and the tangent line of the boundary curve. Then, since the boundary curve again satisfies

$$\Im\left(\int_{\lambda_G(x,t)}^{\lambda} R(\eta;x,t)Y(\eta;x,t)\,d\eta\right) = 0, \tag{5.95}$$

the fixed-point theory predicts smooth deformation of this curve with respect to x and t near $x = x_0$ and $t = t_0$, which in conjunction with the continuity of the near-identity map $w(s) \to w(s; x, t)$ gives the necessary inequality in V_1. This concludes the analysis near $s = 1$ corresponding to the origin in the λ-plane.

With the endpoints taken care of in this way, the argument that the inequality holds on parts of $\Gamma^+_{G/2+1}(x, t)$ that are bounded away from the two endpoints λ_G and 0 is analogous to the corresponding argument we used in proving the persistence of the "internal" gaps. Therefore, for all (x, t) in the sufficiently small disk D', the "final" gap contour $\Gamma^+_{G/2+1}(x, t)$ exists as well. This completes the proof. □

Passing from the local to the global, these continuation arguments can be developed into a partial characterization of the boundary of the region of existence of a successful genus-G ansatz in the (x, t)-plane. Given (x_0, t_0) and a continuous branch of the collection of endpoint functions $\Lambda(x, t)$ such that the conditions of Lemma 5.2.1 and Lemma 5.2.2 are met, let U be the intersection of the largest open set in the (x, t)-plane, where the selected branch of $\Lambda(x, t)$ is differentiable, and the largest open set containing (x_0, t_0), where the genus-G ansatz corresponding to these endpoints satisfies all of the inequalities. Let $(x_{\mathrm{crit}}, t_{\mathrm{crit}})$ be a boundary point of U. It is necessary that at this boundary point at least one of the conditions of either Lemma 5.2.1 or Lemma 5.2.2 fails. Otherwise, the open set D' guaranteed to exist by these results would contain $(x_{\mathrm{crit}}, x_{\mathrm{crit}})$ and be contained in U—a contradiction.

To catalog the possible modes of failure of the ansatz at the boundary of U is a task complicated by the geometry of the cut upper half-plane $\mathbb{H}$. It is possible for a point on the boundary of U to correspond to an ansatz for which one of the band contours meets $\partial\mathbb{H}$ at a point or for which a gap contour is "forced" to meet $\partial\mathbb{H}$ because the boundary of the region where $\Re(\tilde{\phi}^\sigma) < 0$ does so. However, there are also modes of failure that do not involve the contour C meeting $\partial\mathbb{H}$. These modes can be characterized by equations for curves in the (x, t)-plane.

The onset of failure of the inequality for the bands can correspond to a point λ on one of the bands I_k^+ (including endpoints) for which the function $\rho^\sigma(\lambda; x_{\mathrm{crit}}, t_{\mathrm{crit}})/R(\lambda; x_{\mathrm{crit}}, t_{\mathrm{crit}})$, analytic on the closure of each band, has a zero. Here, ρ^σ is given by the formula (5.34) valid in the bands. Therefore, if the ansatz fails by this mechanism at the point $(x_{\mathrm{crit}}, t_{\mathrm{crit}})$, then for some $k = 0, \ldots, G/2$ the following conditions hold for some $\lambda \in \mathbb{H}$:

$$\Im\left(\int_{\lambda_{2k}(x_{\mathrm{crit}}, t_{\mathrm{crit}})}^{\lambda} \rho^\sigma(\eta; x_{\mathrm{crit}}, t_{\mathrm{crit}})\, d\eta\right) = 0, \qquad \frac{\rho^\sigma(\lambda; x_{\mathrm{crit}}, t_{\mathrm{crit}})}{R(\lambda; x_{\mathrm{crit}}, t_{\mathrm{crit}})} = 0, \tag{5.96}$$

and λ is on the band I_k^+. We note here that, neglecting the topological condition that $\lambda \in I_k^+$ (which amounts to the selection of a particular branch of the first relation above) and upon elimination of λ, these relations imply one real relation satisfied by x_{crit} and t_{crit}, a curve in the (x, t)-plane. If, in addition, λ is actually on the band I_k^+, then these conditions imply that the band I_k^+ has the interpretation of a chain of (at least) two connected heteroclinic orbits of the vector field (5.56).

The onset of failure of inequality for the gaps can correspond to the pinching off of a narrow "isthmus" in the region $\Re(\tilde{\phi}^\sigma) < 0$ in the λ-plane through which a gap curve is forced to pass. Exactly at onset, when the inequality first fails, the boundary curve where $\tilde{\phi}^\sigma(\lambda)$ is purely imaginary becomes singular. The existence of a singular point on the imaginary level can be expressed by the equations

$$\frac{\partial \tilde{\phi}^\sigma}{\partial \lambda}(\lambda; x_{\rm crit}, t_{\rm crit}) = 0, \qquad \Re(\tilde{\phi}^\sigma(\lambda; x_{\rm crit}, t_{\rm crit})) = 0. \tag{5.97}$$

Here, $\tilde{\phi}^\sigma$ refers to the expression valid in the gaps. Again, observe that if $\lambda \in \mathbb{H}$ may be eliminated between these two equations, then what remains is a single real equation in the two unknowns $x_{\rm crit}$ and $t_{\rm crit}$. These relations thus describe a union of curves in the real (x, t)-plane.

Remark. It is a consequence of the duality of the function $\tilde{\phi}^\sigma(\lambda)$ evaluated in the gaps with the function $\theta^\sigma(\lambda)$ evaluated in the bands (cf. equations (4.32) and (4.33)) that the conditions (5.96) and (5.97) are essentially *equivalent*. They result in the same curves in the (x, t)-plane.

Remark. The point $(x_{\rm crit}, t_{\rm crit})$ being a solution of either (5.96) or (5.97) is neither necessary nor sufficient for $(x_{\rm crit}, t_{\rm crit})$ to lie on the boundary of U. Even if $(x_{\rm crit}, t_{\rm crit})$ satisfies (5.96), the value of λ establishing the consistency might not lie on the band I_k^+, instead being contained in another of the three curves emanating from the band endpoint (see figure 5.3) or even in a curve branch that is not connected to the endpoint at all. Similarly if $(x_{\rm crit}, t_{\rm crit})$ satisfies (5.97), the bottleneck that is created might not actually constrain any gap contours to pass through the point λ; the pinching might occur in an irrelevant part of the region where $\Re(\tilde{\phi}^\sigma(\lambda)) < 0$. On the other hand, even if it is known that $(x_{\rm crit}, t_{\rm crit})$ is on the boundary of U, the failure of the ansatz may correspond to contact of the contour with $\partial\mathbb{H}$, a mode of failure that is not captured by the condition (5.96) or (5.97). What may be said with precision is this: *If the point $(x_{\rm crit}, t_{\rm crit})$ is known to be a point of failure of the genus-G ansatz and the contour may be taken to avoid $\partial\mathbb{H}$, then $(x_{\rm crit}, t_{\rm crit})$ is contained in the union of solution curves of (5.96) and (5.97) in the real (x, t)-plane.*

One expects that for points in the (x, t)-plane on the other side of the boundary of U, the inequalities can be satisfied by choosing an ansatz corresponding to a different genus G. For example, in chapter 7 we prove that when the condition (5.97) holds for a genus zero-ansatz, the curve defined by (5.97) in the (x, t)-plane is a boundary between values of x and t where the genus-zero ansatz is valid and values of x and t where a genus-two ansatz is valid. It then follows from the analysis in chapter 4 that the asymptotic behavior of the solution $\psi(x, t)$ of the nonlinear Schrödinger equation will be qualitatively different for (x, t) on opposite sides of the boundary of U, being described by Riemann theta functions of different genera. The boundary curve may thus be given the physical interpretation of a *phase transition*. Such sharp transitions are indeed clearly visible in computer reconstructions of the Satsuma-Yajima ensemble [MK98], for example. They have also been seen in recent simulations of (1.1) [BK99, CM02] for more general initial data.

5.3 MODULATION EQUATIONS

Here, we show that if the endpoints $\lambda_0, \ldots, \lambda_G$ satisfy the $2G + 2$ real equations contained in (5.25), (5.52), and (5.58)—equations in which x and t appear analytically as explicit parameters—then it turns out that the endpoints considered as functions of the independent variables x and t also satisfy a quasilinear system of partial differential equations. This system has no explicit dependence on x and t in its coefficients, and also the system takes the same form regardless of the function $A(x)$ that approximates the initial data for (1.1). The equations making up this quasilinear system are the Whitham or modulation equations associated with genus-G wavetrain solutions of the focusing nonlinear Schrödinger equation. They are elliptic, which makes the initial-value problem for them ill-posed.

We will begin by returning to the function $F(\lambda)$ guaranteed to exist by Lemma 5.1.2 because the moment conditions (5.25) are among those satisfied by the endpoints. The first observation that we make about the function $F(\lambda)$ is the following.

Lemma 5.3.1 *Whenever the endpoints satisfy the measure reality conditions* (5.58), *the function* $F(\lambda)$ *satisfies* $F(\lambda) = O(\lambda^{-2})$ *as* $\lambda \to \infty$.

Proof. From the Cauchy integral representation (5.16) of $F(\lambda)$, the result will follow if it is true that

$$\int_{C_I} \bar{\rho}^\sigma(\eta)\,d\eta + \int_{C_I^*} \bar{\rho}^\sigma(\eta^*)^*\,d\eta = 0. \tag{5.98}$$

Now, using the conjugation symmetry of the contours, the definition (5.9), and analyticity of $\rho^0(\eta)$, this is equivalent to the condition

$$\Im\left(\int_0^{iA} \rho^0(\eta)\,d\eta\right) - \Im\left(\int_{C_I} \rho^\sigma(\eta)\,d\eta\right) = 0. \tag{5.99}$$

The first term then vanishes because the given asymptotic eigenvalue measure is real on the imaginary axis, and the second term is equivalent to a sum of integrals of $\rho^\sigma(\eta)\,d\eta$ over the bands I_k^+ of C_I. The reality of each of these integrals is exactly the content of the equations (5.58), which proves the lemma. □

By assumption, the endpoints satisfy the moment conditions $M_p = 0$ for $p = 0, \ldots, G$. If the endpoints also satisfy the measure reality conditions, then slightly more is true.

Lemma 5.3.2 *Whenever the endpoints satisfy the moment conditions* (5.25) *for* $p = 0, \ldots, G$ *and the measure reality conditions* (5.58), *then*

$$M_{G+1} = 0 \tag{5.100}$$

as well.

Proof. This follows immediately from the series expansion (5.30) for the function $H(\lambda)$, along with the fact that $F(\lambda) = R(\lambda)H(\lambda)$, where $R(\lambda) \sim -\lambda^G$ for large λ and the large λ asymptotic behavior of $F(\lambda)$ guaranteed by Lemma 5.3.1. □

Remark. Lemma 5.3.2 means that with the use of the measure reality conditions (5.58) we can deduce one additional moment condition. We now make the correspondence between the reality conditions and the moment conditions more precise. Summing up the integrals R_ℓ, we find

$$\sum_{\ell=0}^{G/2} R_\ell = \Im\left(\int_{\cup_k I_k^+} \rho^\sigma(\eta)\,d\eta\right). \tag{5.101}$$

Now for $\eta \in I_k^+$, $\rho^\sigma(\eta)$ can be expressed in terms of $\rho^0(\eta)$ and the difference of boundary values of $F(\eta)$. Therefore, (5.101) becomes

$$\sum_{\ell=0}^{G/2} R_\ell = \Im\left(\int_{\cup_k I_k^+} \rho^0(\eta)\,d\eta\right) - \frac{1}{2\pi}\Re\left(\int_{\cup_k I_k^+} (F_+(\eta) - F_-(\eta))\,d\eta\right). \quad (5.102)$$

Now for $\eta \in \Gamma_I \cap C_I$, we have that $F_+(\eta) - F_-(\eta) = -2\pi i \rho^0(\eta)$. Thus we may rewrite (5.102) as

$$\sum_{\ell=0}^{G/2} R_\ell = \Im\left(\int_{C_I} \rho^0(\eta)\,d\eta\right) - \frac{1}{2\pi}\Re\left(\int_{C_I} (F_+(\eta) - F_-(\eta))\,d\eta\right) \quad (5.103)$$

or, using analyticity to deform the path in the first integral to the imaginary interval $[0, iA]$ and exploiting reality of the asymptotic eigenvalue measure $\rho^0(\eta)\,d\eta$ on that path,

$$\sum_{\ell=0}^{G/2} R_\ell = -\frac{1}{2\pi}\Re\left(\int_{C_I} (F_+(\eta) - F_-(\eta))\,d\eta\right). \quad (5.104)$$

Finally, since $F(\lambda)$ satisfies the symmetry (5.19), we can write this as

$$\sum_{\ell=0}^{G/2} R_\ell = -\frac{1}{4\pi}\int_{C_I \cup C_I^*} (F_+(\eta) - F_-(\eta))\,d\eta. \quad (5.105)$$

Now since $F(\lambda)$ is analytic in $\mathbb{C}\backslash(C_I \cup C_I^*)$, we may express this integral as a contour integral on any counterclockwise-oriented loop L completely encircling the contour $C_I \cup C_I^*$:

$$\sum_{\ell=0}^{G/2} R_\ell = \frac{1}{4\pi}\oint_L F(\eta)\,d\eta. \quad (5.106)$$

Using the residue theorem to evaluate the integral, assuming only that $F(\lambda)$ decays like $1/\lambda$ at infinity, we find at last

$$\sum_{\ell=0}^{G/2} R_\ell = \frac{i}{2}\lim_{\lambda\to\infty} \lambda F(\lambda) = -\frac{M_{G+1}}{2\pi}. \quad (5.107)$$

The last equality follows from the formula $F(\lambda) = H(\lambda)R(\lambda)$, the asymptotic behavior $R(\lambda) \sim -\lambda^{G+1}$, and the fact that the moments M_0 through M_G are presumed to vanish. This establishes the fact that any one of the reality conditions (5.58) may be replaced with the additional moment condition $M_{G+1} = 0$.

Next, we consider computing derivatives of the moments. We begin with the following lemma.

LEMMA 5.3.3 *The moments M_j, for $j = 1, 2, \ldots$, satisfy the following differential equations:*

$$\frac{\partial M_j}{\partial \lambda_k} = \frac{1}{2}M_{j-1} + \lambda_k \frac{\partial M_{j-1}}{\partial \lambda_k}, \quad (5.108)$$

$$\frac{\partial M_j}{\partial \lambda_k^*} = \frac{1}{2}M_{j-1} + \lambda_k^* \frac{\partial M_{j-1}}{\partial \lambda_k^*}. \quad (5.109)$$

Furthermore, the function $F(\lambda)$ satisfies the following equations, valid for $\lambda \in \mathbb{C}\backslash(C_I \cup C_I^*)$, with appropriate boundary values taken on $C_I \cup C_I^*$:

$$\frac{\partial F}{\partial \lambda_j} = \frac{1}{\pi i} \cdot \frac{R(\lambda)}{\lambda - \lambda_j} \cdot \frac{\partial M_0}{\partial \lambda_j}, \tag{5.110}$$

$$\frac{\partial F}{\partial \lambda_j^*} = \frac{1}{\pi i} \cdot \frac{R(\lambda)}{\lambda - \lambda_j^*} \cdot \frac{\partial M_0}{\partial \lambda_j^*}. \tag{5.111}$$

Proof. To prove (5.108), note that

$$M_j - \lambda_k M_{j-1} = J \int_{\cup_\ell I_\ell^\pm} \frac{2ix + 4i\eta t}{R_+(\eta)} \eta^{j-1}(\eta - \lambda_k)\, d\eta + \int_{\Gamma_I \cap C_I} \frac{i\pi\rho^0(\eta)}{R(\eta)} \eta^{j-1}(\eta - \lambda_k)\, d\eta$$
$$+ \int_{\Gamma_I \cap C_I^*} \frac{i\pi\rho^0(\eta^*)^*}{R(\eta)} \eta^{j-1}(\eta - \lambda_k)\, d\eta. \tag{5.112}$$

Differentiating (5.112) with respect to λ_k, we find

$$\frac{\partial M_j}{\partial \lambda_k} - M_{j-1} - \lambda_k \frac{\partial M_{j-1}}{\partial \lambda_k} = -\frac{1}{2} M_{j-1}, \tag{5.113}$$

and we have proved (5.108). To prove (5.109), replace λ_k with λ_k^* in (5.112) and differentiate with respect to λ_k^*.

To prove (5.110), we use the formula $F(\lambda) = H(\lambda)R(\lambda)$, the Laurent series representation (5.30) for $H(\lambda)$, and differentiate with respect to λ_k:

$$\frac{\partial F}{\partial \lambda_k} = \frac{R(\lambda)}{\pi i \lambda} \sum_{j=0}^{\infty} \left(-\frac{1}{\lambda - \lambda_k} \cdot \frac{M_j}{2} + \frac{\partial M_j}{\partial \lambda_k} \right) \lambda^{-j}$$
$$= -\frac{R(\lambda)}{2\pi i \lambda} \cdot \frac{1}{\lambda - \lambda_k} \left[-2\lambda \frac{\partial M_0}{\partial \lambda_k} + \sum_{j=0}^{\infty} \left(-2 \frac{\partial M_{j+1}}{\partial \lambda_k} + 2\lambda_k \frac{\partial M_j}{\partial \lambda_k} + M_j \right) \lambda^{-j} \right], \tag{5.114}$$

where the second equality results from factoring out $(\lambda - \lambda_k)^{-1}$ and rearranging the sum. The relation (5.110) then follows by using (5.108). To prove (5.111), one differentiates with respect to λ_k^* and uses (5.109). □

Now, we show that we can use the equations $M_p = 0$ taken for $p = 0, \ldots, G+1$ together with the gap conditions (5.52) and the measure reality conditions $R_k = 0$ taken for $k = 1, \ldots, G/2$ to derive the elliptic modulation equations. First, observe that if we evaluate (5.108) and (5.109) for a set of endpoints $\lambda_0, \ldots, \lambda_G$ chosen to satisfy these $2G + 2$ real conditions, then we have for all $j = 1, \ldots, G + 1$ and $k = 0, \ldots, G$,

$$\frac{\partial M_j}{\partial \lambda_k} = \lambda_k^j \frac{\partial M_0}{\partial \lambda_k}, \qquad \frac{\partial M_j}{\partial \lambda_k^*} = \lambda_k^{*j} \frac{\partial M_0}{\partial \lambda_k^*}. \tag{5.115}$$

Second, the formulae (5.110) and (5.111) yield rather simple representations for the derivatives of the functions V_j defined in (5.52) and R_ℓ defined in (5.58), with respect to λ_k and λ_k^*. Let us first consider the function V_j for any $j, 0 \le j \le G/2-1$.

From the formula (5.52), and the symmetry (5.19) of the function $F(\lambda)$, we have the following representation of the function V_j:

$$V_j = \frac{1}{2}\int_{\Gamma_{j+1}^+ \cup \Gamma_{j+1}^-}\left[2iJx + 4iJ\eta t + \frac{1}{2}(F_+(\eta) + F_-(\eta))\right] d\eta. \tag{5.116}$$

Recall that by definition Γ_{j+1}^+ is oriented from λ_{2j} to λ_{2j+1} and Γ_{j+1}^- is oriented from λ_{2j+1}^* to λ_{2j}^*. Since the boundary values of $F(\eta)$ are Hölder-continuous with exponent $1/2$ and hence uniformly continuous, it follows from the boundary conditions satisfied by $F(\eta)$ on $C_I \cup C_I^*$ that the integrand in (5.116) vanishes at the endpoints of the two gaps of integration. Therefore, differentiating (5.116) with respect to λ_k, one finds simply

$$\frac{\partial V_j}{\partial \lambda_k} = \frac{1}{4}\int_{\Gamma_{j+1}^+ \cup \Gamma_{j+1}^-}\frac{\partial}{\partial \lambda_k}(F_+(\eta) + F_-(\eta))\, d\eta. \tag{5.117}$$

Now inserting (5.110) into (5.117), we find

$$\frac{\partial V_j}{\partial \lambda_k} = \frac{1}{2\pi i}\left(\frac{\partial M_0}{\partial \lambda_k}\right)\int_{\Gamma_{j+1}^+ \cup \Gamma_{j+1}^-}\frac{R(\eta)}{\eta - \lambda_k}\, d\eta. \tag{5.118}$$

Repeating the above calculations but differentiating with respect to λ_k^*, one may easily verify

$$\frac{\partial V_j}{\partial \lambda_k^*} = \frac{1}{2\pi i}\left(\frac{\partial M_0}{\partial \lambda_k^*}\right)\int_{\Gamma_{j+1}^+ \cup \Gamma_{j+1}^-}\frac{R(\eta)}{\eta - \lambda_k^*}\, d\eta. \tag{5.119}$$

To obtain the derivatives of the functions R_j, $1 \le j \le G/2$, with respect to $\lambda_0, \dots, \lambda_G$ and $\lambda_0^*, \dots, \lambda_G^*$, we start with the following formula for R_j:

$$\begin{aligned} R_j = {}& \frac{1}{2i}\int_{I_j^+}\left[\rho^0(\eta) + \frac{1}{2\pi i}(F_+(\eta) - F_-(\eta))\right] d\eta \\ &+ \frac{1}{2i}\int_{I_j^-}\left[\rho^0(\eta^*)^* + \frac{1}{2\pi i}(F_+(\eta) - F_-(\eta))\right] d\eta, \end{aligned} \tag{5.120}$$

obtained by representing $\rho^\sigma(\eta)$ in the bands I_j^+ in terms of the asymptotic eigenvalue density $\rho^0(\eta)$ and $F(\eta)$ and using the symmetry property (5.19). Recall that by definition for $j > 0$, the band I_j^+ is oriented from λ_{2j-1} to λ_{2j} while the conjugate band I_j^- is oriented from λ_{2j}^* to λ_{2j-1}^*. Also, by the same arguments as in our discussion of the quantities V_j, the integrand vanishes at the endpoints. Therefore, differentiating (5.120) with respect to λ_k, we find

$$\frac{\partial R_j}{\partial \lambda_k} = -\frac{1}{4\pi}\int_{I_j^+ \cup I_j^-}\frac{\partial}{\partial \lambda_k}(F_+(\eta) - F_-(\eta))\, d\eta. \tag{5.121}$$

Now inserting (5.110) into (5.121), we find

$$\frac{\partial R_j}{\partial \lambda_k} = \frac{i}{2\pi^2}\left(\frac{\partial M_0}{\partial \lambda_k}\right)\int_{I_j^+ \cup I_j^-}\frac{R_+(\eta)}{\eta - \lambda_k}\, d\eta. \tag{5.122}$$

Repeating the above calculations but differentiating with respect to λ_k^*, one may easily derive

$$\frac{\partial R_j}{\partial \lambda_k^*} = \frac{i}{2\pi^2}\left(\frac{\partial M_0}{\partial \lambda_k^*}\right)\int_{I_j^+ \cup I_j^-}\frac{R_+(\eta)}{\eta - \lambda_k^*}\, d\eta. \tag{5.123}$$

We have proved the following lemma.

LEMMA 5.3.4 *The partial derivatives of the quantities M_j, V_j, and R_j with respect to the endpoints $\lambda_0, \ldots, \lambda_G$ and their complex conjugates satisfy a set of canonical formulae whenever the endpoints solve the equations*

$$\begin{aligned} M_j &= 0, \quad j = 0, \ldots, G+1, \\ V_j &= 0, \quad j = 0, \ldots, G/2 - 1, \\ R_j &= 0, \quad j = 1, \ldots, G/2. \end{aligned} \tag{5.124}$$

These formulae are

$$\frac{\partial M_j}{\partial \lambda_k} = \lambda_k^j \frac{\partial M_0}{\partial \lambda_k}, \qquad \frac{\partial M_j}{\partial \lambda_k^*} = \lambda_k^{*j} \frac{\partial M_0}{\partial \lambda_k^*}, \tag{5.125}$$

for $j = 1, \ldots, G+1$ and $k = 0, \ldots, G$,

$$\frac{\partial V_j}{\partial \lambda_k} = \frac{1}{2\pi i} \frac{\partial M_0}{\partial \lambda_k} \int_{\Gamma_{j+1}^+ \cup \Gamma_{j+1}^-} \frac{R(\eta)}{\eta - \lambda_k} d\eta, \quad \frac{\partial V_j}{\partial \lambda_k^*} = \frac{1}{2\pi i} \frac{\partial M_0}{\partial \lambda_k^*} \int_{\Gamma_{j+1}^+ \cup \Gamma_{j+1}^-} \frac{R(\eta)}{\eta - \lambda_k^*} d\eta, \tag{5.126}$$

for $j = 0, \ldots, G/2 - 1$ and $k = 0, \ldots, G$, and

$$\frac{\partial R_j}{\partial \lambda_k} = \frac{i}{2\pi^2} \frac{\partial M_0}{\partial \lambda_k} \int_{I_j^+ \cup I_j^-} \frac{R_+(\eta)}{\eta - \lambda_k} d\eta, \quad \frac{\partial R_k}{\partial \lambda_k^*} = \frac{i}{2\pi^2} \frac{\partial M_0}{\partial \lambda_k^*} \int_{I_j^+ \cup I_j^-} \frac{R_+(\eta)}{\eta - \lambda_k^*} d\eta, \tag{5.127}$$

for $j = 1, \ldots, G/2$ and $k = 0, \ldots, G$.

Third, we compute the partial derivatives of M_p, V_k, and R_ℓ with respect to x and t. For fixed endpoints, a simple residue calculation shows that M_p satisfies

$$\frac{\partial M_p}{\partial x} = 0, \qquad p = 0, \ldots, G-1, \tag{5.128}$$

while

$$\frac{\partial M_G}{\partial x} = -2J\pi, \qquad \frac{\partial M_{G+1}}{\partial x} = -J\pi \sum_{k=0}^{G} (\lambda_k + \lambda_k^*). \tag{5.129}$$

Similarly, one finds

$$\frac{\partial M_p}{\partial t} = 0, \qquad p = 0, \ldots, G-2, \tag{5.130}$$

while

$$\begin{aligned} \frac{\partial M_{G-1}}{\partial t} &= -4\pi J, \qquad \frac{\partial M_G}{\partial t} = -2\pi J \sum_{k=0}^{G} (\lambda_k + \lambda_k^*), \\ \frac{\partial M_{G+1}}{\partial t} &= -\pi J \left[\sum_{0 \le j < k \le G} (\lambda_j + \lambda_j^*)(\lambda_k + \lambda_k^*) - \frac{1}{2} \sum_{j=0}^{G} (\lambda_j - \lambda_j^*)^2 \right]. \end{aligned} \tag{5.131}$$

To compute the partial derivatives of the functions V_j and R_j with respect to x and t, we first observe that from the representation $F(\lambda) = H(\lambda)R(\lambda)$ and the explicit formula (5.29) for $H(\lambda)$, we find

$$\left. \frac{\partial F}{\partial x}(\lambda) \right|_{\text{endpoints fixed}} = \frac{2J}{\pi} R(\lambda) \int_{\cup_k I_k^\pm} \frac{d\eta}{(\lambda - \eta) R_+(\eta)} = -2iJ \tag{5.132}$$

and

$$\left.\frac{\partial F}{\partial t}(\lambda)\right|_{\text{endpoints fixed}} = \frac{4J}{\pi}R(\lambda)\int_{\cup_k I_k^{\pm}} \frac{\eta\, d\eta}{(\lambda-\eta)R_+(\eta)} = -4iJ\lambda - 4iR(\lambda)\cdot\delta_{G,0}, \tag{5.133}$$

where the integral is evaluated explicitly by residues and for the t-derivative there is a residue at infinity only for $G = 0$, which explains the Kronecker delta.

Remark. These partial derivatives of $F(\lambda)$ are computed holding the endpoints $\lambda_0, \ldots, \lambda_G$ and their complex conjugates fixed. These formulae therefore do not contradict the discussion in §5.1.2, which concerned the total variations of $F(\lambda)$ with respect to x and t when the endpoints are constrained by the moment conditions (5.25).

Combining these with (5.116), we find that V_j satisfies simply

$$\frac{\partial V_j}{\partial x} = 0, \qquad \frac{\partial V_j}{\partial t} = 0, \tag{5.134}$$

for $j = 0, \ldots, G/2-1$. Observe that for $G = 0$ there are no gap conditions, and in this case the Kronecker delta term in (5.133) plays no role. Similarly, from (5.120) and the fact that $\rho^0(\eta)$ is independent of x and t, we find simply

$$\frac{\partial R_j}{\partial x} = 0, \qquad \frac{\partial R_j}{\partial t} = 0, \tag{5.135}$$

for $j = 1, \ldots, G/2$. Again, note that for $G = 0$, the Kronecker delta term in (5.133) plays no role because we are considering the only measure reality condition $R_0 = 0$ present to be absorbed into the additional moment condition $M_1 = 0$.

Finally, we indicate how the $2G + 2$ real conditions (5.124) imply the elliptic Whitham modulation equations. Define the column vector $\vec{\lambda} := [\lambda_0, \lambda_0^*, \lambda_1, \lambda_1^*, \ldots, \lambda_G, \lambda_G^*]^T$ and the vector-valued function $\mathcal{G}(\vec{\lambda})$ via

$$\mathcal{G}(\vec{\lambda})^T = [\mathcal{G}_1, \ldots, \mathcal{G}_{2G+2}] := [M_0, \ldots, M_{G+1}, V_0, \ldots, V_{G/2-1}, R_1, \ldots, R_{G/2}]. \tag{5.136}$$

Then the equations (5.124) are written compactly as

$$\mathcal{G}(\vec{\lambda}) = \vec{0}. \tag{5.137}$$

Differentiating with respect to x and t, we find

$$\mathbf{M}(\vec{\lambda})\,\frac{\partial\vec{\lambda}}{\partial x} = -\frac{\partial\mathcal{G}}{\partial x}, \qquad \mathbf{M}(\vec{\lambda})\,\frac{\partial\vec{\lambda}}{\partial t} = -\frac{\partial\mathcal{G}}{\partial t}, \tag{5.138}$$

where

$$\frac{\partial\mathcal{G}_j}{\partial x} = 0, \qquad \text{for } j = 1, \ldots, G \text{ and } j = G+3, \ldots, 2G+2, \tag{5.139}$$

while

$$\frac{\partial\mathcal{G}_{G+1}}{\partial x} = -2J\pi, \qquad \frac{\partial\mathcal{G}_{G+2}}{\partial x} = -J\pi\sum_{k=0}^{G}(\lambda_k + \lambda_k^*) \tag{5.140}$$

and

$$\frac{\partial \mathcal{G}_j}{\partial t} = 0, \qquad \text{for } j = 1, \ldots, G-1 \text{ and } j = G+3, \ldots, 2G+2, \tag{5.141}$$

while

$$\begin{aligned} \frac{\partial \mathcal{G}_G}{\partial t} &= -4J\pi, \qquad \frac{\partial \mathcal{G}_{G+1}}{\partial t} = -2J\pi \sum_{k=0}^{G} (\lambda_k + \lambda_k^*), \\ \frac{\partial \mathcal{G}_{G+2}}{\partial t} &= -J\pi \left(\sum_{0 \le j < k \le G} (\lambda_j + \lambda_j^*)(\lambda_k + \lambda_k^*) - \frac{1}{2} \sum_{j=0}^{G} (\lambda_j - \lambda_j^*)^2 \right), \end{aligned} \tag{5.142}$$

and the (Jacobian) matrix $\mathbf{M}(\vec{\lambda})$ is defined by

$$\mathbf{M}(\vec{\lambda}) := \frac{\partial \mathcal{G}}{\partial \vec{\lambda}} = \begin{bmatrix} \frac{\partial M_0}{\partial \lambda_0} & \frac{\partial M_0}{\partial \lambda_0^*} & \cdots & \frac{\partial M_0}{\partial \lambda_G^*} \\ \vdots & \vdots & \vdots & \vdots \\ \frac{\partial M_{G+1}}{\partial \lambda_0} & \frac{\partial M_{G+1}}{\partial \lambda_0^*} & \cdots & \frac{\partial M_{G+1}}{\partial \lambda_G^*} \\ \frac{\partial V_0}{\partial \lambda_0} & \frac{\partial V_0}{\partial \lambda_0^*} & \cdots & \frac{\partial V_0}{\partial \lambda_G^*} \\ \vdots & \vdots & \vdots & \vdots \\ \frac{\partial V_{G/2-1}}{\partial \lambda_0} & \frac{\partial V_{G/2-1}}{\partial \lambda_0^*} & \cdots & \frac{\partial V_{G/2-1}}{\partial \lambda_G^*} \\ \frac{\partial R_1}{\partial \lambda_0} & \frac{\partial R_1}{\partial \lambda_0^*} & \cdots & \frac{\partial R_1}{\partial \lambda_G^*} \\ \vdots & \vdots & \vdots & \vdots \\ \frac{\partial R_{G/2}}{\partial \lambda_0} & \frac{\partial R_{G/2}}{\partial \lambda_0^*} & \cdots & \frac{\partial R_{G/2}}{\partial \lambda_G^*} \end{bmatrix}. \tag{5.143}$$

Now using the relations (5.125), (5.126), and (5.127), we find that, miraculously, the Jacobian $\mathbf{M}(\vec{\lambda})$ factors:

$$\mathbf{M}(\vec{\lambda}) = \operatorname{diag}\left(1, \ldots, 1, \frac{1}{2\pi i}, \ldots, \frac{1}{2\pi i}, \frac{i}{2\pi^2}, \ldots, \frac{i}{2\pi^2}\right) \cdot \widetilde{\mathbf{M}}(\vec{\lambda}) \cdot \frac{\partial M_0}{\partial \vec{\lambda}}, \tag{5.144}$$

where

$$\frac{\partial M_0}{\partial \vec{\lambda}} := \operatorname{diag}\left(\frac{\partial M_0}{\partial \lambda_0}, \frac{\partial M_0}{\partial \lambda_0^*}, \ldots, \frac{\partial M_0}{\partial \lambda_G}, \frac{\partial M_0}{\partial \lambda_G^*}\right), \tag{5.145}$$

and where

$$\widetilde{\mathbf{M}}(\vec{\lambda}) := \begin{bmatrix} 1 & 1 & \cdots & 1 \\ \lambda_0 & \lambda_0^* & \cdots & \lambda_G^* \\ \vdots & \vdots & \vdots & \vdots \\ \lambda_0^{G+1} & \lambda_0^{*G+1} & \cdots & \lambda_G^{*G+1} \\ \int_{\Gamma_1^+\cup\Gamma_1^-} \frac{R(\eta)\,d\eta}{\eta-\lambda_0} & \int_{\Gamma_1^+\cup\Gamma_1^-} \frac{R(\eta)\,d\eta}{\eta-\lambda_0^*} & \cdots & \int_{\Gamma_1^+\cup\Gamma_1^-} \frac{R(\eta)\,d\eta}{\eta-\lambda_G^*} \\ \vdots & \vdots & \vdots & \vdots \\ \int_{\Gamma_{G/2}^+\cup\Gamma_{G/2}^-} \frac{R(\eta)\,d\eta}{\eta-\lambda_0} & \int_{\Gamma_{G/2}^+\cup\Gamma_{G/2}^-} \frac{R(\eta)\,d\eta}{\eta-\lambda_0^*} & \cdots & \int_{\Gamma_{G/2}^+\cup\Gamma_{G/2}^-} \frac{R(\eta)\,d\eta}{\eta-\lambda_G^*} \\ \int_{I_1^+\cup I_1^-} \frac{R_+(\eta)\,d\eta}{\eta-\lambda_0} & \int_{I_1^+\cup I_1^-} \frac{R_+(\eta)\,d\eta}{\eta-\lambda_0^*} & \cdots & \int_{I_1^+\cup I_1^-} \frac{R_+(\eta)\,d\eta}{\eta-\lambda_G^*} \\ \vdots & \vdots & \vdots & \vdots \\ \int_{I_{G/2}^+\cup I_{G/2}^-} \frac{R_+(\eta)\,d\eta}{\eta-\lambda_0} & \int_{I_{G/2}^+\cup I_{G/2}^-} \frac{R_+(\eta)\,d\eta}{\eta-\lambda_0^*} & \cdots & \int_{I_{G/2}^+\cup I_{G/2}^-} \frac{R_+(\eta)\,d\eta}{\eta-\lambda_G^*} \end{bmatrix}. \tag{5.146}$$

The determinant of $\widetilde{\mathbf{M}}(\vec{\lambda})$ can be calculated explicitly. First, one uses the linearity of the determinant in each row to write

$$\det \widetilde{\mathbf{M}}(\vec{\lambda}) = \int_{\Gamma_1^+\cup\Gamma_1^-} \cdots \int_{\Gamma_{G/2}^+\cup\Gamma_{G/2}^-} \prod_{j=1}^{G/2} R(\eta_j)\,d\eta_j \int_{I_1^+\cup I_1^-} \cdots \int_{I_{G/2}^+\cup I_{G/2}^-} \prod_{k=G/2+1}^{G} R_+(\eta_k)\,d\eta_k \times \det \mathbf{S}(\vec{\lambda}, \vec{\eta}), \tag{5.147}$$

where

$$\mathbf{S}(\vec{\lambda}, \vec{\eta}) = \begin{bmatrix} 1 & 1 & \cdots & 1 \\ \lambda_0 & \lambda_0^* & \cdots & \lambda_G^* \\ \vdots & \vdots & \vdots & \vdots \\ \lambda_0^{G+1} & \lambda_0^{*G+1} & \cdots & \lambda_G^{*G+1} \\ \frac{1}{\eta_1-\lambda_0} & \frac{1}{\eta_1-\lambda_0^*} & \cdots & \frac{1}{\eta_1-\lambda_G^*} \\ \vdots & \vdots & \vdots & \vdots \\ \frac{1}{\eta_G-\lambda_0} & \frac{1}{\eta_G-\lambda_0^*} & \cdots & \frac{1}{\eta_G-\lambda_G^*} \end{bmatrix}. \tag{5.148}$$

This matrix is a combination of a Vandermonde matrix and a Cauchy matrix. The determinant $\det \mathbf{S}(\vec{\lambda}, \vec{\eta})$ can be computed by observing that it is a rational function in each variable with obvious singularities and with the same number of explicit zeros. For example, as a function of λ_0, the determinant has G simple poles at $\eta_1, \ldots, \eta_G$ and behaves like λ_0^{G+1} near infinity. Therefore, it has exactly $2G+1$ zeros, and it is easy to see that these occur exactly for $\lambda_0 = \lambda_0^*, \lambda_1, \ldots, \lambda_G^*$, since each of these choices makes two columns identical. By Liouville's theorem, this fixes the determinant up to a constant factor, which may be obtained by similar considerations viewing the determinant as a function of the other variables. In any case, we find

$$\det\,\mathbf{S}(\vec{\lambda},\vec{\eta}) = \frac{\prod_{j=0}^{G}\prod_{k=0}^{G}(\lambda_k^*-\lambda_j)\prod_{0\leq j<k\leq G}(\lambda_k-\lambda_j)(\lambda_k^*-\lambda_j^*)\prod_{1\leq j<k\leq G}(\eta_k-\eta_j)}{(-1)^G\prod_{j=1}^{G}\prod_{k=0}^{G}(\eta_j-\lambda_k)(\eta_j-\lambda_k^*)}. \tag{5.149}$$

It is straightforward to solve (5.138) for $\partial\vec{\lambda}/\partial x$ and $\partial\vec{\lambda}/\partial t$ by Cramer's rule. In doing so, one first inverts the diagonal prefactor and notes that from the positions of the only nonzero entries on the right-hand side, (5.138) is really just

$$\widetilde{\mathbf{M}}(\vec{\lambda})\cdot\frac{\partial M_0}{\partial\vec{\lambda}}\cdot\frac{\partial\vec{\lambda}}{\partial x} = -\frac{\partial\mathcal{G}}{\partial x}, \qquad \widetilde{\mathbf{M}}(\vec{\lambda})\cdot\frac{\partial M_0}{\partial\vec{\lambda}}\cdot\frac{\partial\vec{\lambda}}{\partial t} = -\frac{\partial\mathcal{G}}{\partial t}. \tag{5.150}$$

By a direct calculation, one can see that none of the partial derivatives of M_0 with respect to an endpoint vanishes identically. Therefore, inverting the diagonal matrix $\partial M_0/\partial\vec{\lambda}$ explicitly, one finds that for $k = 1,\ldots,2G+2$,

$$\left(\frac{\partial\vec{\lambda}}{\partial t}\right)_k + c_k(\vec{\lambda})\left(\frac{\partial\vec{\lambda}}{\partial x}\right)_k = 0, \tag{5.151}$$

where

$$c_k(\vec{\lambda}) := -\frac{\det\widetilde{\mathbf{M}}^{(k,t)}}{\det\widetilde{\mathbf{M}}^{(k,x)}} \tag{5.152}$$

and $\widetilde{\mathbf{M}}^{(k,x)}$ is the matrix obtained from $\widetilde{\mathbf{M}}$ by replacing the kth column with $\partial\mathcal{G}/\partial x$, while $\widetilde{\mathbf{M}}^{(k,t)}$ is the matrix obtained from $\widetilde{\mathbf{M}}$ by replacing the kth column with $\partial\mathcal{G}/\partial t$. Note that (5.151) is a first-order system of quasilinear partial differential equations in x and t that is explicitly written in Riemann-invariant form regardless of the size of the system (value of G). Also, it is clear from the definition (5.152) that the characteristic velocities $c_k(\vec{\lambda})$ have no explicit dependence on x and t.

Without belaboring the point, let us observe in passing that in computing these determinants, we have established that the $2G+2$ conditions contained in (5.25), (5.52), and (5.58) actually imply that the endpoints $\lambda_0,\ldots,\lambda_G$ and their complex conjugates solve the partial differential equations (5.151). The characteristic velocities (5.152) are explicitly expressed in terms of ratios of determinants of matrices whose entries are all hyperelliptic integrals. Also, as seen in §4.3, the dependent variables $\lambda_0,\ldots,\lambda_G$ and their complex conjugates have the interpretation of moduli of a hyperelliptic Riemann surface used in the reconstruction of the asymptotic semiclassical solution in the vicinity of fixed x and t. The system (5.151) is therefore just the set of Whitham or modulation equations for genus-G wavetrain solutions of the focusing nonlinear Schrödinger equation, expressed in Riemann-invariant form. See, for example, [FL86] for a formal derivation of these equations from the starting point of the assumption of an approximate solution of the focusing nonlinear Schrödinger equation in the form of a slowly modulated wavetrain.

Our formula (5.152) for the characteristic velocities is not written in exactly the same form as in Forest and Lee's paper [FL86]. Making the identification requires identifying the ratios of determinants in (5.152) with those obtained in [FL86] by the normalization of the pair of canonical meromorphic differentials by adding appropriate holomorphic differentials to achieve zero a-cycles. We do not concern ourselves further with the aforementioned equivalence, leaving this to the interested reader.

We want to emphasize that certain steps in obtaining the equations (5.151) from the solution of (5.25), (5.52), and (5.58), such as the nontrivial issue of proving that the matrix $\widetilde{\mathbf{M}}$ possesses an inverse, require very delicate analysis. In any case, from the point of view of computing rigorous semiclassical asymptotics, the nondifferential relations (5.25), (5.52), and (5.58) form a complete characterization of the endpoints, always containing information about the approximate initial data $A(x)$ encoded in the asymptotic eigenvalue density $\rho^0(\eta)$. From this point of view, the fact that the system (5.151) does not contain any reference to the initial data via $\rho^0(\eta)$ and yet is satisfied by solutions of (5.25), (5.52), and (5.58), which do depend on $\rho^0(\eta)$, is a happy coincidence.

To make the derivation of the Whitham equations more concrete, let us now carry out the above program for the case of genus $G = 0$. We have the following matrix equation satisfied by $\{\lambda_0, \lambda_0^*\}$:

$$\begin{bmatrix} \frac{\partial M_0}{\partial \lambda_0} & \frac{\partial M_0}{\partial \lambda_0^*} \\ \frac{\partial M_1}{\partial \lambda_0} & \frac{\partial M_1}{\partial \lambda_0^*} \end{bmatrix} \begin{bmatrix} \frac{\partial \lambda_0}{\partial x} \\ \frac{\partial \lambda_0^*}{\partial x} \end{bmatrix} = - \begin{bmatrix} \frac{\partial M_0}{\partial x} \\ \frac{\partial M_1}{\partial x} \end{bmatrix}. \tag{5.153}$$

Now from (5.129) we find that

$$\frac{\partial M_0}{\partial x} = -2\pi J, \qquad \frac{\partial M_1}{\partial x} = -2\pi J a_0, \tag{5.154}$$

where $a_0 := (\lambda_0 + \lambda_0^*)/2$, and so equation (5.153) becomes

$$\begin{bmatrix} \frac{\partial M_0}{\partial \lambda_0} & \frac{\partial M_0}{\partial \lambda_0^*} \\ \lambda_0 \frac{\partial M_0}{\partial \lambda_0} & \lambda_0^* \frac{\partial M_0}{\partial \lambda_0^*} \end{bmatrix} \begin{bmatrix} \frac{\partial \lambda_0}{\partial x} \\ \frac{\partial \lambda_0^*}{\partial x} \end{bmatrix} = 2\pi J \begin{bmatrix} 1 \\ a_0 \end{bmatrix}. \tag{5.155}$$

Here we have also used (5.115). We may simplify this as follows:

$$\begin{bmatrix} 1 & 1 \\ \lambda_0 & \lambda_0^* \end{bmatrix} \begin{bmatrix} \frac{\partial M_0}{\partial \lambda_0} \cdot \frac{\partial \lambda_0}{\partial x} \\ \frac{\partial M_0}{\partial \lambda_0^*} \cdot \frac{\partial \lambda_0^*}{\partial x} \end{bmatrix} = 2\pi J \begin{bmatrix} 1 \\ a_0 \end{bmatrix}, \tag{5.156}$$

whose solution is given by

$$\begin{bmatrix} \frac{\partial M_0}{\partial \lambda_0} \cdot \frac{\partial \lambda_0}{\partial x} \\ \frac{\partial M_0}{\partial \lambda_0^*} \cdot \frac{\partial \lambda_0^*}{\partial x} \end{bmatrix} = \pi J \begin{bmatrix} 1 \\ 1 \end{bmatrix}. \tag{5.157}$$

Similarly, for the t-derivatives, we find

$$\begin{bmatrix} \frac{\partial M_0}{\partial \lambda_0} & \frac{\partial M_0}{\partial \lambda_0^*} \\ \lambda_0 \frac{\partial M_0}{\partial \lambda_0} & \lambda_0^* \frac{\partial M_0}{\partial \lambda_0^*} \end{bmatrix} \begin{bmatrix} \frac{\partial \lambda_0}{\partial t} \\ \frac{\partial \lambda_0^*}{\partial t} \end{bmatrix} = 2\pi J \begin{bmatrix} 2a_0 \\ 2a_0^2 - b_0^2 \end{bmatrix}, \tag{5.158}$$

where $b_0 := (\lambda_0 - \lambda_0^*)/(2i)$. This uses (5.131), the fact that $\partial M_0/\partial t = -4\pi a_0$, and $\partial M_1/\partial t = -2\pi(2a_0^2 - b_0^2)$. From this we find that

$$\begin{bmatrix} \frac{\partial M_0}{\partial \lambda_0} \cdot \frac{\partial \lambda_0}{\partial t} \\ \frac{\partial M_0}{\partial \lambda_0^*} \cdot \frac{\partial \lambda_0^*}{\partial t} \end{bmatrix} = \pi J \begin{bmatrix} 2a_0 + ib_0 \\ 2a_0 - ib_0 \end{bmatrix}. \tag{5.159}$$

Combining (5.157) and (5.159) gives at last the following theorem.

Theorem 5.3.1 *Let $G = 0$, and let $\lambda_0(x,t)$ be any solution of the moment equations $M_0 = 0$ and $M_1 = 0$ that is differentiable with respect to x and t in some open set in the (x,t)-plane. Then the function $\lambda_0(x,t)$ satisfies the system of partial differential equations*

$$\frac{\partial \lambda_0}{\partial t} + (-2a_0 - ib_0)\frac{\partial \lambda_0}{\partial x} = 0, \qquad \frac{\partial \lambda_0^*}{\partial t} + (-2a_0 + ib_0)\frac{\partial \lambda_0^*}{\partial x} = 0, \tag{5.160}$$

where $\lambda_0(x,t) = a_0(x,t) + ib_0(x,t)$. This system is exactly the complex form of the elliptic modulation equations for genus $G = 0$.

5.4 SYMMETRIES OF THE ENDPOINT EQUATIONS

The relations that determine the endpoints as functions of x and t for an ansatz of a given even genus G involve contour integrals over paths that are not known a priori. In §5.1, it was shown by elementary contour deformation arguments that given an ordered sequence of complex endpoints $\lambda_0, \dots, \lambda_G$, the moment conditions (5.25), vanishing conditions (5.52), and measure reality conditions (5.58) have the same value for all contours C_I in the cut upper half-plane $\mathbb{H}$ connecting the origin to iA via this sequence of points that can be smoothly deformed into each other while holding the intermediate points $\lambda_0, \dots, \lambda_G$ fixed.

But this fact alone does not provide sufficient invariance. One would really like to know that the determination of the endpoints is completely insensitive to the choice of integration contour and even the ordering of the endpoints along the contour (i.e., which intervals between the endpoints constitute bands and which constitute gaps). For example, if the configuration on the upper left in figure 5.5 satisfies the endpoint relations for genus $G = 2$, then it should follow that the other three configurations do as well. Note that these invariance issues are nontrivial compared with inverse problems such as the zero-dispersion limit of the Korteweg–de Vries equation [LL83] and the continuum limit of the Toda lattice [DM98]. In these selfadjoint problems, the endpoints are totally ordered because they are necessarily real and similarly there is no ambiguity whatsoever about paths of integration.

In this section, we explore the symmetries of the equations (5.25), (5.52), and (5.58) in more detail. According to the calculations presented in §5.3, we are free to replace the condition $R_0 = 0$ with $M_{G+1} = 0$, and we do this here. We begin with the following lemma.

Lemma 5.4.1 *Each moment M_p defined by (5.25) depends only on the endpoints $\lambda_0, \dots, \lambda_G$. Considered as a function of the independent complex variables $\lambda_0, \dots,$*

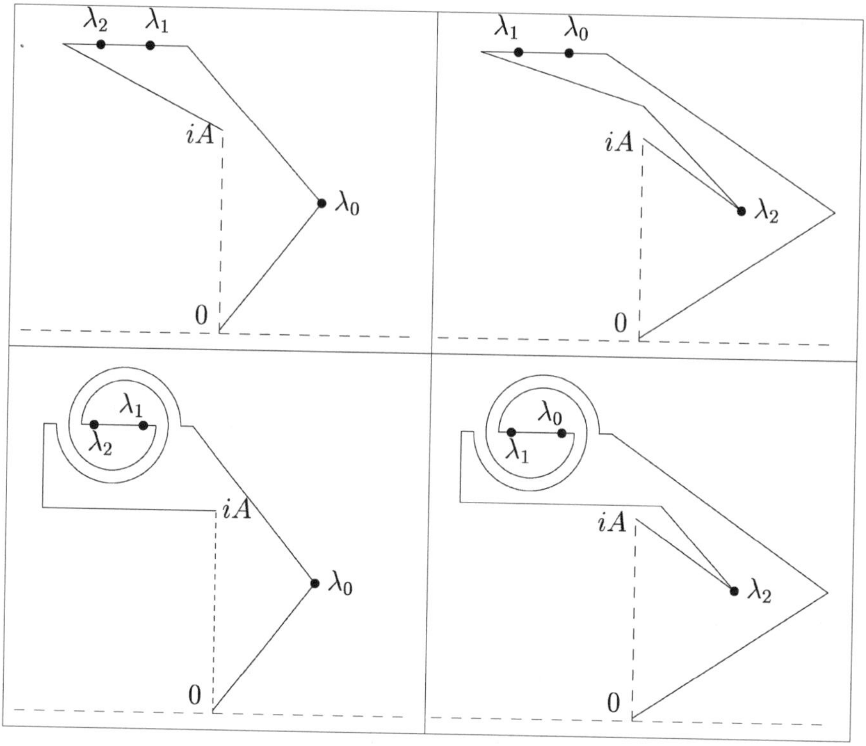

Figure 5.5 Four ways to thread a contour C_I through the same three points in $\mathbb{H}$. Upper left: An apparently simple way. Upper right: Different ordering of the same endpoints. Lower left: Same ordering of points but a topologically distinct contour C_I; the contour contains a Dehn twist, which is a generator of braid groups. Lower right: Combining the two effects.

$\lambda_G \in \mathbb{H}$ and $\lambda_0^*, \ldots, \lambda_G^* \in \mathbb{H}^*$, it is analytic in $(\mathbb{H} \times \mathbb{H}^*)^{G+1}$ and is symmetric under any permutation among the endpoints $(\lambda_0, \ldots, \lambda_G)$ or, independently, among $(\lambda_0^*, \ldots, \lambda_G^*)$.

Proof. The complexification of the moment M_p is the formula

$$M_p = J \int_{\cup_k I_k^{\pm}} \frac{2ix + 4i\eta t}{R_+(\eta)} \eta^p \, d\eta + \int_{\Gamma_I} \frac{\pi i \rho^0(\eta)}{R(\eta)} \eta^p \, d\eta + \int_{\Gamma_I^*} \frac{\pi i \rho^0(\eta^*)^*}{R(\eta)} \eta^p \, d\eta; \tag{5.161}$$

that is, when λ_k^* is taken to be the complex conjugate of λ_k, this formula agrees with (5.25). Using now-familiar contour deformations and the paths C_{I+} and C_{I-} introduced in the proof of Lemma 5.1.3 to represent the function $Y(\lambda)$, one rewrites the moment as

$$M_p = -\frac{J}{2}\oint_L \frac{2ix + 4i\eta t}{R(\eta)}\eta^p\,d\eta + \frac{1}{2}\int_{C_{I+}\cup C_{I-}} \frac{\pi i\rho^0(\eta)}{R(\eta)}\eta^p\,d\eta$$
$$+ \frac{1}{2}\int_{C_{I+}^*\cup C_{I-}^*} \frac{\pi i\rho^0(\eta^*)^*}{R(\eta)}\eta^p\,d\eta\,, \tag{5.162}$$

where L is an arbitrarily large, positively oriented loop contour and where the conjugate paths $C_{I\pm}^*$ are taken to be oriented from $-iA$ toward the origin. Note that the paths $C_{I\pm}$ may be taken to be the same for all choices of the path C_I, the path $C_{I\sigma}$ may be taken as the imaginary interval $[0, iA]$, and the path $C_{I(-\sigma)}$ may be deformed toward infinity with the only obstruction being any points of nonanalyticity of $\rho^0(\eta)$.

With the moment M_p rewritten in this way, the only dependence on the endpoints enters through the function $R(\eta)$. Since all cuts of this function are contained inside the closed contour L and also between the contours C_{I+} and C_{I-} or between C_{I+}^* and C_{I-}^* and since the branch of the square root is determined by asymptotic behavior at infinity, M_p is easily seen to be completely independent of C_I and an analytic function of the $2G+2$ independent complex variables $\lambda_0, \ldots, \lambda_G, \lambda_0^*, \ldots, \lambda_G^*$. The permutation symmetry of swapping any pair of endpoints $\lambda_j \leftrightarrow \lambda_k$ or any pair $\lambda_j^* \leftrightarrow \lambda_k^*$ follows from similar considerations. □

Next, we consider the vanishing conditions (5.52) and the reality conditions (5.58). These two apparently different kinds of conditions are essentially equivalent. This is because it follows from differentiating the relation (4.32) with respect to λ that in the gaps of C_I,

$$\frac{d\tilde{\phi}^\sigma}{d\lambda}(\lambda) = i\pi\rho^\sigma(\lambda), \tag{5.163}$$

where on the right-hand side the function ρ^σ is analytically continued from the "+" side of any band. Using the formula (5.35), we see that while the reality functions R_k can be written for $k = 1, \ldots, G/2$ as

$$R_k = \frac{1}{2i}\left[\int_{I_k^+} R_+(\eta)Y(\eta)\,d\eta + \int_{I_k^-} R_+(\eta)Y(\eta)\,d\eta\right], \tag{5.164}$$

the vanishing functions V_k can be similarly written for $k = 0, \ldots, G/2-1$ as

$$V_k = -\frac{\pi}{2i}\left[\int_{\Gamma_{k+1}^+} R(\eta)Y(\eta)\,d\eta + \int_{\Gamma_{k+1}^-} R(\eta)Y(\eta)\,d\eta\right]. \tag{5.165}$$

By passing to the Riemann surface of the square root function $R(\lambda)$, the functions πR_k for $k = 1, \ldots, G/2$ and V_j for $j = 0, \ldots, G/2-1$ can be reinterpreted as periods (integrals over complete homology cycles) *of the same differential*. Fix a set of $G+1$ complex endpoints in the cut upper half-plane $\mathbb{H}$, and consider two different paths C_I and C_I' interpolating these points, possibly in different order. Let $\mathbf{v} := [R_1, \ldots, R_{G/2}, V_0, \ldots, V_{G/2-1}]^T$ be the vector of functions corresponding to the path C_I, and likewise let $\mathbf{v}'$ correspond to the path C_I'. Then it is possible to show using homology arguments that $\mathbf{v}' = \mathbf{G}_{\mathbf{v}\to\mathbf{v}'}\mathbf{v}$, where $\mathbf{G}_{\mathbf{v}\to\mathbf{v}'}$ is an invertible matrix with integer entries. Thus, while each separate function R_k and V_j undergoes nontrivial monodromy when the path C_I is changed by adding a cycle or the branch points are reordered, the zero locus of the full set of equations is invariant. This leads us to state the following.

Lemma 5.4.2 *The common zero locus of the vanishing conditions* (5.52) *and the reality conditions* (5.58) *is independent of the ordering of the endpoints* $\lambda_0, \ldots, \lambda_G$ *and of the contour* C_I.

Remark. These statements about the vanishing conditions and reality conditions are only valid in the real subspace of $\mathbb{C}^{2G+2}$ when the variables λ_k and λ_k^* are linked by complex conjugation.

Remark. Unlike the moments M_p, the functions V_j and R_k are multivalued functions. They are branched when either of the two endpoints of the corresponding integral coalesces with another λ_k different from the opposite endpoint.

Together, these two results imply the main symmetry result.

Theorem 5.4.1 *Consider the equations* $M_p = 0$ *for* $p = 0, \ldots, G+1$, $V_j = 0$ *for* $j = 0, \ldots, G/2 - 1$, *and* $R_k = 0$ *for* $k = 1, \ldots, G/2$. *Then the set of real solutions (that is, where* λ_k^* *is the complex conjugate of* λ_k*) of this system is invariant under permutations of the endpoints and arbitrary redefinitions of the interpolating contour* C_I.

Chapter Six

The Genus-Zero Ansatz

6.1 LOCATION OF THE ENDPOINTS FOR GENERAL DATA

For $G = 0$, there is only one complex endpoint to determine, λ_0. This endpoint is constrained by one moment condition and one measure reality condition. Both conditions are real and, taken together, are expected to determine the endpoint up to a discrete multiplicity of solutions. The equations that constrain the endpoint for $G = 0$ are

$$M_0 = J \int_{I_0} \frac{2ix + 4i\eta t}{R_+(\eta)}\, d\eta + \int_{\Gamma_I \cap C_I} \frac{\pi i \rho^0(\eta)}{R(\eta)}\, d\eta + \int_{\Gamma_I \cap C_I^*} \frac{\pi i \rho^0(\eta^*)^*}{R(\eta)}\, d\eta = 0 \tag{6.1}$$

and

$$R_0 = \Im\left(\int_0^{\lambda_0} \rho^\sigma(\eta)\, d\eta\right) = 0. \tag{6.2}$$

In the measure reality condition $R_0 = 0$, we use the formula (5.34) for the candidate measure $\rho^\sigma(\eta)$ valid in the band I_0^+:

$$\begin{aligned}\rho^\sigma(\lambda) = \rho^0(\lambda) - \frac{4Jt}{\pi} R_+(\lambda) + \frac{R_+(\lambda)}{\pi i} \int_{\Gamma_I \cap C_I} \frac{\rho^0(\eta)\, d\eta}{(\lambda - \eta) R(\eta)} \\ + \frac{R_+(\lambda)}{\pi i} \int_{\Gamma_I \cap C_I^*} \frac{\rho^0(\eta^*)^*\, d\eta}{(\lambda - \eta) R(\eta)}.\end{aligned} \tag{6.3}$$

In these formulas, $I_0 = I_0^+ \cup I_0^-$ is the unknown band connecting λ_0^* in the lower half-plane to λ_0 via the origin. Also, $\Gamma_I \cap C_I$ denotes a path from λ_0 to iA and $\Gamma_I \cap C_I^*$ denotes a path from $-iA$ to λ_0^*, both in the complex plane cut at I_0.

It is useful to simplify somewhat the two conditions $M_0 = 0$ and $R_0 = 0$. We begin with the moment condition $M_0 = 0$, evaluating the first term by residues by rewriting the integral as a closed loop around the band I_0 as described in chapter 5. Thus,

$$M_0 = -2J\pi(x + 2a_0 t) + \int_{\Gamma_I \cap C_I} \frac{\pi i \rho^0(\eta)}{R(\eta)}\, d\eta + \int_{\Gamma_I \cap C_I^*} \frac{\pi i \rho^0(\eta^*)^*}{R(\eta)}\, d\eta, \tag{6.4}$$

where $a_0 = \Re(\lambda_0)$. Continuing with the reality condition $R_0 = 0$, we use similar reasoning as in §5.3 to obtain the representation

$$R_0 = \frac{1}{8\pi} \oint_L F(\eta)\, d\eta, \tag{6.5}$$

where L is an arbitrarily large counterclockwise circular loop. Using the relation $F(\eta) = H(\eta)R(\eta)$ and the Laurent series expansion (5.30) for $H(\eta)$ along with the expansion of $R(\eta)$ for genus $G = 0$,

$$R(\eta) = -\eta + a_0 - \frac{b_0^2}{2\eta} + O(\eta^{-2}), \quad \text{as } \eta \to \infty, \tag{6.6}$$

one finds simply

$$R_0 = \frac{1}{4\pi}(a_0 M_0 - M_1). \tag{6.7}$$

Computing the first term in the moment M_1 by residues as done for M_0 above, we obtain

$$2iR_0 = -2iJtb_0^2 + \int_{\Gamma_I \cap C_I} \frac{(\eta - a_0)\rho^0(\eta)}{R(\eta)}\, d\eta + \int_{\Gamma_I \cap C_I^*} \frac{(\eta - a_0)\rho^0(\eta^*)^*}{R(\eta)}\, d\eta. \tag{6.8}$$

Finally, since for $G = 0$,

$$\frac{\eta - a_0}{R(\eta)} = \frac{\partial R}{\partial \eta}(\eta), \tag{6.9}$$

the measure reality condition becomes

$$2iR_0 = -2iJtb_0^2 + \int_{\Gamma_I \cap C_I} \rho^0(\eta) \frac{\partial R}{\partial \eta}(\eta)\, d\eta + \int_{\Gamma_I \cap C_I^*} \rho^0(\eta^*)^* \frac{\partial R}{\partial \eta}(\eta)\, d\eta. \tag{6.10}$$

Further analysis of these conditions on the endpoint $\lambda_0(x, t)$ requires either detailed knowledge of the function $\rho^0(\eta)$ or a simplifying assumption such as $t = 0$ or $x = 0$.

6.2 SUCCESS OF THE ANSATZ FOR GENERAL DATA AND SMALL TIME. RIGOROUS SMALL-TIME ASYMPTOTICS FOR SEMICLASSICAL SOLITON ENSEMBLES

6.2.1 The Genus-Zero Ansatz for $t = 0$. Success of the Ansatz and Recovery of the Initial Data

When $t = 0$, it follows from the fact that the function $\rho^0(\eta)$ is purely imaginary for η on the imaginary axis between the origin and iA that the measure reality condition $R_0 = 0$ is satisfied by assuming that the endpoint λ_0 is purely imaginary and lies below $\lambda = iA$. We write $\lambda_0 = ib_0$ for $0 < b_0 < A$. Using this information, the moment condition $M_0 = 0$ becomes for $t = 0$

$$-\int_{b_0}^{A} \frac{i\rho^0(iv)}{\sqrt{v^2 - b_0^2}}\, dv = Jx. \tag{6.11}$$

Here, the square root symbol refers to the principal branch. Since the measure $i\rho^0(iv)\, dv$ is strictly negative (cf. the WKB formula (3.2)), this formula is inconsistent unless we choose the Jost function normalization index J to satisfy

$$J := \text{sign}(x). \tag{6.12}$$

Inserting the WKB formula (3.2) for even, single-maximum initial data $A(x)$ (in which case the symmetry $x_-(\eta) = -x_+(\eta)$ holds) into (6.11) subject to (6.12) gives

$$\frac{2}{\pi}\int_{b_0}^{A} d\nu \int_0^{x_+(i\nu)} dx \frac{\nu}{\sqrt{\nu^2 - b_0^2}\sqrt{A(x)^2-\nu^2}} = |x|. \tag{6.13}$$

Exchanging the order of integration, using the fact that $x_+(i\nu)$ is an inverse function to $A(x)$, (i.e., $A(x_+(i\nu)) = \nu$ for $0 \le \nu \le A$, one finds

$$\frac{2}{\pi}\int_0^{x_+(ib_0)} dx \int_{b_0}^{A(x)} d\nu \frac{\nu}{\sqrt{\nu^2 - b_0^2}\sqrt{A(x)^2-\nu^2}} = |x|. \tag{6.14}$$

Let $S(\nu)$ denote the square root function satisfying $S(\nu)^2 = (\nu^2 - b_0^2)(A(x)^2 - \nu^2)$, defined in the ν-plane cut in the real intervals $[-A(x), -b_0]$ and $[b_0, A(x)]$ and normalized so that for $\mu \in (b_0, A(x))$

$$\lim_{\epsilon\downarrow 0} S(\mu + i\epsilon) > 0. \tag{6.15}$$

Then, the inner integral can be written as

$$\begin{aligned}\int_{b_0}^{A(x)} \frac{\nu\, d\nu}{\sqrt{\nu^2 - b_0^2}\sqrt{A(x)^2-\nu^2}} &= \frac{1}{2}\lim_{\epsilon\downarrow 0}\left(\int_{-A(x)}^{-b_0} + \int_{b_0}^{A(x)}\right)\frac{\nu\, d\nu}{S(\nu+i\epsilon)} \\ &= -\frac{1}{4}\oint_L \frac{\nu\, d\nu}{S(\nu)},\end{aligned} \tag{6.16}$$

where L is an arbitrarily large counterclockwise-oriented closed loop. This integral can be evaluated exactly by residues. Since $S(\nu) = -i\nu^2 + O(\nu)$ for large ν, we obtain simply

$$\int_{b_0}^{A(x)} \frac{\nu\, d\nu}{\sqrt{\nu^2 - b_0^2}\sqrt{A(x)^2-\nu^2}} = \frac{\pi}{2}. \tag{6.17}$$

Therefore, the moment condition at $t = 0$ becomes

$$x_+(ib_0) = |x|, \text{ which implies } b_0 = b_0(x) := A(x). \tag{6.18}$$

Combining this information about the endpoint with the final remark at the end of §5.1.2, we can obtain a useful formula for the candidate density $\rho^\sigma(\eta)$ for $t = 0$ by expressing it as the integral of its derivative with respect to x. Using the fact that $\rho^0(\eta)$ is independent of x and the formula for the total derivative $\partial F/\partial x$ (i.e., the derivative including dependence of the endpoints on x and t), one finds that for any x_0

$$\rho^\sigma(\eta; x) = \rho^\sigma(\eta; x_0) - \frac{2J}{\pi}\int_{x_0}^{x} \frac{\partial R_+}{\partial \eta}(\eta; x')\, dx'. \tag{6.19}$$

In particular, if η is on the imaginary axis between the origin and $ib_0(x)$, we may choose $x_0 = Jx_+(\eta)$. For this choice, we obtain

$$\rho^\sigma(\eta; x) = -\frac{2J}{\pi}\int_{Jx_+(\eta)}^{x} \frac{\partial}{\partial \eta}\sqrt{\eta^2 + b_0(x')^2}\, dx', \tag{6.20}$$

using the explicit formula for R_+ written in terms of the standard branch of the square root function, valid for $t = 0$ and $\eta \in (0, ib_0(x))$. Thus, we see that the integrand is always positive imaginary (the square root function is real and decreasing in the positive imaginary direction). From the relation (6.12), it then follows that $\rho^\sigma(\eta; x)$ is positive imaginary for all $\eta \in (0, ib_0(x))$. We have proven the following lemma.

Lemma 6.2.1 *For $t = 0$ and a genus-zero ansatz, the oriented contour band I_0^+ may be taken to coincide with the vertical segment $[0, ib_0(x)] = [0, iA(x)]$. That is, on this segment the differential $\rho^\sigma(\eta; x)\, d\eta$ is real and negative. Moreover, for all x such that the function $x_+(\eta)$ is differentiable at $ib_0 = ib_0(x) = iA(x)$, the function $\rho^\sigma(\eta; x)$ vanishes exactly like a square root at $\eta = ib_0(x)$ and not to higher order.*

Note that our monotonicity assumption on the initial data $A(x)$ implies that $x_+(\eta)$ fails to be differentiable only when $\eta = iA$, and therefore the only value of x, where $\rho^\sigma(\eta; x)$ fails to vanish exactly like a square root at $\eta = ib_0(x)$, is $x = 0$. In fact when $x = 0$, we have $\rho^\sigma(\eta) \equiv \rho^0(\eta)$, and the latter does not generally vanish at all in the limit as η approaches iA from below.

This result requires some clarification in the context of the fundamental assumption that the contour C should be a loop encircling the imaginary interval $(0, iA)$. We choose to imagine the band I_0^+ lying infinitesimally either to the right of $(0, iA)$ (for $\sigma = +1$) or to the left of $(0, iA)$ (for $\sigma = -1$). Using the relations (4.32) and (4.33), it is easy to find a formula for the boundary value of the function $\tilde{\phi}^\sigma(\lambda; x)$ on the same side of the imaginary interval $[0, iA]$ above the endpoint. Let $\lambda \in (ib_0(x), iA)$. Then,

$$\lim_{\epsilon \downarrow 0} \tilde{\phi}^\sigma(\lambda + \sigma\epsilon; x) = \tilde{\phi}^\sigma|_{\lambda \in I_0^+}(x) + 2iJ \int_{ib_0(x)}^{\lambda} d\eta \int_{Jx_+(\eta)}^{x} dx' \frac{\partial R}{\partial \eta}(\eta; x'). \tag{6.21}$$

Since by construction $\tilde{\phi}^\sigma$ evaluates to an imaginary constant in the band I_0^+ for each x, using the relation (6.12) along with the fact that $\partial R/\partial \eta$ is negative real over the range of integration, we obtain the following result.

Lemma 6.2.2 *For $t = 0$ and a genus zero ansatz and for all $x \neq 0$ and all $\lambda \in (ib_0(x), iA]$,*

$$\lim_{\epsilon \downarrow 0} \Re(\tilde{\phi}^\sigma(\lambda + \sigma\epsilon; x)) < 0. \tag{6.22}$$

The boundary values of the analytic function $\tilde{\phi}^\sigma(\lambda; x)$ on the interval $[0, iA]$ are related by (cf. definition (4.13))

$$\lim_{\epsilon \downarrow 0} \tilde{\phi}^\sigma(\lambda + \sigma\epsilon; x) - \lim_{\epsilon \downarrow 0} \tilde{\phi}^\sigma(\lambda - \sigma\epsilon; x) = -2\pi i\sigma \int_{\lambda}^{iA} \rho^0(\eta)\, d\eta. \tag{6.23}$$

Because this quantity is purely imaginary, we can immediately extend the previous result to the boundary value on the other side of the interval $[0, iA]$ *above* the endpoint $\lambda = ib_0(x)$. This proves the following.

Lemma 6.2.3 *For $t = 0$ and a genus-zero ansatz and for all $x \neq 0$ and all $\lambda \in (ib_0(x), iA]$,*

$$\lim_{\epsilon \downarrow 0} \Re(\tilde{\phi}^\sigma(\lambda - \sigma\epsilon; x)) < 0. \tag{6.24}$$

To complete the analysis on this side of the asymptotic spectral interval $[0, iA]$, we must obtain a similar inequality valid *below* the endpoint $ib_0(x)$. Using (6.23) and the relations (4.32) and (4.33), we find that for $\lambda \in (0, ib_0(x))$,

$$\lim_{\epsilon\downarrow 0}\tilde{\phi}^\sigma(\lambda - \sigma\epsilon; x) = -i\pi\sigma\left[\int_\lambda^{ib_0(x)} \rho^\sigma(\eta; x)\, d\eta - 2\int_\lambda^{iA} \rho^0(\eta)\, d\eta\right]$$
$$= i\pi\sigma\left[\int_\lambda^{iA} \bar{\rho}^\sigma(\eta; x)\, d\eta + \int_\lambda^{iA} \rho^0(\eta)\, d\eta\right], \qquad (6.25)$$

where we have used the definition (5.9) of the complementary density $\bar{\rho}^\sigma(\eta; x)$. This boundary value is purely imaginary. Now, we show that for $t = 0$ the genus-zero ansatz yields a measure $\bar{\rho}^\sigma(\eta; x)\, d\eta$, which like $\rho^0(\eta)\, d\eta$ is negative real for all $\eta \in (0, iA)$ taken with upward orientation. Since for η above the endpoint $ib_0(x)$ on C (against the imaginary axis) the function $\rho^\sigma(\eta; x)$ vanishes identically, it is only necessary to check this for $\eta \in (0, ib_0(x))$. For this purpose, we note that for $t = 0$ and genus zero, with the endpoint $\lambda_0 = ib_0(x)$, the formula (5.32) that holds for $\eta \in (0, ib_0(x))$ can be written as

$$\bar{\rho}^\sigma(\eta; x) = \frac{iR_+(\eta; x)}{\pi}\int_{ib_0(x)}^{iA}\left(\frac{1}{|\xi| - |\eta|} + \frac{1}{|\xi| + |\eta|}\right)\frac{i\rho^0(\xi)}{R(\xi; x)}\, d\xi. \qquad (6.26)$$

Since $R_+(\eta; x)$ is positive real and $R(\xi; x)$ is negative imaginary over the range of integration and since the measure $\rho^0(\xi)\, d\xi$ is negative real, this formula shows that $\bar{\rho}^\sigma(\eta; x)$ is strictly positive imaginary for $\eta \in (0, ib_0(x))$.

Remark. This result shows that at $t = 0$ the measure $-\rho^\sigma(\eta; x)\, d\eta$ on the contour satisfies an *upper constraint* as well as a positivity condition, since $0 < -\rho^\sigma(\eta; x)\, d\eta < -\rho^0(\eta)\, d\eta$. In this respect, the analysis at $t = 0$ on the imaginary axis is very similar to the analysis of Lax and Levermore [LL83].

For $\lambda \in (0, ib_0(x))$, it therefore follows that $\Re(\tilde{\phi}^\sigma(\lambda + \sigma 0; x)) \equiv 0$ and at the same time that $\Im(\tilde{\phi}^\sigma(\lambda + \sigma 0; x))$ is increasing (respectively, decreasing) in the positive imaginary direction for $\sigma = +1$ (respectively, for $\sigma = -1$). An application of the Cauchy-Riemann equations for the analytic function $\tilde{\phi}^\sigma(\lambda; x)$ then yields the following.

Lemma 6.2.4 *For $\sigma = +1$ (respectively, $\sigma = -1$), there exists a lens-shaped region to the left (respectively, right) of the imaginary interval $(0, ib_0(x))$ in which $\Re(\tilde{\phi}^\sigma(\lambda; x)) < 0$.*

At this point we have proved that at $t = 0$ and for each nonzero x, there is a genus-zero ansatz for each choice of $\sigma = \pm 1$ that satisfies the inequalities $\rho^\sigma(\eta; x)\, d\eta < 0$ for $\eta \in I_0^+$ and $\Re(\tilde{\phi}^\sigma(\lambda; x)) \le 0$ for $\lambda \in \Gamma_1^+$, where I_0^+ and Γ_1^+ are the band and gap components of a degenerate contour C that barely encircles the imaginary interval $[0, iA]$. See figure 6.1. The inequality $\Re(\tilde{\phi}^\sigma(\lambda; x)) \le 0$ is in fact strict at all points interior to Γ_1^+ *except* for $\lambda = ib_0(x) - \sigma 0$. However, using the fact that $\rho^\sigma(\eta; x) = O(|\eta - ib_0(x)|^{1/2})$ for η near $ib_0(x)$, the formula (6.25) gives

$$\lim_{\epsilon\downarrow 0}\tilde{\phi}^\sigma(\lambda - \sigma\epsilon; x) = 2\pi i\sigma\int_{ib_0(x)}^{iA} \rho^0(\eta)\, d\eta - 2\pi i\sigma\rho^0(ib_0(x))$$
$$\cdot(\lambda - ib_0(x)) + O(|\lambda - ib_0(x)|^{3/2}) \qquad (6.27)$$

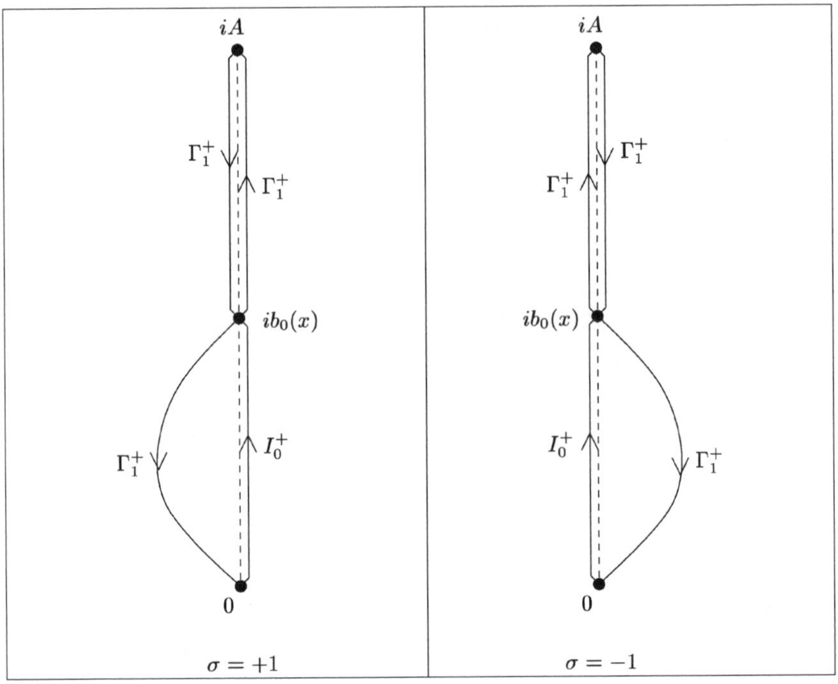

Figure 6.1 The degenerate contours for $t = 0$. For each x there are two possibilities corresponding to $\sigma = +1$ and $\sigma = -1$.

in the limit $\lambda \to ib_0(x)$. Therefore, since the function $\rho^0(\eta)$ defined by (3.2) never vanishes for $\eta \in (0, iA)$, the linear terms dominate, and for all λ in a sufficiently small half-disk centered at $\lambda = ib_0(x)$ and lying in the left (respectively, right) half-plane for $\sigma = +1$ (respectively, $\sigma = -1$), the inequality $\Re(\tilde{\phi}^\sigma(\lambda; x)) < 0$ holds. This means that the gap contour Γ_1^+ can be pulled slightly away from the interval $[0, iA]$ except at the endpoints $\lambda = ib_0(x) + \sigma 0$ and $\lambda = -\sigma 0$ while achieving *strict* inequality on the interior. This proves the following theorem.

Theorem 6.2.1 *For $t = 0$, the endpoint function may be taken as $\lambda_0 = iA(x)$, and then for all $x \neq 0$ and both signs of σ, the genus-zero ansatz corresponding to a contour C for which I_0^+ is the imaginary interval $[\sigma 0, iA(x) + \sigma 0]$, and Γ_1^+ has endpoints $\lambda = iA(x) + \sigma 0$ and $-\sigma 0$ and lies in the slit half-plane $\mathbb{H}$, satisfies the following:*

- *The differential $\rho^\sigma(\eta; x)\, d\eta$ is strictly negative in the interior of I_0^+ and vanishes exactly like a square root at the endpoint $\eta = iA(x) + \sigma 0$;*
- *The inequality $\Re(\tilde{\phi}^\sigma(\lambda; x)) < 0$ holds strictly in the interior of the contour Γ_1^+.*

The contour C is illustrated in figure 6.2.

Finally, we observe that the genus-zero ansatz for $t = 0$ formally reconstructs the initial data for (1.1) in the semiclassical limit $\hbar \downarrow 0$.

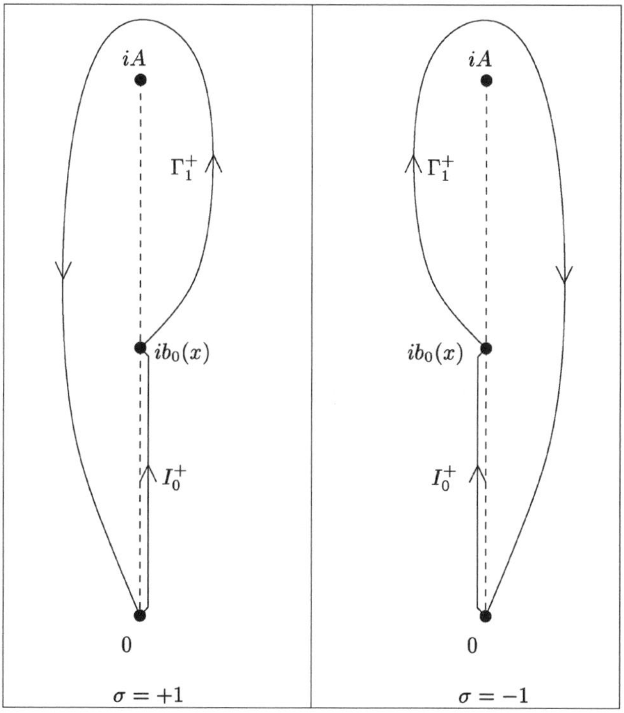

Figure 6.2 The contours on which Theorem 6.2.1 holds. Once again, both signs of σ are available for each $x \neq 0$.

Theorem 6.2.2 *The function* $\tilde{\psi}$ *corresponding to the genus-zero ansatz for* $t = 0$ *and defined by* (4.108) *is given by* $\tilde{\psi} \equiv A(x)$.

Proof. Since we have already shown that $\Im(\lambda_0) = A(x)$, it remains to show that α_0, the constant value of the function $-iJ\tilde{\phi}^\sigma(\lambda)$ in the band I_0^+, is identically zero as a function of x.

To do this, we first establish a general formula, holding for an arbitrary genus-zero ansatz (i.e., not only for $t = 0$) for the derivative of $-\alpha_0$ with respect to x. In §4.3 we observed that this quantity had the interpretation of a local wavenumber k. Let us calculate the wavenumber k in terms of the endpoints λ_0 and λ_0^*. First, from the definition of the constant α_0, we have

$$k = iJ\frac{\partial \tilde{\phi}^\sigma}{\partial x}\bigg|_{\lambda=0}. \tag{6.28}$$

Next, using (4.13), we find

$$k = iJ\frac{\partial}{\partial x}\int_{C_I}\overline{L_\eta^{C,\sigma}}(0)\bar{\rho}^\sigma(\eta)\,d\eta + iJ\frac{\partial}{\partial x}\int_{C_I^*}\overline{L_\eta^{C,\sigma}}(0)\bar{\rho}^\sigma(\eta^*)^*\,d\eta, \tag{6.29}$$

where we recall that the overbar on the logarithm indicates the average of the two boundary values taken on the contour C. We now show that the derivative with respect to x can be moved inside the integral and put onto the complementary density $\bar{\rho}^\sigma(\eta)$. Recalling the definition (5.9) and using the facts that the logarithmic integral of the function $\rho^0(\eta)$ is independent of x and that $\rho^\sigma(\eta) \equiv 0$ for $\eta \in \Gamma_I \cap C_I$, (6.29) implies

$$\begin{aligned} k &= -iJ\frac{\partial}{\partial x}\int_{I_0^+}\overline{L_\eta^{C,\sigma}}(0)\rho^\sigma(\eta)\,d\eta - iJ\frac{\partial}{\partial x}\int_{I_0^-}\overline{L_\eta^{C,\sigma}}(0)\rho^\sigma(\eta^*)^*\,d\eta \\ &= -iJ\int_{I_0^+}\overline{L_\eta^{C,\sigma}}(0)\frac{\partial}{\partial x}\rho^\sigma(\eta)\,d\eta - iJ\int_{I_0^-}\overline{L_\eta^{C,\sigma}}(0)\frac{\partial}{\partial x}\rho^\sigma(\eta^*)^*\,d\eta \\ &= iJ\int_{I_0^+}\overline{L_\eta^{C,\sigma}}(0)\frac{\partial}{\partial x}\bar{\rho}^\sigma(\eta)\,d\eta + iJ\int_{I_0^-}\overline{L_\eta^{C,\sigma}}(0)\frac{\partial}{\partial x}\bar{\rho}^\sigma(\eta^*)^*\,d\eta. \end{aligned} \quad (6.30)$$

Here, the x derivative can be brought inside the integral since $\rho^\sigma(\eta)$ vanishes at the moving endpoints, and the last step follows because the density function $\rho^0(\eta)$ is independent of x. In terms of the function $F(\lambda)$, we then have

$$k = -\frac{J}{2\pi}\int_{I_0}\overline{L_\eta^{C,\sigma}}(0)\left(\frac{\partial F_+}{\partial x}(\eta) - \frac{\partial F_-}{\partial x}(\eta)\right)d\eta. \quad (6.31)$$

Using the expression for $\partial F/\partial x$ obtained in the final remark in §5.1.2, we find that for $\eta \in I_0$,

$$\frac{\partial F_+}{\partial x}(\eta) - \frac{\partial F_-}{\partial x}(\eta) = -4iJ\frac{\partial R_+}{\partial \eta}(\eta), \quad (6.32)$$

and therefore

$$\begin{aligned} k &= \frac{2i}{\pi}\int_{I_0}\overline{L_\eta^{C,\sigma}}(0)\,\frac{\partial R_+}{\partial \eta}(\eta)\,d\eta \\ &= \frac{i}{\pi}\int_{I_0}L_{\eta+}^{C,\sigma}(0)\,\frac{\partial R_+}{\partial \eta}(\eta)\,d\eta + \frac{i}{\pi}\int_{I_0}L_{\eta-}^{C,\sigma}(0)\,\frac{\partial R_+}{\partial \eta}(\eta)\,d\eta, \end{aligned} \quad (6.33)$$

where we have used the definition of $\overline{L_\eta^{C,\sigma}}(\lambda)$ as an average of boundary values. Now the function $L_{\eta+}^{C,\sigma}(0)$ is the boundary value of $L_\eta^{C,\sigma}(\lambda)$ as λ approaches the origin in the oriented contour $[C \cup C^*]_\sigma$ from the "+" side. This means that for $\eta \in [C \cup C^*]_\sigma$, $L_{\eta+}^{C,\sigma}(0)$ has an analytic extension as a function of η to the "−" side of $[C \cup C^*]_\sigma$. A similar argument shows that $L_{\eta-}^{C,\sigma}(0)$ is analytic in η on the "+" side of $[C \cup C^*]_\sigma$. And of course $R_+(\lambda)$ extends analytically to the "+" side of $[C \cup C^*]_\sigma$, while $R_-(\lambda)$ extends to the "−" side. These observations allow us to move the path of integration away from the integrable singularity at the origin in each integral. Namely, if we let C_a be a path from λ_0^* to λ_0 lying to the right of I_0 and C_b be a path from λ_0 to λ_0^* lying to the left of I_0, then it is easy to see that

$$k = -\frac{i}{\pi}\int_{C_a}L_{\eta+}^{C,\sigma}(0)\,\frac{\partial R}{\partial \eta}(\eta)\,d\eta - \frac{i}{\pi}\int_{C_b}L_{\eta-}^{C,\sigma}(0)\,\frac{\partial R}{\partial \eta}(\eta)\,d\eta. \quad (6.34)$$

Now, integrate by parts in each integral using the fact that R vanishes at the endpoints to find

$$k = \frac{i}{\pi}\int_{C_a}\frac{R(\eta)\,d\eta}{\eta} + \frac{i}{\pi}\int_{C_b}\frac{R(\eta)\,d\eta}{\eta}. \quad (6.35)$$

The paths of integration may now be combined into a single counterclockwise loop surrounding I_0 and the singularity at the origin. Calculating this loop integral by a residue at infinity (again using detailed information about the form of $R(\lambda)$ valid for genus $G = 0$), one finds at last that

$$k = -(\lambda_0 + \lambda_0^*) = -2\Re(\lambda_0) = -2a_0. \tag{6.36}$$

Using the fact that for the genus-zero ansatz at $t = 0$ the endpoint is purely imaginary, we see from this general formula that α_0 is independent of x. We now show that with $t = 0$,

$$\lim_{x \to 0} \alpha_0 = 0. \tag{6.37}$$

This follows from two observations. First, for η fixed on the imaginary axis below iA, the function $\bar{\rho}^\sigma(\eta)$ converges pointwise to zero as $x \to 0$. This can be seen by noting that for $|x|$ small enough η lies in the band I_0^+; a direct estimate of the boundary values of the functions $H(\eta)$ and $R_+(\eta)$ that vanishes as $x \to 0$ is then easily obtained from the exact formula (5.29). Next, since as noted above there is an effective upper constraint for $t = 0$ on the measure $\bar{\rho}^\sigma(\eta)\,d\eta$ on the imaginary axis, it follows from a dominated convergence argument that the function $\tilde{\phi}^\sigma(\lambda)$ converges pointwise to zero for $t = 0$ as $x \to 0$. These results imply that for all x at $t = 0$, $\alpha_0 = 0$, and the proof is complete. □

Remark. Although the inequalities are all strict, the fact that the band I_0^+ lies against the imaginary axis when $t = 0$ precludes the application of the asymptotic inverse theory in chapter 4 to establish the recovery of the initial data. In other words, the fact that our conditions on the complex phase function select at $t = 0$ a contour that coincides with polar singularities of the matrix $\mathbf{m}(\lambda)$ solving the meromorphic Riemann-Hilbert Problem 2.0.1 means that the strong $O(\hbar_N^{1/3})$ error estimate we obtained in Theorem 4.5.2 is not uniformly valid in any neighborhood that should happen to include $t = 0$ (at least not without a modification of the methods we have presented here; see [M02]). On the other hand, we know from the Lax-Levermore-type analysis carried out by Ercolani, Jin, Levermore, and MacEvoy [EJLM93] that it is possible to prove $L^2(\mathbb{R})$ convergence of ψ to $\tilde{\psi} \equiv A(x)$ exactly at $t = 0$. Note however, that at least for the special case of the Satsuma-Yajima ensemble, this strange situation is no real obstruction to our analysis since there is no adjustment of the initial data (e.g., neglect of a reflection coefficient) for $\hbar$ values in the "quantum" sequence $\hbar = \hbar_N$ (cf. (3.9)) and consequently nothing to prove at $t = 0$.

6.2.2 Perturbation Theory for Small Time

We begin this section by establishing the existence of the endpoint $\lambda_0(x, t)$ for small time.

Lemma 6.2.5 *Let the initial data $A(x) > 0$ be real-analytic, even, and monotone decreasing in $|x|$. Then, for each fixed $x \neq 0$, the equations $M_0 = 0$ and $R_0 = 0$ have a solution for $|t|$ sufficiently small that is differentiable in t and agrees with the purely imaginary solution obtained in §6.2.1 upon setting $t = 0$.*

Proof. We need to show that the Jacobian determinant $\partial(M_0, R_0)/\partial(\lambda_0, \lambda_0^*)$ does not vanish. Since R_0 is a linear combination of M_0 and M_1, it is equivalent to show that $\partial(M_0, M_1)/\partial(\lambda_0, \lambda_0^*) \neq 0$. In §5.3, it is shown that

$$\frac{\partial(M_0, M_1)}{\partial(\lambda_0, \lambda_0^*)} := \det \begin{bmatrix} \dfrac{\partial M_0}{\partial \lambda_0} & \dfrac{\partial M_0}{\partial \lambda_0^*} \\ \dfrac{\partial M_1}{\partial \lambda_0} & \dfrac{\partial M_1}{\partial \lambda_0^*} \end{bmatrix} = \det \begin{bmatrix} \dfrac{\partial M_0}{\partial \lambda_0} & \dfrac{\partial M_0}{\partial \lambda_0^*} \\ \lambda_0 \dfrac{\partial M_0}{\partial \lambda_0} & \lambda_0^* \dfrac{\partial M_0}{\partial \lambda_0^*} \end{bmatrix}$$
$$= -(\lambda_0 - \lambda_0^*) \frac{\partial M_0}{\partial \lambda_0} \frac{\partial M_0}{\partial \lambda_0^*}. \tag{6.38}$$

To calculate the partial derivatives, we first establish a simple formula for M_0 that involves the initial data $A(x)$. For this purpose, we define a quantity $I(\lambda_0, \lambda_0^*)$ from (6.4) by writing $M_0 = -2J\pi(x + (\lambda_0 + \lambda_0^*)t) + I(\lambda_0, \lambda_0^*)$, and to calculate $I(\lambda_0, \lambda_0^*)$ we momentarily suppose that $\lambda_0 = i\alpha$ and $\lambda_0^* = i\beta$ with α and β being independent real numbers with $0 < \alpha < A$ and $-A < \beta < 0$. Then, substituting the formula (3.2) for $\rho^0(\eta)$ into the formula (6.4), one exchanges the order of integration to find

$$I(i\alpha, i\beta) = \int_0^{x_+(i\alpha)} dx \int_\alpha^{A(x)} d\nu \, \frac{2\nu}{\sqrt{(\nu - \alpha)(\nu - \beta)}\sqrt{A(x)^2 - \nu^2}}$$
$$- \int_0^{x_+(-i\beta)} dx \int_{-A(x)}^{\beta} d\nu \, \frac{2\nu}{\sqrt{(\nu - \alpha)(\nu - \beta)}\sqrt{A(x)^2 - \nu^2}}. \tag{6.39}$$

Now, define a square root function $T(\nu)$ satisfying $T(\nu)^2 = (\nu - \alpha)(\nu - \beta)(A(x)^2 - \nu^2)$ and defined in the ν-plane slit along the real intervals $[-A(x), \beta]$ and $[\alpha, A(x)]$ with the normalization

$$\lim_{\epsilon \downarrow 0} T(\mu + i\epsilon) > 0, \qquad \mu \in (\alpha, A(x)). \tag{6.40}$$

Letting L_α (respectively, L_β) be a small counterclockwise-oriented loop encircling $[\alpha, A]$ (respectively, encircling $[-A, \beta]$) as shown in figure 6.3, we may therefore write

$$I(i\alpha, i\beta) = -\int_0^{x_+(i\alpha)} dx \oint_{L_\alpha} \frac{\nu \, d\nu}{T(\nu)} - \int_0^{x_+(-i\beta)} dx \oint_{L_\beta} \frac{\nu \, d\nu}{T(\nu)}. \tag{6.41}$$

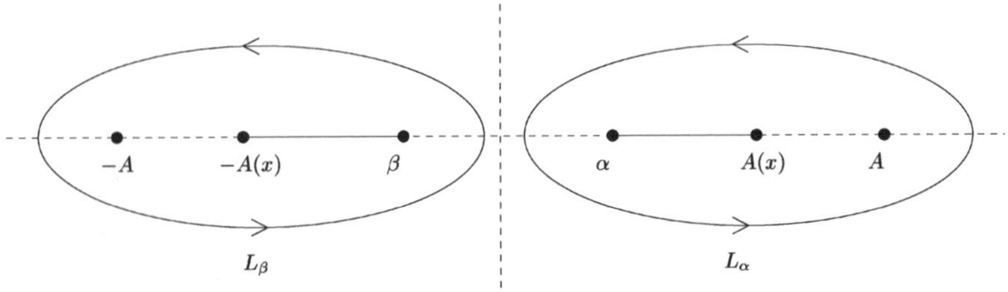

Figure 6.3 The integration contours L_α and L_β surrounding the cuts of $T(\nu)$.

This formula will be analytic in $\lambda_0 = -i\alpha$ and $\lambda_0^* = -i\beta$ for λ_0 in a complex neighborhood of the imaginary interval $(0, iA)$ if the initial data $A(x)$ is analytic.

With the help of the formula (6.41), we may now easily compute derivatives of $I(\lambda_0, \lambda_0^*)$ with respect to λ_0 and λ_0^* at $\lambda_0 = iA(x)$ and $\lambda_0^* = -iA(x)$. First, note that

$$\frac{\partial}{\partial\alpha} I(i\alpha, i\beta) = -\frac{d}{d\alpha} x_+(i\alpha) \cdot \oint_{L_\alpha} \frac{\nu\, d\nu}{T(\nu)}\bigg|_{x=x_+(i\alpha)} - \int_0^{x_+(i\alpha)} dx \oint_{L_\alpha} \frac{\nu\, d\nu}{2(\nu-\alpha)T(\nu)} - \int_0^{x_+(-i\beta)} dx \oint_{L_\beta} \frac{\nu\, d\nu}{2(\nu-\alpha)T(\nu)} \tag{6.42}$$

and

$$\frac{\partial}{\partial\beta} I(i\alpha, i\beta) = \frac{d}{d\beta} x_+(-i\beta) \cdot \oint_{L_\beta} \frac{\nu\, d\nu}{T(\nu)}\bigg|_{x=x_+(-i\beta)} - \int_0^{x_+(i\alpha)} dx \oint_{L_\alpha} \frac{\nu\, d\nu}{2(\nu-\beta)T(\nu)} - \int_0^{x_+(-i\beta)} dx \oint_{L_\beta} \frac{\nu\, d\nu}{2(\nu-\beta)T(\nu)}. \tag{6.43}$$

Setting $\alpha = A(x)$ and $\beta = -A(x)$, the final two terms in each of the above formulae can be combined, and the integrand of the x integral then can be calculated by a residue at $\nu = \infty$, which vanishes identically as a function of x. Thus, only the first term survives in each case, and these too can be computed by residues. Using $\partial/\partial\alpha = i\,\partial/\partial\lambda_0$ and $\partial/\partial\beta = i\partial/\partial\lambda_0^*$ and expressing the derivatives of the inverse function $x_+(\cdot)$ in terms of derivatives of $A(\cdot)$, one obtains

$$\frac{\partial I}{\partial\lambda_0}(iA(x), -iA(x)) = \frac{i\pi}{A'(x)}, \qquad \frac{\partial I}{\partial\lambda_0^*}(iA(x), -iA(x)) = -\frac{i\pi}{A'(x)}. \tag{6.44}$$

Finally, we use these formulas to evaluate the Jacobian for $t = 0$. We find

$$\frac{\partial(M_0, M_1)}{\partial(\lambda_0, \lambda_0^*)}\bigg|_{t=0} (iA(x), -iA(x)) = -2i\pi^2 \frac{A(x)}{A'(x)^2}. \tag{6.45}$$

By our monotonicity assumptions on the initial data, this Jacobian is finite and strictly nonzero for all $x \neq 0$. Thus, the lemma is proved by appealing to the implicit function theorem. □

Theorem 6.2.3 *Let the initial data $A(x)$ be real-analytic, even, and monotone decreasing in $|x|$. Then for each fixed $x \neq 0$, there is a $\tau_x > 0$ such that for all positive $t < \tau_x$ (respectively, negative $t > -\tau_x$) the genus-zero ansatz holds with $\sigma = \operatorname{sign}(x)$ (respectively, for $\sigma = -\operatorname{sign}(x)$) for a loop contour C lying in the cut upper half-plane $\mathbb{H}$. The value $t = 0$ is excluded only in the sense that as explained in §6.2.1 the loop C cannot be taken to lie completely in $\mathbb{H}$ for either value of σ.*

Proof. From Lemma 6.2.5, the endpoint function $\lambda_0(x, t)$ is differentiable in a neighborhood of $t = 0$ for each nonzero x. The proofs of the local continuation results given in §5.2 can easily be applied here to show that, if the endpoint leaves the imaginary axis by moving into the right half-plane for a $\sigma = +1$ ansatz or by moving into the left half-plane for a $\sigma = -1$ ansatz, then a gap contour $\Gamma_1^+(x,t)$ will exist in $\mathbb{H}$ connecting $\lambda_0(x,t)$ to $\lambda = -\sigma 0$, on the interior points of which the inequality $\Re(\tilde{\phi}^\sigma(\lambda; x, t)) < 0$ holds strictly. Similarly, these proofs show that for small $|t|$ a band contour $I_0^+(x,t)$ will exist on which the differential $\rho^\sigma(\eta; x, t)\, d\eta$ is real and strictly negative. However, in this case the difficulty is that for $t = 0$ the band $I_0^+(x, 0)$ lies on the boundary of $\mathbb{H}$, and we must therefore prove that the band $I_0^+(x,t)$ lies entirely on one side or the other of the imaginary axis for small time.

Note that as λ moves along the contour $I_0^+(x,t)$ (whose existence for small time is guaranteed by the arguments in §5.2) from $\lambda = 0$ to $\lambda = \lambda_0(x,t)$, the function $B_0(\lambda)$ defined in (5.57) is real and strictly decreasing. Therefore, by the Cauchy-Riemann equations, $\Im(B_0(\lambda))$ is negative (respectively, positive) for all λ in a small lens-shaped region just to the left (respectively, right) of $I_0^+(x,t)$. We use the expression for the total derivative of the function $F(\lambda)$ with respect to t obtained in the final remark in §5.1.2 to calculate

$$\begin{aligned}\frac{\partial}{\partial t}\Im(B_0(\lambda)) &= \Im\left(\frac{1}{2\pi i}\int_0^\lambda \left[\frac{\partial F_+}{\partial t}(\eta) - \frac{\partial F_-}{\partial t}(\eta)\right] d\eta\right)\\ &= -\frac{2J}{\pi}\Im\left(\int_0^\lambda \frac{\partial}{\partial \eta}\,[\eta R_+(\eta)]\, d\eta\right)\\ &= -\frac{2J}{\pi}\Im(\lambda R_+(\lambda)). \end{aligned} \tag{6.46}$$

For the purely imaginary endpoint configuration at $t = 0$, this quantity is strictly nonzero with sign $-J$ for all λ on the positive imaginary axis below the endpoint $\lambda_0(x, 0) = iA(x)$.

Using the relation (6.12), we therefore see that for $x > 0$ the interior points of the band $I_0^+(x,t)$ all move into the right half-plane for t small and positive and into the left half-plane for t small and negative. Similarly, for $x < 0$ the band moves to the left for $t > 0$ and to the right for $t < 0$. This shows that the ansatz corresponding to $\sigma = \operatorname{sign}(xt)$ always deforms for small time so that all inequalities remain valid, which proves the theorem. □

Combining Lemma 6.2.5 with Theorem 5.3.1 leads to a representation of the solution of the analytic Cauchy problem for the elliptic genus-zero Whitham equations (5.160).

Theorem 6.2.4 (Solution of the analytic Cauchy problem for the Whitham equations). *The algebraic equations $M_0 = 0$ and $R_0 = 0$, with $\rho^0(\eta)$ given in terms of the even, single-maximum, real-analytic function $A(x)$ by* (3.2), *implicitly defines for small t and all $x \neq 0$ the solution $(\lambda_0(x,t), \lambda_0^*(x,t))$ of the Cauchy (initial-value) problem for the elliptic Whitham system* (5.160) *corresponding to the initial data $\lambda_0(x,0) = iA(x)$ and $\lambda_0^*(x,0) = -iA(x)$.*

So the genus-$G = 0$ ansatz is sufficient to enable the error analysis of §4.5 to be valid, as long as t is different from zero but is sufficiently small for any given x. Combining Theorem 6.2.3 with Theorem 4.5.2 yields one of our most important results, the rigorous description of the small-time semiclassical limit of WKB soliton ensembles for the focusing nonlinear Schrödinger equation.

THEOREM 6.2.5 (Rigorous small-time asymptotics for semiclassical soliton ensembles). *Let $\psi(x,t)$ be for each $\hbar = \hbar_N$ the solution of the focusing nonlinear Schrödinger equation that is the WKB soliton ensemble corresponding to the approximate initial data $\psi(x,0) = A(x)$. Then, for each $x \neq 0$ there is an open time interval T_x containing $t = 0$ and independent of $\hbar$ such that the formula* (4.108) *for $\tilde{\psi}(x,t)$ built from the genus-zero ansatz satisfies $|\psi - \tilde{\psi}| \leq K_{x,t}\hbar_N^{1/3}$ for $\hbar_N$ sufficiently small as long as $t \in T_x \backslash \{0\}$. The constant $K_{x,t}$ may vary in x and t.* q

For the special case of the Satsuma-Yajima ensemble, where the semiclassical soliton ensemble coincides with the solution of the initial-value problem (1.1) because there is no modification of the initial data by the WKB approximation of the spectrum for $\hbar = \hbar_N$, we obtain a uniform description of the semiclassical limit for the focusing nonlinear Schrödinger equation.

COROLLARY 6.2.1 (The semiclassical limit for the Satsuma-Yajima initial data). *Let $\psi(x,t)$ be the solution of the focusing nonlinear Schrödinger equation with initial data $\psi(x,0) = A\,\text{sech}(x)$. Then, for each $x \neq 0$ there is an open-time interval T_x containing $t = 0$ and independent of $\hbar$ such that the formula* (4.108) *for $\tilde{\psi}(x,t)$ built from the genus-zero ansatz satisfies $|\psi - \tilde{\psi}| \leq K_{x,t}\hbar_N^{1/3}$ whenever $t \in T_x$. The error is uniformly small in compact subsets of the (x,t)-plane where the approximation is valid.*

6.3 LARGER-TIME ANALYSIS FOR SOLITON ENSEMBLES

Here, we consider the genus-zero ansatz for larger times. First, we establish a simple formula for the solution of the analytic Cauchy problem for the genus-zero Whitham modulation equations (5.160) that holds when $x = 0$, that is, in the center of the symmetric evolution. Then, we use the concrete example of the Satsuma-Yajima soliton ensemble; that is, we assume $\rho^0(\eta) = \rho^0_{\text{SY}}(\eta) \equiv i$ to show how to determine the boundary of the genus-zero region of the (x,t)-plane in practice. In this case, with the help of numerical calculations, we are able to indicate the success of the ansatz in regard to satisfying the relevant inequalities in a certain region of the (x,t)-plane and to make concrete the mechanism of failure of the ansatz, as described in general terms above, at the boundary of this region.

Given values of x and t, we may choose the discrete parameters $J = \pm 1$ and $\sigma = \pm 1$. This choice turns out to be essential in order to treat the whole (x,t)-plane; in fact, different values of J and σ are needed for different signs of x and t. Of course, for all real-valued, even initial data, we have the symmetries $\psi(-x,t;\hbar) = \psi(x,t;\hbar)$ and $\psi(x,-t;\hbar) = \psi(x,t;\hbar)^*$ that allow the solution for all x and t to be obtained from the solution for x and t positive. Therefore, for the semiclassical soliton ensembles we consider in this book, it is not necessary strictly speaking to carry out any more analysis for other signs of x and t. Nonetheless, it is useful to document how the other signs of x and t break symmetry for more general future applications.

6.3.1 The Explicit Solution of the Analytic Cauchy Problem for the Genus-Zero Whitham Equations along the Symmetry Axis $x = 0$

We now study the equations $M_0 = 0$ and $R_0 = 0$ for the endpoint $\lambda_0(x, t)$ under the assumption that $x = 0$ with $|t|$ sufficiently small. Our main result is contained in the following theorem.

THEOREM 6.3.1 (Explicit location of the endpoint for $x = 0$). *Assume that $A(x)$ is a real-analytic, even, bell-shaped function satisfying $A''(0) < 0$. Then, if $x = 0$ and $|t|$ is sufficiently small, the equations $M_0 = 0$ and $R_0 = 0$ are satisfied by a point $\lambda_0(t) = ib_0(t)$ with $b_0 > A$. The relation between b_0 and t is simply*

$$|t| = \frac{2}{\pi b_0} \int_0^{-iA^{-1}(b_0)} E(1 - A(iy)^2/b_0^2)\, dy, \tag{6.47}$$

where the upper limit of integration is the number $y > 0$ for which $A(iy) = b_0 > A$ and where $E(m)$ is the complete elliptic integral of the second kind. The upper limit of integration makes sense for $b_0 > A$ because $A''(0) < 0$ implies that $A(iy)$ is an increasing function of $|y|$ for y real and small enough.

Proof. To prove this, we need to examine the equations $M_0 = 0$ and $R_0 = 0$ for such endpoint configurations, which requires in particular that $\rho^0(\eta)$ can be defined for η on the imaginary axis above $\eta = iA$. We now show that under the condition $A''(0) < 0$ the function $\rho^0(\eta)$ is analytic at $\eta = iA$ and therefore has a unique analytic continuation for some distance along the imaginary axis above $\eta = iA$. We begin with the WKB formula (3.1) that defines $\rho^0(\eta)$ for η on the imaginary axis between 0 and iA. In this formula, $x_-(\eta)$ and $x_+(\eta)$ are, respectively, the negative and positive real roots of the equation $A(x)^2 + \eta^2 = 0$. For even bell-shaped functions $A(x)$, $x_-(\eta) = -x_+(\eta)$ and both functions are well defined for η in the imaginary interval in question. To show the analyticity at $\eta = iA$, we use the fact that $A(x)$ is real-analytic and for η just below iA define a function $B(x, \eta)$ satisfying $B(x, \eta)^2 = A(x)^2 + \eta^2$ in a neighborhood U of $x = 0$ containing only $x_\pm(\eta)$ as roots of $B(x, \eta)^2 = 0$. $B(x, \eta)$ is taken to be cut on the real axis between $x_-(\eta)$ and $x_+(\eta)$ and has positive boundary values on the the upper half-plane side of the cut. With this normalization, we also have

$$\lim_{x \to \pm\infty} B(x, \eta) = \mp\eta. \tag{6.48}$$

Then, the WKB density (3.1) is rewritten as

$$\rho^0(\eta) = \frac{\eta}{2\pi} \oint_L \frac{dx}{B(x, \eta)}, \tag{6.49}$$

where L is a clockwise-oriented loop surrounding the cut of $B(x, \eta)$ and lying in U. Because we are assuming that $A''(0) \neq 0$, we can choose the neighborhood U and the loop contour L to be independent of η below but sufficiently near $\eta = iA$ such that for all such η the contour L only ever encloses the two roots $x_\pm(\eta)$. If $A''(0) = 0$, then more than two roots would have to coalesce at $x = 0$ when $\eta = iA = iA(0)$, and the contour L would have to shrink as η approaches iA in order to exclude the unwanted roots. Now, with $A''(0) < 0$, the two roots $x_\pm(\eta)$ coalesce as η moves up the axis through iA, and reemerge as a purely imaginary complex-conjugate pair for η just above iA. For η just above iA, we still have only two roots within U and enclosed by L, and we define $B(x, \eta)$ to be cut along the imaginary axis between these two roots and to be normalized by the same relation as before (6.48). With this choice, it is then clear that the formula (6.49) defines the analytic continuation of the original formula (3.1) for $\rho^0(\eta)$ through the point $\eta = iA$.

For η on the imaginary axis above iA, the function $B(x, \eta)$ takes positive imaginary boundary values on the left of the vertical cut. Thus, for such η we can write (6.49) in the form

$$\rho^0(\eta) = \frac{2\eta}{\pi} \int_0^{A^{-1}(-i\eta)} \frac{dx}{i\sqrt{-(A(x)^2 + \eta^2)}}, \tag{6.50}$$

where the positive square root is taken and where $A^{-1}(\eta)$ is the positive imaginary number x that satisfies $A(x) = -i\eta$. Or, changing variables to $x = iy$,

$$\rho^0(\eta) = \frac{2\eta}{\pi} \int_0^{-iA^{-1}(-i\eta)} \frac{dy}{\sqrt{-(A(iy)^2 + \eta^2)}}. \tag{6.51}$$

This formula, representing the analytic continuation of $\rho^0(\eta)$, is positive imaginary for η above iA on the imaginary axis.

The moment condition $M_0 = 0$ is satisfied automatically for $x = 0$ by any $G = 0$ configuration with endpoint $\lambda_0 = ib_0$ on the imaginary axis with $b_0 > A$. In this situation, the moment M_0 explicitly takes the form

$$M_0 := \int_{\Gamma_I \cap C_I} \frac{\pi i \rho^0(\eta)}{R(\eta)}\, d\eta + \int_{\Gamma_I \cap C_I^*} \frac{\pi i \rho^0(\eta^*)^*}{R(\eta)}\, d\eta. \tag{6.52}$$

Now, with the band I_0^+ connecting the origin to ib_0 in the first quadrant of the η-plane for $\sigma = +1$ and the second quadrant of the η-plane for $\sigma = -1$ (we are assuming that I_0^+ does not coincide identically with an interval of the positive imaginary axis), it is easy to see that the function $R(\eta)$, satisfying $R(\eta)^2 = \eta^2 + b_0^2$, cut on the band I_0, and normalized to $-\eta$ for large η, is purely real for η in the imaginary interval $[-ib_0, ib_0]$, and in fact

$$R(\eta) = \sigma\sqrt{\eta^2 + b_0^2} \tag{6.53}$$

for such η, where the positive square root is intended. The contour $\Gamma_I \cap C_I$ may be taken to coincide with the interval $[ib_0, iA]$ oriented from ib_0 down to iA, and correspondingly, $\Gamma_I \cap C_I^*$ coincides with the interval $[-iA, -ib_0]$ oriented from $-iA$ down to $-ib_0$. So, the moment becomes

$$M_0 := \int_{ib_0}^{iA} \frac{\pi i \rho^0(\eta)}{\sigma\sqrt{\eta^2 + b_0^2}}\, d\eta + \int_{-iA}^{-ib_0} \frac{\pi i \rho^0(\eta^*)^*}{\sigma\sqrt{\eta^2 + b_0^2}}\, d\eta. \tag{6.54}$$

Using $\rho^0(\eta^*)^* = -\rho^0(-\eta)$ and changing variables $\eta \to -\eta$ in the second term, one sees that $M_0 = 0$ holds identically for all $b_0 > A$.

We now show that the reality condition $R_0 = 0$ determines the endpoint $\lambda_0 = ib_0$ at $x = 0$ as a function of t. Using the formula (6.53) for $R(\eta)$ and the fact that $\partial R/\partial\eta = \eta/R(\eta)$, the relevant quantity to consider is

$$R_0 := -Jtb_0^2 + \frac{1}{2i}\int_{ib_0}^{iA} \frac{\eta\rho^0(\eta)}{\sigma\sqrt{\eta^2+b_0^2}}\,d\eta + \frac{1}{2i}\int_{-iA}^{-ib_0} \frac{\eta\rho^0(\eta^*)^*}{\sigma\sqrt{\eta^2+b_0^2}}\,d\eta \qquad (6.55)$$

or, with $\rho^0(\eta^*)^* = -\rho^0(-\eta)$ and a change of variables $\eta \to -\eta$,

$$R_0 := -Jtb_0^2 + \int_{ib_0}^{iA} \frac{\eta\rho^0(\eta)}{i\sigma\sqrt{\eta^2+b_0^2}}\,d\eta. \qquad (6.56)$$

Using (6.51) and $\eta = iz$ with z real and positive, we get

$$R_0 := -Jtb_0^2 + \int_A^{b_0} \frac{2z^2}{\pi\sigma\sqrt{b_0^2-z^2}} \int_0^{-iA^{-1}(z)} \frac{dy}{\sqrt{z^2-A(iy)^2}}\,dz. \qquad (6.57)$$

The equation $R_0 = 0$ can evidently only have solutions if $\sigma Jt \geq 0$. In this case we have

$$R_0 := -|t|b_0^2 + \int_A^{b_0} \frac{2z^2}{\pi\sqrt{b_0^2-z^2}} \int_0^{-iA^{-1}(z)} \frac{dy}{\sqrt{z^2-A(iy)^2}}\,dz. \qquad (6.58)$$

We simplify further by exchanging the order of integration using

$$\int_A^{b_0}\left[\int_0^{-iA^{-1}(z)} f(y,z)\,dy\right]dz = \int_0^{-iA^{-1}(b_0)}\left[\int_{A(iy)}^{b_0} f(y,z)dz\right]dy, \qquad (6.59)$$

and thus find

$$R_0 := -|t|b_0^2 + \frac{2}{\pi}\int_0^{-iA^{-1}(b_0)}\left[\int_{A(iy)}^{b_0} \frac{z^2\,dz}{\sqrt{(b_0^2-z^2)(z^2-A(iy)^2)}}\right]dy, \qquad (6.60)$$

where the positive square root is meant. The inner integral is identified as (cf. page 596 of [AS65])

$$\int_{A(iy)}^{b_0} \frac{z^2\,dz}{\sqrt{(b_0^2-z^2)(z^2-A(iy)^2)}} = b_0E(1-A(iy)^2/b_0^2), \qquad (6.61)$$

where $E(m)$ denotes the complete elliptic integral of the second kind with modulus m. Thus, the condition $R_0 = 0$ becomes the relation (6.47), which completes the proof of our claim. □

Remark. If $A''(0) = 0$, then the argument used in the proof to show that $\rho^0(\eta)$ defined by (3.1) is analytic at $\eta = iA$ does not apply, and $\rho^0(\eta)$ simply may not continue through $\eta = iA$. We can understand this qualitatively as follows. From the perspective of one-dimensional quantum mechanics (i.e., the theory of Schrödinger operators in one dimension), the formula (3.1) gives the density of energy levels of the potential well $-A(x)^2$ in the vicinity of the negative energy $E = \eta^2$. If the potential well is too flat near the bottom, then for energies just above the bottom the well will look like a constant potential that supports a continuous spectrum of scattering states at this energy. So intuitively, one expects the density of states $\rho^0(\eta)$ to be infinite at $\eta = iA$ if $A''(0) = 0$.

The solution formula (6.47) is explicit enough to be very useful in applications for quite general initial data $A(x)$. In particular, using (6.47) it is easy to calculate the earliest time when the endpoint function $b_0(t)$ fails to be analytic (at $x = 0$). This gives an elementary upper bound on the breaking time, which is the earliest time for which the genus-zero ansatz fails, and more complicated behavior takes over in the semiclassical solution. This sort of calculation is carried out using the formula (6.47) in [CM02], where it is also shown that the formula (6.47) provides a very accurate approximation to the square modulus of the numerical solution of (1.1) for small $\hbar$ with initial data $\psi_0(x) = A(x)$. The formula used in [CM02] is simply $|\psi(0,t)|^2 \sim b_0(t)^2$, where $b_0(t)$ satisfies (6.47).

6.3.2 Determination of the Endpoint for the Satsuma-Yajima Ensemble and General x and t

Using the explicit formula for the eigenvalue density function $\rho^0(\eta) \equiv \rho^0_{\rm SY}(\eta) \equiv i$ appropriate for the Satsuma-Yajima ensemble, let us obtain more detailed information about the endpoint $\lambda_0(x,t)$ for finite times. First, consider the moment condition $M_0 = 0$. To evaluate the integral term in M_0 (cf. (6.4)) in terms of standard functions, we observe that on the paths of integration Γ_I we have

$$R(\eta) = \sigma\sqrt{(\eta - a_0)^2 + b_0^2}, \tag{6.62}$$

where $a_0 = \Re(\lambda_0)$, $b_0 = \Im(\lambda_0)$, and the function $\sqrt{z}$ refers to the principal branch whose cut is the negative real z-axis. We consider only positive values of b_0 and therefore find that the integrals on the right-hand side of (6.4) can be written as

$$\begin{aligned}\int_{\Gamma_I\cap C_I} &\frac{\pi i\rho^0_{\rm SY}(\eta)}{R(\eta)}\,d\eta + \int_{\Gamma_I\cap C_I^*}\frac{\pi i\rho^0_{\rm SY}(\eta^*)^*}{R(\eta)}\,d\eta \\ &= -\pi\sigma\int_{a_0+ib_0}^{iA}\frac{d\eta}{\sqrt{(\eta-a_0)^2+b_0^2}} + \pi\sigma\int_{-iA}^{a_0-ib_0}\frac{d\eta}{\sqrt{(\eta-a_0)^2+b_0^2}}.\end{aligned} \tag{6.63}$$

Performing the quadrature puts the moment condition $M_0 = 0$ in the form

$$M_0 = -J\pi(2x + 4a_0t) + \pi\sigma\left(\operatorname{arcsinh}\left(\frac{a_0 + iA}{b_0}\right) + \operatorname{arcsinh}\left(\frac{a_0 - iA}{b_0}\right)\right) = 0, \tag{6.64}$$

where $\operatorname{arcsinh}(z)$ is the principal branch whose cuts are on the imaginary z axis for $|z| \geq 1$.

Next, consider the reality condition $R_0 = 0$ with R_0 given by (6.10). Since the function $\rho^0_{\rm SY}(\eta)$ is constant on $\Gamma_I \cap C_I$, the integrals on the right-hand side of (6.10) can be evaluated explicitly:

$$\begin{aligned}2iR_0 &= -2iJtb_0^2 + \int_{\Gamma_I\cap C_I}\rho^0_{\rm SY}(\xi)\frac{\partial R}{\partial \xi}(\xi)\,d\xi + \int_{\Gamma_I\cap C_I^*}\rho^0_{\rm SY}(\xi^*)^*\frac{\partial R}{\partial\xi}(\xi)\,d\xi \\ &= -2iJtb_0^2 + iR(iA) + iR(-iA).\end{aligned} \tag{6.65}$$

With the observation that, in terms of the principal branch of the square root $\sqrt{z}$,

$$R(\pm iA) = \sigma\sqrt{(a_0 \mp iA)^2 + b_0^2}, \tag{6.66}$$

the reality condition is therefore expressed in terms of standard functions as

$$2R_0 = \sigma\sqrt{(a_0 + iA)^2 + b_0^2} + \sigma\sqrt{(a_0 - iA)^2 + b_0^2} - 2Jtb_0^2 = 0. \tag{6.67}$$

Let us now investigate the degree to which the conditions (6.64) and (6.67) determine the endpoint λ_0 as a function of x and t. The first observation is the following lemma.

Lemma 6.3.1 *The reality condition* (6.67) *is consistent only if*

$$\sigma Jt \geq 0. \tag{6.68}$$

Thus, our options in choosing the parameters $J = \pm 1$ and $\sigma = \pm 1$ are limited by the sign of t.

If J and σ are chosen so that the reality condition is consistent, then it may be solved explicitly for a_0:

$$a_0 = \pm tb_0^2\sqrt{\frac{A^2 - b_0^2 + t^2b_0^4}{A^2 + t^2b_0^4}}. \tag{6.69}$$

For each fixed t, the graph of (6.69) is a curve in the real (a_0, b_0)-plane that contains the endpoint $\lambda_0 = a_0 + ib_0$. Several of these curves are plotted in figure 6.4. The curve becomes singular when $|t| = 1/(2A)$, developing a double point at $a_0 = 0$ and $b_0 = A\sqrt{2}$.

Remark. The endpoint λ_0 must of course lie on the graph. It is important to note that the bounded component of the graph of (6.69), namely, the loop encircling the

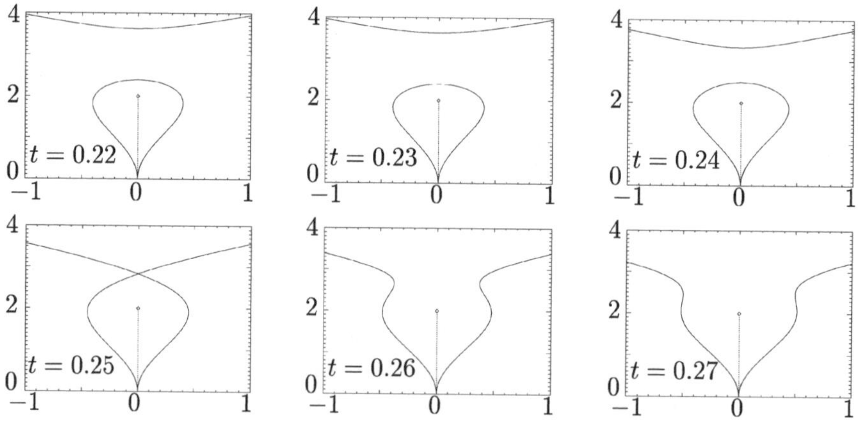

Figure 6.4 Plots of the real graph of the measure reality condition $R_0 = 0$ for $A = 2$. Note the double point appearing when $t = 1/4$. In each plot, the dashed vertical line indicates the imaginary interval $[0, iA]$.

imaginary interval $[0, iA]$ for $|t| < 1/(2A)$, turns out *not* to be directly related to the loop contour C. In particular, the contour C may be different for different values of x, while the graph of (6.69) is the same for all x.

We now return to the moment condition $M_0 = 0$ in which we set $\sigma J = \text{sgn}(t)$ for consistency. Solving (6.64) for x gives

$$x = -2a_0 t + \frac{\text{sgn}(t)}{2}\left(\text{arcsinh}\left(\frac{a_0 + iA}{b_0}\right) + \text{arcsinh}\left(\frac{a_0 - iA}{b_0}\right)\right). \tag{6.70}$$

For each fixed $|t| < 1/(2A)$, this transformation continuously and invertibly maps the loop enclosing $[0, iA]$ onto the whole real-x line, with the point of the loop on the imaginary axis being mapped to $x = 0$. For $t > 0$, the left (respectively, right) half of the loop is mapped to $x < 0$ (respectively, $x > 0$), while for $t < 0$, the situation is reversed. On this bounded component of the graph, the point $\lambda = 0$ corresponds to $x = \pm\infty$. The unbounded component of the graph for $|t| < 1/(2A)$ is also placed continuously into one-to-one correspondence with the real-x line. For $t > 0$, the left (respectively, right) half of the graph is mapped to $x > 0$ (respectively, $x < 0$), and for $t < 0$ the situation is again reversed. For $|t| > 1/(2A)$, there are two branches of the graph, left and right, each one unbounded. Each branch of the graph is placed into one-to-one correspondence with the real-x line.

We therefore arrive at the result that, given x and t (nonzero), the equations for the endpoint $\lambda_0 = a_0 + ib_0$ are only consistent if we choose $\sigma J = \text{sgn}(t)$. This leaves us free to choose, say, $\sigma = \pm 1$ with J being then determined. And for each case $\sigma = +1$ and $\sigma = -1$, there are *two distinct solutions of the constraint equations in the upper half-plane*. For $|t| < 1/(2A)$, there is one solution on the bounded branch of the graph and one solution on the unbounded branch. For $|t| > 1/(2A)$, there is one solution on each of the left and right branches. Thus, there is always one solution $\lambda_0(x, t)$ in the right half-plane and one in the left half-plane. So for each x and t, we still have four possibilities to investigate: $\sigma = \pm 1$ and $\text{sgn}(a_0) = \pm 1$.

6.3.3 Numerical Determination of the Contour Band for the Satsuma-Yajima Ensemble

At this point, we turn to numerical computations in order to determine whether there exists a connected component of the graph of $\Im(B_0(\lambda)) = 0$ containing $\lambda = 0$ and $\lambda = \lambda_0 = a_0 + ib_0$, and if so, whether the candidate measure $\rho^\sigma(\eta)d\eta$ supported there is of the correct sign. Later, we will exploit numerics further to verify the possibility of satisfying the inequality $\Re(\tilde{\phi}^\sigma(\lambda)) < 0$ in the gap. We search for the band I_0 by integrating numerically the differential equation (5.56) for the bands. We used a Simpson's rule integrator in conjunction with local changes of variables to remove integrable inverse square-root singularities and the formula (5.34) with the desingularization afforded by the representation (5.35) to compute the function $\rho^\sigma(\lambda)$. Then, we used a fourth-order Runge-Kutta scheme to integrate numerically the ordinary differential equations (5.56) using arc length as a parameter. The integration proceeded from $\lambda = 0$ in the direction where $\rho^\sigma(\eta)d\eta$ was negative.

The desired result of this numerical procedure is a path of integration in the complex plane that emerges in the half-plane determined by the choice of orientation σ, avoids the imaginary interval $[-iA, iA]$ that is the support of the measure $\rho^0_{SY}(\eta)\,d\eta$, and ultimately meets the endpoint $\lambda = \lambda_0$. However, quite often the numerical integration revealed other possibilities. Sometimes, the integral curve starting at $\lambda = 0$ either fails to emerge in the correct half-plane as determined by the choice of orientation $\sigma = \pm 1$ or intersects the support of the measure $\rho^0_{SY}(\eta)\,d\eta$ (the imaginary interval $[-iA, iA]$). And even if this is not the case, sometimes the integral curve misses the endpoint λ_0 altogether, being deflected off to infinity in the left or right half-planes.

What actually is observed for given choices of $\sigma = \pm 1$ and $\text{sgn}(a_0) = \pm 1$ depends on the values of x and t. Our numerical experiments indicate that the (x, t)-plane can be partitioned into twenty regions, in each of which the integral curve displays qualitatively uniform behavior. The regions are illustrated for $A = 2$ in figure 6.5, and the meaning of each region for the integration contour is given

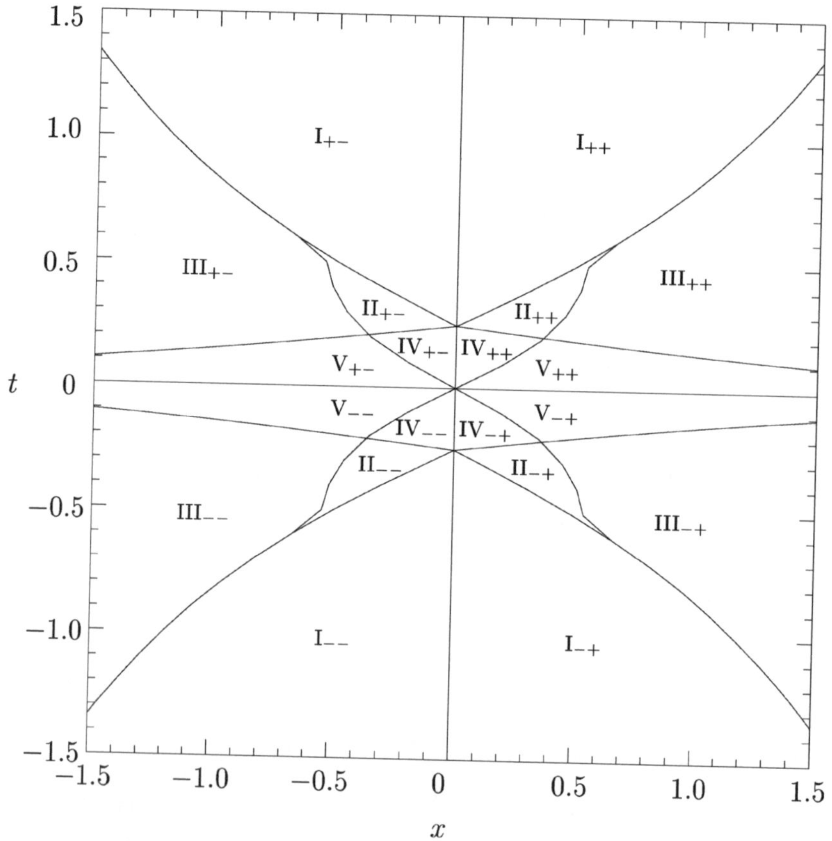

Figure 6.5 Regions of the (x, t)-plane in which different assumptions regarding $\sigma = \pm 1$ and $\text{sgn}(a_0) = \pm 1$ lead to a compact band I_0^+ supporting a measure $\rho^\sigma(\eta)\,d\eta$ of the correct sign. This picture is for $A = 2$. See table 6.1 and the text for a full explanation.

Table 6.1 Regions of the (x, t)-plane in which there exist compactly supported candidate measures of the appropriate sign. See figure 6.5.

	$\sigma = +1$, $a_0 > 0$	$\sigma = -1$, $a_0 > 0$	$\sigma = +1$, $a_0 < 0$	$\sigma = -1$, $a_0 < 0$
I_{++} and I_{--}	connection	left deflection	right deflection	connection
II_{++} and II_{--}	connection	connection	right deflection	connection
III_{++} and III_{--}	connection	intersection	right deflection	connection
IV_{++} and IV_{--}	connection	connection	left deflection	right deflection
V_{++} and V_{--}	connection	intersection	left deflection	right deflection
I_{+-} and I_{-+}	connection	left deflection	right deflection	connection
II_{+-} and II_{-+}	connection	left deflection	connection	connection
III_{+-} and III_{-+}	connection	left deflection	intersection	connection
IV_{+-} and IV_{-+}	left deflection	right deflection	connection	connection
V_{+-} and V_{-+}	left deflection	right deflection	intersection	connection

in table 6.1. Note that in figure 6.5, the regions I_{**}, II_{**}, and IV_{**} all meet on the t-axis at $t = \pm 1/(2A)$. The meanings of the various scenarios listed in table 6.1 are as follows:

- *Intersection* means that the contour of integration meets the imaginary interval $[0, iA]$ either immediately or after some finite arc length of integration.
- *Left/Right deflection* means that the integration contour emerges from the origin in the correct quadrant but does not terminate at the endpoint. Instead, it misses, and the orbit goes off to infinity in the left or right half-plane.
- *Connection* means that the path of integration lies completely in the cut upper half-plane $\mathbb{H}$ and terminates at the endpoint λ_0 with finite arc length.

Therefore, only the cases labeled as "connection" are admissible for the asymptotic analysis described in chapter 4 to succeed.

For each x and t, we see that there is at least one choice of $\sigma = \pm 1$ and $\mathrm{sgn}(a_0) = \pm 1$ for which there exists an a band I_0 that connects $\lambda = 0$ to $\lambda = \lambda_0$. In particular, in each quadrant of the (x, t)-plane there is one choice that always works uniformly throughout all five subregions. Note also that there are subregions where more than one choice yields a connecting band I_0: in the regions II_{**} there are three possible choices, each of which yields an admissible band I_0. In order to distinguish further among these, we need to continue by checking whether each possible band I_0 admits a gap contour connecting the endpoint to $\lambda = 0$ in such a way that the band and the gap together make a loop encircling the imaginary interval $[0, iA]$ and such that everywhere on the gap $\tilde{\phi}^\sigma(\lambda)$ has a strictly negative real part.

6.3.4 Seeking a Gap Contour on Which $\Re(\tilde{\phi}^\sigma(\lambda)) < 0$. The Primary Caustic for the Satsuma-Yajima Ensemble

For given values of x and t, one can choose σ and the sign of a_0 in one or more ways such that the genus-zero ansatz results in a negative candidate measure $\rho^\sigma(\eta)\, d\eta$

on I_0^+, a contour connecting the origin to λ_0 that can be determined numerically as described in §6.3.3. Given this measure, it is then possible to compute numerically the real part of the corresponding function $\tilde{\phi}^\sigma(\lambda)$ as given by the formula (5.59). This is computationally very efficient, since the first two terms of (5.59) can be integrated explicitly for the special case $\rho^0(\eta) = \rho^0_{\mathrm{SY}}(\eta)$ (cf. §??). Similarly, the real part of the integral involving $\rho^\sigma(\eta)\,d\eta$ is easy to evaluate numerically because the candidate measure has already been computed in the process of finding its support (cf. §6.3.3) and is real by construction. Thus, to calculate this term one simply replaces $\overline{L_\eta^{C,\sigma}}(\lambda)$ by $\log|\lambda - \eta|$ and sums over the support weighted by the measure.

In our numerical investigations, we of course restrict attention at this point to those cases labeled "connection" in table 6.1. Our first observation is that it appears that there can only exist an appropriate gap contour on which $\Re(\tilde{\phi}^\sigma(\lambda)) < 0$ everywhere if $x \cdot t \cdot a_0 > 0$. This means that the connections given in the final column of table 6.1 for $x \cdot t > 0$ and in the first column of table 6.1 for $x \cdot t < 0$ do not appear to admit *any* connected path from λ_0 to zero that closes the loop and on which the relevant inequality is satisfied everywhere. In a given quadrant of the (x, t)-plane, the behavior of this ansatz is the same in all three regions I_*, II_*, and III_*. Representative figures showing the region where $\Re(\tilde{\phi}^\sigma(\lambda)) < 0$ and thus where a gap contour might live are shown in figures 6.6 and 6.7.

If we accept these numerical results, we see that for each x and t at most one solution of the equations for the endpoint is relevant for constructing a genus-zero ansatz that satisfies all necessary inequalities. We now concern ourselves exclusively with the unique solution λ_0 that is in the right half-plane for x and t of the same sign and in the left half-plane for x and t of opposite signs.

In studying this case, we first consider those cases when the band I_0^+ must "wrap around" the imaginary interval $[0, iA]$ because σ and a_0 are of opposite signs. The possible connections are listed in the second column of table 6.1 for x and t of the same sign and in the third column of table 6.1 for x and t of opposite sign. These connections are only possible in the small regions labeled II_* and IV_* of figure 6.5. Based on our numerical experiments, the main observation we want to make for these cases is that *a gap contour may always be found.* Two examples of such cases are shown in figures 6.8 and 6.9. In each of these figures, we note that there exists a shaded connected region where $\Re(\tilde{\phi}^\sigma(\lambda)) < 0$ that contains many paths connecting the endpoint to zero and completing the loop around the imaginary interval $[0, iA]$. For future reference, we also record a case of this type when the point (x, t) is very close to the boundary of the region I_*. Such a case is shown in figure 6.10. In this figure, the ansatz is close to failure because a complex zero of the function $\rho^\sigma(\eta)$ is about to move onto the band contour I_0^+. This is evidenced by the nearly square angle made by the integral curve of the vector field of (5.56) as it makes a close approach to this fixed point.

We now study the case represented in the first column of table 6.1 for x and t of the same sign and in the last column for x and t of opposite sign. In this case, the band I_0^+ connecting zero to the endpoint exists for x and t in the entire quadrant, and we can use plots of the type presented so far to try and distinguish any regions within a given quadrant where it appears that the inequality $\Re(\tilde{\phi}^\sigma(\lambda)) < 0$ can or cannot be

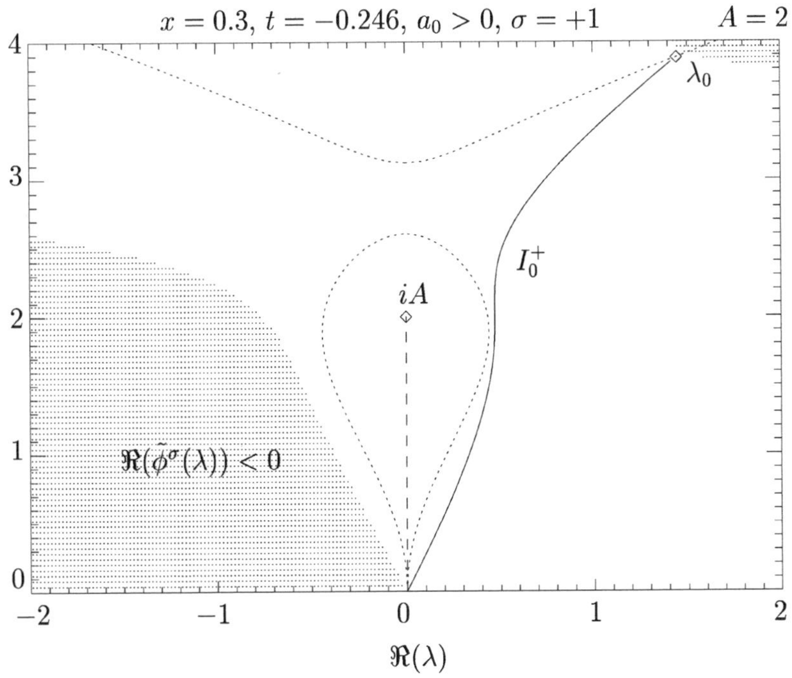

Figure 6.6 The impossibility of satisfying the gap inequality for an incorrectly chosen genus-zero ansatz. The dashed curves are the components of the graph of the reality relation $R_0 = 0$, one of which contains the endpoint $\lambda_0 = a_0 + ib_0$ as indicated. The band I_0^+ as shown is found by Runge-Kutta integration of the differential equation (5.56). The regions of the plane where the real part of the associated function $\tilde{\phi}^\sigma(\lambda)$ is negative are shaded. Note that it does not appear possible to find any path from the endpoint to zero that lies entirely within the shaded region.

satisfied on an appropriate gap contour closing the loop around $[0, iA]$. The results appear to be that *an ansatz of this type is successful everywhere in the* (x, t)*-plane except in the regions* I_* *as shown in figure* 6.5. To illustrate the mechanism for the breakdown of the ansatz in this case, we first look at two plots corresponding to points just on either side of the boundary between the regions I_{--} and II_{--}. The first plot, shown in figure 6.11, corresponds to exactly the same values of x and t as in figure 6.10. We see that a path representing the gap by completing the loop C surrounding the imaginary interval $[0, iA]$ while satisfying $\Re(\tilde{\phi}^\sigma(\lambda)) < 0$ everywhere can indeed be found, albeit barely. For x and t just on the other side of the boundary, in region I_{--} the picture that we obtain is given in figure 6.12.

Figures 6.10, 6.11, and 6.12 clearly demonstrate the *duality* of the successful ansätze corresponding to two orientations $\sigma = +1$ and $\sigma = -1$ in the small regions II_* and III_* of the (x, t)-plane where they coexist. The band and gap are dual to each other and interchangeable by reversing orientation while maintaining the same endpoint. Also, the band I_0^+ for one ansatz coincides with a connected component

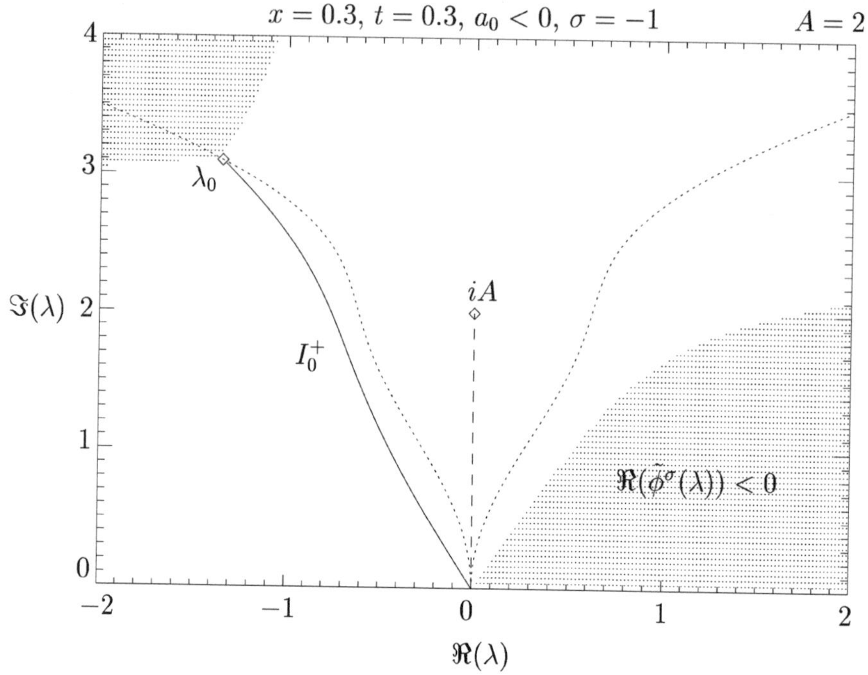

Figure 6.7 The impossibility of satisfying the gap inequality. As in figure 6.6 it appears to be impossible to find a gap contour on which the relevant inequality is satisfied everywhere.

of the graph of $\Re(\tilde{\phi}^\sigma(\lambda)) = 0$ for the dual ansatz. Note that for fixed x and t, this interchange requires changing from $J = +1$ to $J = -1$ and vice versa, so each choice corresponds to the asymptotic simplification of a different Riemann-Hilbert problem. When the mutually dual ansätze break down at the boundary with region I_*, they fail *at the same point in the complex λ-plane*. When the failure occurs in the gap, the situation is exactly as described in chapter 5; the "isthmus" through which the gap contour must pass becomes singular at a certain point in the λ-plane and pinches off when x and/or t are tuned into the region I_*. When the failure occurs in the band, a complex zero of the candidate density $\rho^\sigma(\lambda)$ approaches the band I_0^+ and ultimately meets it at a definite point in the λ-plane—*exactly the same point where the isthmus pinches off in the dual case*. When x and t are tuned into the region I_* from the region II_*, the zero has crossed the contour and there is no longer any possibility of finding a band I_0^+ connecting zero to the endpoint supporting $\rho^\sigma(\eta)\,d\eta$ as a negative real measure.

At the boundary between the regions III_* and I_*, the mechanism of breakdown in the ansatz is similar, although the interpretation in terms of duality is no longer viable because the "wraparound" ansatz is not valid as described above (it fails because the integration of (5.56) from zero intersects the imaginary interval $[0, iA]$). Representative diagrams are shown in figures 6.13 and 6.14. Figure 6.13 concerns

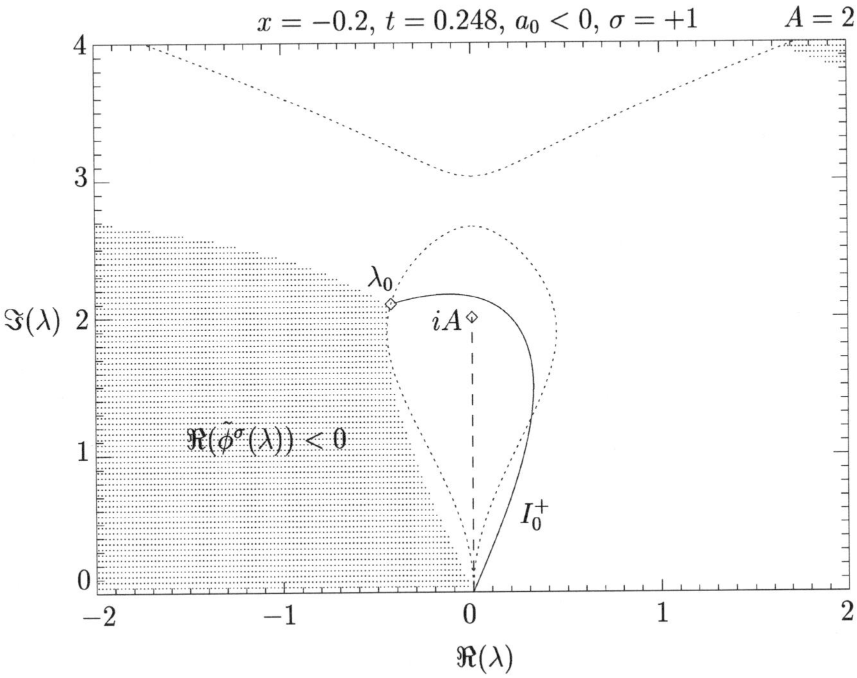

Figure 6.8 A case where the inequalities can be satisfied.

a point in the (x, t)-plane in region III_{+-} very close to the boundary with region I_{+-}. The isthmus through which the gap must pass is very close to pinch-off. In figure 6.14, the value of x has been tuned so that the point (x, t) is just barely in region I_{+-}. The pinch-off has occurred, and it is no longer possible to find an admissible gap contour.

The results we have obtained with our numerical computations are summarized in table 6.2. For all x and t outside the region $\mathrm{I} = \mathrm{I}_{++} \cup \mathrm{I}_{+-} \cup \mathrm{I}_{-+} \cup \mathrm{I}_{--}$, we have at least one genus-zero ansatz (and sometimes there is also the dual ansatz) that satisfies all of the conditions required of a complex phase function $g^{\sigma}(\lambda)$. For such x and t, it follows from the rigorous analysis carried out in chapter 4 that the corresponding complex phase function therefore correctly captures the behavior of the solution of the Riemann-Hilbert problem in the limit $\hbar \downarrow 0$. In light of the exact solution of the outer model problem given in §4.3, we see that that the Satsuma-Yajima semiclassical soliton ensemble behaves like a modulated exponential plane-wave solution of the nonlinear Schrödinger equation for all x and t outside of the boundary of region I. This is in agreement with the observations made in [MK98], where the curve in the (x, t)-plane at which the modulated plane-wave behavior was seen to break down was called the *primary caustic* (a secondary caustic was also observed). Comparing with the figures in [MK98], it is clear that *the boundary of the region* I *is exactly the primary caustic.*

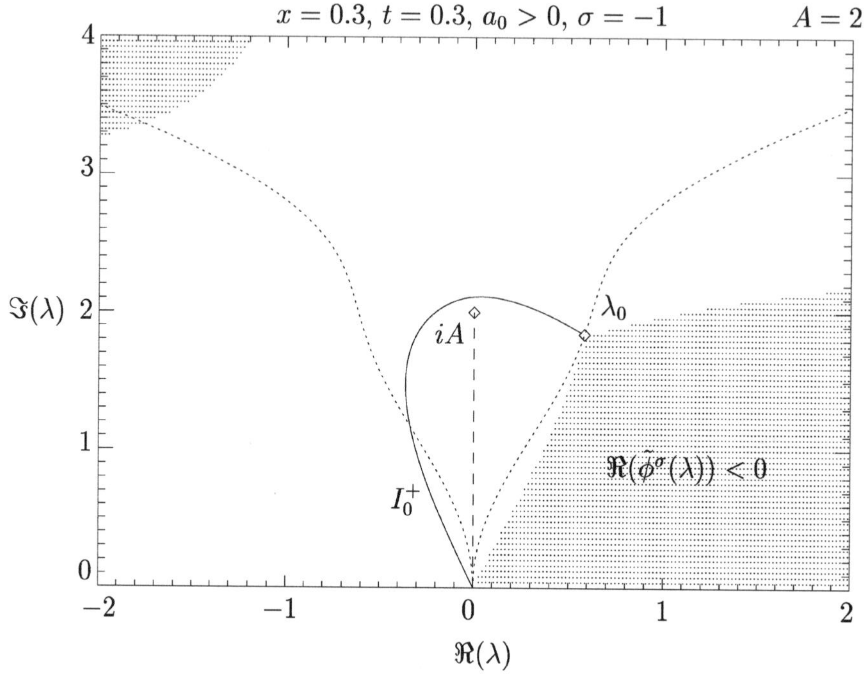

Figure 6.9 The inequalities can also be satisfied for this case.

Remark. These numerical experiments indicate that for the Satsuma-Yajima soliton ensemble, the failure of the genus-zero ansatz can be essentially captured by the two conditions (5.96) and (5.97). Indeed, if we examine the boundary of region I, we see from table 6.1 that whenever for a particular choice of σ and the sign of a_0 a band fails to exist upon crossing this boundary, it is due to a transition from "connection" to "left/right deflection." In this sense, the "intersection" scenario plays no real role in the breakdown of the ansatz. The transition from "connection" to "deflection" corresponds to the passage of a zero of the candidate density $\rho^\sigma(\eta)$ through the band I_0^+, and therefore such a transition point $(x_{\text{crit}}, t_{\text{crit}})$ satisfies the conditions (5.96). This phenomenon is illustrated in figure 6.10; since from the relations (4.32) and (4.33) the curves satisfying $\Re(\tilde{\phi}^\sigma(\lambda)) = 0$ are also orbits of the differential equation (5.56), the boundary of the shaded region in the second quadrant is another orbit of (5.56). The mechanism of failure of the band to exist at $(x_{\text{crit}}, t_{\text{crit}})$ is therefore the meeting of these two orbits at a mutual analytic fixed point. Similarly, at each point $(x_{\text{crit}}, t_{\text{crit}})$ on the boundary of region I, where for a particular choice of σ and the sign of a_0 a band I_0^+ exists on both sides of the boundary, the failure of a gap to exist is brought on by the pinching off at some point in the cut upper half-plane $\mathbb{H}$ of the region where $\Re(\tilde{\phi}^\sigma(\lambda)) < 0$. This phenomenon is clearly illustrated in, for example, figures 6.11 and 6.12, or figures 6.13 and 6.14. For such a transition, the boundary point $(x_{\text{crit}}, t_{\text{crit}})$ satisfies the gap failure criterion (5.97).

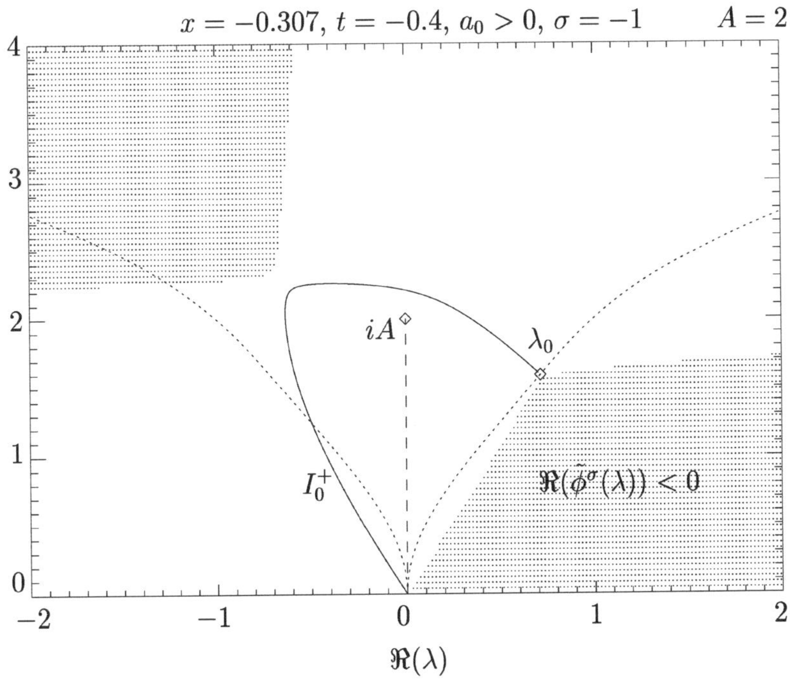

Figure 6.10 A case of the successful "wraparound" genus-zero ansatz very close to the boundary between region II_{--} and region I_{--}.

Although we have verified the existence of a complex phase function corresponding to the genus-zero ansatz all the way to the boundary of region I by verifying all inequalities numerically (i.e., using the data leading to pictures of the sort that have been presented in this section for a fine grid of values of x and t), we do not have at this point a direct method to verify these inequalities analytically. We are, however, compelled to state the following.

THEOREM 6.3.2 *The conclusion of Corollary* 6.2.1, *giving the rigorous semiclassical limit of the initial-value problem* (1.1) *for the focusing nonlinear Schrödinger equation with the Satsuma-Yajima initial data* $\psi_0(x) = A\,\mathrm{sech}(x)$, *extends from* $t = 0$ *all the way to the boundary of region* I. *This boundary is explicitly characterized by the gap failure criterion* (5.97).

An analytical proof of this theorem awaits the development of tools that generalize the "integration in x" methods that have proven so useful for real-line problems like the zero-dispersion limit of the Korteweg–de Vries equation [LL83] and the continuum limit of the Toda lattice [DM98] to the complex plane. Formal WKB theory [M01] may be of some assistance in this connection, as it identifies certain significant paths in the complex x-plane that may play the role usually played simply by the real x-axis.

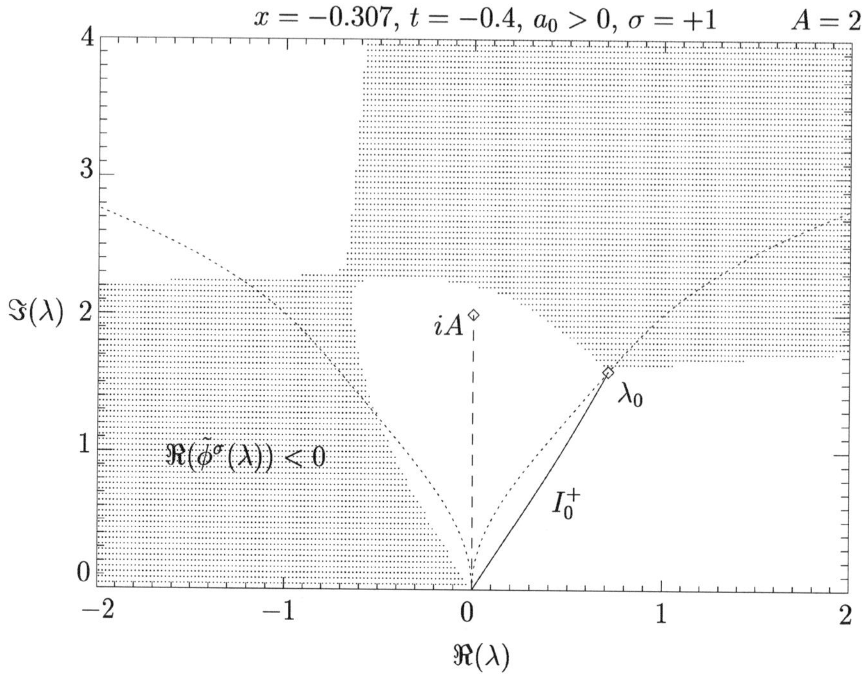

Figure 6.11 A case of a barely successful genus-zero ansatz. The values of x and t are the same as in figure 6.10.

Remark. Note that while the genus-zero ansatz breaks down at the boundary of region I, the endpoint $\lambda_0(x, t)$ is analytic in x and t at the boundary and indeed continues into region I. As is shown in §6.4, the only singular point of $\lambda_0(x, t)$ in common with the boundary of region I is for $x = 0$.

Remark. The numerics suggest that the point $x = 0$ is not as singular as one might expect from the small-time analysis. Indeed, the ansatz for x small and positive appears to smoothly continue around $x = 0$ to negative values of x on any path with $t \neq 0$. In fact, for $t \neq 0$ and $x = 0$, the endpoint lies on the imaginary axis *above* iA, and the genus-zero ansatz appears to be successful for both $\sigma = +1$ and $\sigma = -1$, although in neither case does the band I_0^+ coincide with the imaginary axis. For $x = 0$ the two workable ansätze are mutually dual and are mapped into each other by reflection through the imaginary axis. Thus, in the regions $\mathrm{IV}_{++} \cup \mathrm{IV}_{+-}$ and $\mathrm{IV}_{--} \cup \mathrm{IV}_{-+}$, *both* workable ansätze are continuous at $x = 0$ for $t \neq 0$ in that as the endpoint crosses the imaginary axis, the ansatz for which the band I_0^+ lies entirely in one quadrant continuously becomes an ansatz for which the band I_0^+ must "wrap around" the point iA in order to meet the endpoint $\lambda_0(x, t)$.

The numerical results on the t-axis are consistent with the exact solution of the endpoint equations in terms of the elliptic integral $E(m)$ given in Theorem 6.3.1.

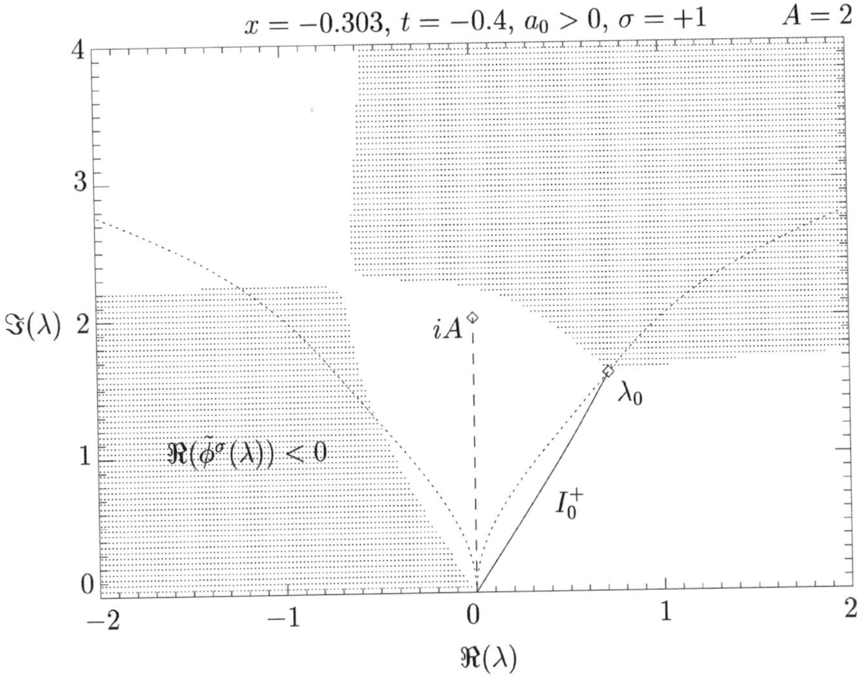

Figure 6.12 A case of a barely unsuccessful genus-zero ansatz. The coordinates x and t lie just on the other side of the boundary between the regions I_{--} and II_{--} in the (x, t)-plane (see figure 6.5) from coordinates used in figures 6.10 and 6.11.

6.4 THE ELLIPTIC MODULATION EQUATIONS AND THE PARTICULAR SOLUTION OF AKHMANOV, SUKHORUKOV, AND KHOKHLOV FOR THE SATSUMA-YAJIMA INITIAL DATA

Recall that in §5.3 it was shown that whenever the endpoints $\lambda_0(x,t), \ldots, \lambda_G(x,t)$ are obtained for a genus-G ansatz as the solution of the three sets of conditions (5.25), (5.52), and (5.58), then they satisfy a first-order quasilinear coupled system of partial differential equations (the Whitham or modulation equations) in x and t and that each endpoint and its complex conjugate is a Riemann invariant of this quasilinear system. In particular, for genus zero, we found that (cf. (5.160))

$$\frac{\partial \lambda_0}{\partial t} + (-2a_0 - ib_0)\frac{\partial \lambda_0}{\partial x} = 0, \qquad \frac{\partial \lambda_0^*}{\partial t} + (-2a_0 + ib_0)\frac{\partial \lambda_0^*}{\partial x} = 0, \tag{6.71}$$

where $a_0 := \Re(\lambda_0)$ and $b_0 := \Im(\lambda_0)$. These equations can be given a more direct physical interpretation in view of the semiclassical solution of the nonlinear Schrödinger equation given by (cf. (4.108))

$$\psi \sim \Im(\lambda_0)e^{-i\alpha_0/\hbar} \tag{6.72}$$

and proved in §4.5 to be uniformly valid in any compact set in the (x, t)-plane where the genus-zero ansatz is valid with all inequalities being strictly satisfied. For fixed

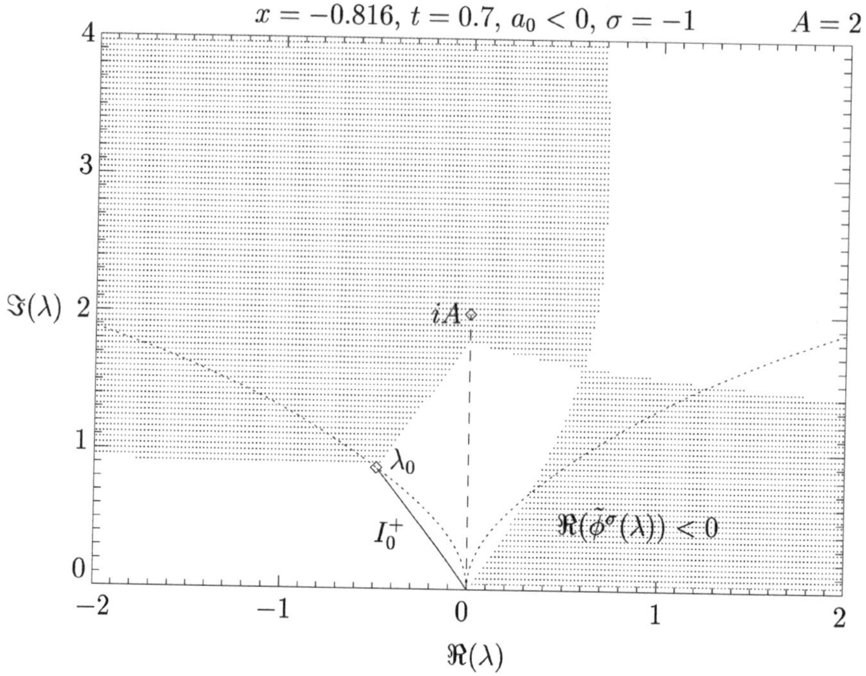

Figure 6.13 A barely successful ansatz. The coordinates x and t are chosen to be within region III_{+-}, but very close to the boundary with region I_{+-}. See figure 6.5.

t and for $x = x_0 + \hbar\hat{x}$, this approximate solution is a modulated plane wave of the form (up to a phase depending on x_0)

$$\psi \sim \sqrt{\rho}e^{ik\hat{x}}, \tag{6.73}$$

where the "fluid density" (not to be confused with the density function $\rho^\sigma(\eta)$ for the complex phase) is defined by $\rho := \Im(\lambda_0)^2$ and the wavenumber is defined by $k := -\partial_x \alpha_0$.

With the help of the formula (6.36) expressing k in terms of the λ_0, we can rewrite the modulation equations in terms of the fluid density ρ and momentum $\mu := \rho k$. Since $\rho = b_0^2$ and $\mu = -2a_0 b_0^2$, we find immediately from (5.160) that

$$\begin{aligned} \frac{\partial \rho}{\partial t} + \frac{\partial \mu}{\partial x} &= 0, \\ \frac{\partial \mu}{\partial t} + \frac{\partial}{\partial x}\left(\frac{\mu^2}{\rho} - \frac{\rho^2}{2}\right) &= 0. \end{aligned} \tag{6.74}$$

This elliptic quasilinear system has been known in this form for some time in connection with the formal semiclassical theory of nonlinear Schrödinger equations. In 1966, an exact solution of (6.74) was obtained in implicit form by Akhmanov, Sukhorukov, and Khokhlov [ASK66] through an application of the hodograph transform method. See also Whitham [W74] for a discussion of their result. They defined

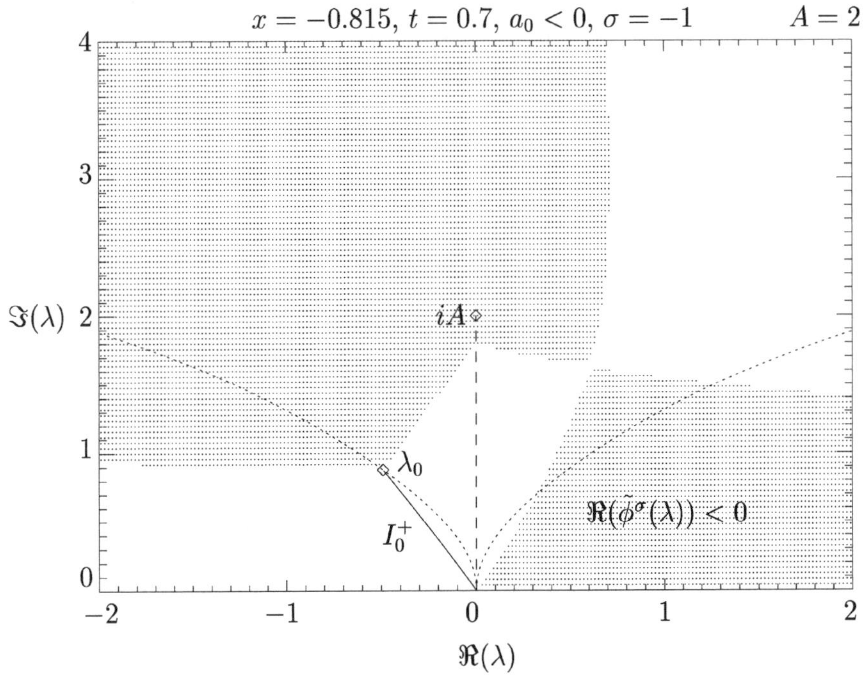

Figure 6.14 A barely unsuccessful ansatz. The coordinates x and t are chosen to be within region I_{+-} but very close to the boundary with region III_{+-}. See figure 6.5.

$\rho(x, t)$ and $\mu(x, t)$ as a branch of the solution of the relations

$$\begin{aligned} \mu &= -2t\rho^2 \tanh\left(\frac{\rho x - \mu t}{\rho}\right), \\ \rho &= (A^2 + t^2\rho^2)\,\mathrm{sech}^2\left(\frac{\rho x - \mu t}{\rho}\right). \end{aligned} \tag{6.75}$$

It is easy to see by setting $t = 0$ that the solution obtained in [ASK66] satisfies the initial condition

$$\rho(x, 0) = A^2\,\mathrm{sech}^2(x), \qquad \mu(x, 0) \equiv 0. \tag{6.76}$$

Clearly, the functions $\rho(x, t)$ and $\mu(x, t)$ obtained from the endpoints $\lambda_0(x, t)$ and $\lambda_0^*(x, t)$ also satisfy these initial conditions. We therefore see that our analysis both reproduces and makes rigorous a formal result that has been in the literature for more than thirty years.

It is instructive to study the implicit relations (6.75) for $x = 0$. In this case, it is easily seen that there is one solution for which $\mu \equiv 0$ as a function of t, and then from the second equation one finds

$$\rho(x = 0, t) = \frac{1 \pm \sqrt{1 - 4A^2t^2}}{2t^2}. \tag{6.77}$$

Table 6.2 Success or failure of the genus-zero ansatz in the regions of the (x, t)-plane in which there exist compactly supported candidate measures of the appropriate sign. See figure 6.5.

	$\sigma = +1$, $a_0 > 0$	$\sigma = -1$, $a_0 > 0$	$\sigma = +1$, $a_0 < 0$	$\sigma = -1$, $a_0 < 0$
I_{++} and I_{--}	inequality violated			inequality violated
II_{++} and II_{--}	gap exists	gap exists		inequality violated
III_{++} and III_{--}	gap exists			inequality violated
IV_{++} and IV_{--}	gap exists	gap exists		
V_{++} and V_{--}	gap exists			
I_{+-} and I_{-+}	inequality violated			inequality violated
II_{+-} and II_{-+}	inequality violated		gap exists	gap exists
III_{+-} and III_{-+}	inequality violated			gap exists
IV_{+-} and IV_{-+}			gap exists	gap exists
V_{+-} and V_{-+}				gap exists

Note: The blank entries in the table correspond to ansätze for which there is no band I_0^+ connecting zero to the endpoint on which the candidate measure $\rho^\sigma(\eta)\, d\eta$ is negative real.

Thus, a branch point singularity develops for $x = 0$ when $t = 1/(2A)$. This singularity corresponds exactly to the intersection of the boundary of region I with the line $x = 0$. The *first* point on the boundary of the region where the genus-zero ansatz holds therefore corresponds to a singularity of the endpoint function $\lambda_0(x, t)$, although as previously remarked the remaining boundary points are no obstruction to the analytic continuation of the function $\lambda_0(x, t)$.

The same formula (6.77) can of course also be obtained from the general formula (6.47) by using $A(x) = A\,\mathrm{sech}(x)$ and setting $\rho(x = 0, t) = b_0(t)^2$.

Chapter Seven

The Transition to Genus Two

Recall that in §6.2.2 it was shown that for each fixed $x \neq 0$ there exists some choice of the parameters σ and J such that the $G = 0$ ansatz holds for $|t|$ sufficiently small. Furthermore, it was shown in §5.2 that if the pair (x_0, t_0) is such that the $G = 0$ ansatz holds and the endpoint functions are differentiable, then there is a small neighborhood of (x_0, t_0) on which the $G = 0$ ansatz holds as well, and this allows us to define a region of the (x, t)-plane containing (x_0, t_0) throughout which the $G = 0$ ansatz satisfies all inequalities necessary for the asymptotic analysis of chapter 4 to be valid.

A point $(x_{\mathrm{crit}}, t_{\mathrm{crit}})$ on the boundary of this region of validity is characterized by one or more of the following six critical events:

1. The endpoint function $\lambda_0(x, t)$ fails to be analytic in x and t at $(x_{\mathrm{crit}}, t_{\mathrm{crit}})$.
2. The endpoint $\lambda_0(x_{\mathrm{crit}}, t_{\mathrm{crit}})$ lies on the boundary of the cut upper half-plane $\mathbb{H}$.
3. The band I_0^+ is a smooth orbit of (5.56) connecting $\lambda = \sigma 0$ to the endpoint $\lambda_0(x_{\mathrm{crit}}, t_{\mathrm{crit}})$, but either I_0^+ has a point of tangency with the boundary of $\mathbb{H}$ on the real axis or the imaginary interval $[0, iA)$ or the point $\lambda = iA$ lies on I_0^+.
4. There is a connected region in the upper half-plane with $\lambda_0(x_{\mathrm{crit}}, t_{\mathrm{crit}})$ and $\lambda = -\sigma 0$ on the boundary where $\Re(\tilde{\phi}^\sigma(\lambda)) < 0$, but $\lambda = iA$ is also on the boundary and the region is bisected by the segment $[0, iA]$.
5. The band I_0^+ passes through an analytic fixed point $\hat{\lambda}$ of the vector field (5.56) on the way to $\lambda_0(x_{\mathrm{crit}}, t_{\mathrm{crit}})$ and thus is not smooth at this point, making an angle of 90 degrees (for a simple zero of $\rho^\sigma(\lambda)$). Strictly speaking, I_0^+ is a union of three orbits of (5.56): two regular orbits and a fixed point. In a degenerate situation, the zero $\hat{\lambda}$ may coincide with an endpoint of the band.
6. The closed region where $\Re(\tilde{\phi}^\sigma(\lambda)) \leq 0$ holds admits a gap contour Γ_1^+ in $\mathbb{H}$ connecting $\lambda_0(x_{\mathrm{crit}}, t_{\mathrm{crit}})$ to $\lambda = -\sigma 0$ but is pinched off at a point $\hat{\lambda}$ through which Γ_1^+ must pass and at which $\Re(\tilde{\phi}^\sigma(\hat{\lambda})) = 0$.

The last two of these critical events may be characterized by the equations (5.96) and (5.97). The computer-assisted analysis of the genus-zero ansatz for the Satsuma-Yajima ensemble carried out in §6.3 indicated that in that particular case the ansatz parameters may be chosen at $(x_{\mathrm{crit}}, t_{\mathrm{crit}})$ such that the failure is indeed due either to the conditions (5.96) or the conditions (5.97). In fact, it is clear from table 6.2 that for particular choices of the ansatz parameters it is sufficient to characterize the boundary of the region where the genus-zero ansatz holds by the gap failure condition (5.97). Also, as pointed out, the band failure condition (5.96) and the gap failure condition (5.97) are essentially equivalent according to relations (4.32) and (4.33); they generate the same set of points in the (x, t)-plane.

In this chapter we assume that $(x_{\text{crit}}, t_{\text{crit}})$ is such that the sixth condition listed above holds, which implies that the gap failure conditions (5.97) hold for $\lambda = \hat{\lambda} \in \mathbb{H}$. We call this situation a *critical genus-zero ansatz*. The problem at hand then is to describe what happens as (x, t) leaves the connected region on which the genus-zero ansatz holds, moving away from $(x_{\text{crit}}, t_{\text{crit}})$. We want to establish that for some genus $G \neq 0$ the ansatz (hopefully with the same parameters J and σ) is valid just beyond the boundary. Following examples from the integrable systems literature (see, for example, [LL83]) and the approximation theory literature (see [DKM98]), we suppose that as (x, t) leaves the connected component of the (x, t)-plane where the genus-zero ansatz holds, the genus jumps from $G = 0$ to $G = 2$.

Remark. The reason for supposing that the genus "skips" the value $G = 1$ is essentially connected to the complex-conjugation symmetry of the contour $C \cup C^*$. Indeed what is expected is that the critical point $\hat{\lambda}$ will open up into a pair of endpoints of a new band. By symmetry, the same thing will happen at the conjugate point $\hat{\lambda}^*$, so the number of bands (and hence the genus) increases by two.

7.1 MATCHING THE CRITICAL $G = 0$ ANSATZ WITH A DEGENERATE $G = 2$ ANSATZ

By a degenerate $G = 2$ ansatz, we simply mean one for which two of the three complex endpoints, say, λ_1 and λ_2, are equal. In general, the three complex endpoints λ_0, λ_1, and λ_2 must satisfy the four real equations $M_p = 0$ for $p = 0, \ldots, 3$, along with the vanishing condition $V_0 = 0$ and the reality condition $R_1 = 0$. Here we are exchanging the reality condition $R_0 = 0$ for the additional moment condition $M_3 = 0$ as described in §5.3. As pointed out in §5.4, the moments are analytic and completely symmetric functions of the endpoints, and are also analytic in x and t. We begin our analysis of the feasibility of a degenerate genus-$G = 2$ ansatz by evaluating the moments on a degenerate set of endpoints satisfying $\lambda_0 = \lambda_0^{\text{crit}}$ and $\lambda_1 = \lambda_2 = \hat{\lambda}$. Let $M_p^{(0)}(x, t, \lambda_0^{\text{crit}})$ for $p = 0, 1$ denote the moment functions for the $G = 0$ ansatz. Then we have the following result.

Lemma 7.1.1 *Evaluating the genus-$G = 2$ moments on a degenerate set of endpoints yields the following four relations:*

$$M_3(x, t, \lambda_0^{\text{crit}}, \hat{\lambda}, \hat{\lambda}) - 2\Re(\hat{\lambda}) M_2(x, t, \lambda_0^{\text{crit}}, \hat{\lambda}, \hat{\lambda}) + |\hat{\lambda}|^2 M_1(x, t, \lambda_0^{\text{crit}}, \hat{\lambda}, \hat{\lambda}) = M_1^{(0)}(x, t, \lambda_0^{\text{crit}}), \tag{7.1}$$

$$M_2(x, t, \lambda_0^{\text{crit}}, \hat{\lambda}, \hat{\lambda}) - 2\Re(\hat{\lambda}) M_1(x, t, \lambda_0^{\text{crit}}, \hat{\lambda}, \hat{\lambda}) + |\hat{\lambda}|^2 M_0(x, t, \lambda_0^{\text{crit}}, \hat{\lambda}, \hat{\lambda}) = M_0^{(0)}(x, t, \lambda_0^{\text{crit}}), \tag{7.2}$$

$$M_1(x, t, \lambda_0^{\text{crit}}, \hat{\lambda}, \hat{\lambda}) - \hat{\lambda}^* M_0(x, t, \lambda_0^{\text{crit}}, \hat{\lambda}, \hat{\lambda}) = \pi^2 Y^{(0)}(\hat{\lambda}), \tag{7.3}$$

$$M_1(x, t, \lambda_0^{\text{crit}}, \hat{\lambda}, \hat{\lambda}) - \hat{\lambda} M_0(x, t, \lambda_0^{\text{crit}}, \hat{\lambda}, \hat{\lambda}) = \pi^2 Y^{(0)}(\hat{\lambda}^*), \tag{7.4}$$

where $Y^{(0)}(\lambda)$ is the function given by (5.41) *for genus-$G = 0$ in terms of the single complex endpoint λ_0^{crit}.*

Proof. The evaluation of the moments on the degenerate solution is completely straightforward using the formula (5.162) that makes clear the analytic dependence of the moments on the endpoints. Thus, for any $p \geq 0$, we find

$$\begin{aligned} M_{p+2} - 2\Re(\hat{\lambda}) M_{p+1} + |\hat{\lambda}|^2 M_p \\ &= -\frac{J}{2}\oint_L \frac{2ix + 4i\eta t}{R(\eta)}\eta^p(\eta - \hat{\lambda})(\eta - \hat{\lambda}^*)\,d\eta \\ &\quad + \frac{1}{2}\int_{C_{I+}\cup C_{I-}} \frac{\pi i \rho^0(\eta)}{R(\eta)}\eta^p(\eta - \hat{\lambda})(\eta - \hat{\lambda}^*)\,d\eta \\ &\quad + \frac{1}{2}\int_{C_{I+}^*\cup C_{I-}^*} \frac{\pi i \rho^0(\eta^*)^*}{R(\eta)}\eta^p(\eta - \hat{\lambda})(\eta - \hat{\lambda}^*)\,d\eta. \end{aligned} \tag{7.5}$$

Since for the degenerate set of endpoints $R(\eta) = (\eta - \hat{\lambda})(\eta - \hat{\lambda})R^{(0)}(\eta)$, where $R^{(0)}(\eta)$ is the square root function for genus-$G = 0$ corresponding to the single complex endpoint $\lambda_0^{\mathrm{crit}}$, we can again use the formula (5.162) to identify the right-hand side as exactly $M_p^{(0)}(x, t, \lambda_0^{\mathrm{crit}})$, which in particular proves (7.1) and (7.2). Similarly, using the formula (5.162) for the moments along with the degenerate form of the square root function for $\lambda_1 = \lambda_2 = \hat{\lambda}$ as described above, we find

$$\begin{aligned} M_1(x, t, \lambda_0^{\mathrm{crit}}, \hat{\lambda}, \hat{\lambda}) - \hat{\lambda}^* M_0(x, t, \lambda_0^{\mathrm{crit}}, \hat{\lambda}, \hat{\lambda}) \\ &= -\frac{J}{2}\oint_L \frac{2ix + 4i\eta t}{(\eta - \hat{\lambda})R^{(0)}(\hat{\lambda})}\,d\eta \\ &\quad + \frac{1}{2}\int_{C_{I+}\cup C_{I-}} \frac{\pi i \rho^0(\eta)}{(\eta - \hat{\lambda})R^{(0)}(\eta)}\,d\eta \\ &\quad + \frac{1}{2}\int_{C_{I+}^*\cup C_{I-}^*} \frac{\pi i \rho^0(\eta^*)^*}{(\eta - \hat{\lambda})R^{(0)}(\eta)}\,d\eta. \end{aligned} \tag{7.6}$$

Evaluating the first integral exactly by residues and identifying the result with the formula (5.41) proves (7.3). Finally, (7.4) is proved either by repeating the above arguments or by simply noting the symmetry $Y^{(0)}(\lambda^*) = Y^{(0)}(\lambda)^*$ and using the reality of the moments. □

Corollary 7.1.1 *Suppose that for some x and t the endpoint $\lambda_0^{\mathrm{crit}}$ satisfies the genus-$G = 0$ equations $M_0^{(0)} = 0$ and $M_1^{(0)} = 0$ and that $\hat{\lambda} \neq \lambda_0^{\mathrm{crit}}$ is chosen so that $\Im(\hat{\lambda}) \neq 0$ and*

$$\frac{d\tilde{\phi}^{(0)\sigma}}{d\lambda}(\hat{\lambda}) = 0, \tag{7.7}$$

where by $\tilde{\phi}^{(0)\sigma}(\lambda)$ we mean the function $\tilde{\phi}^{\sigma}(\lambda)$ constructed for genus-$G = 0$ using the endpoint $\lambda_0^{\mathrm{crit}}$. Then, the genus-$G = 2$ moment equations $M_p(x, t, \lambda_0^{\mathrm{crit}}, \hat{\lambda}, \hat{\lambda}) = 0$ are satisfied for $p = 0, 1, 2, 3$.

Proof. As described in §5.4, it follows from (4.32) and the representation (5.35) of $\rho^{\sigma}(\lambda)$ that

$$\frac{d\tilde{\phi}^{(0)\sigma}}{d\lambda}(\lambda) = i\pi R^{(0)}(\lambda)Y^{(0)}(\lambda). \tag{7.8}$$

Since $\hat{\lambda} \neq \lambda_0^{\rm crit}$, we have $R^{(0)}(\hat{\lambda}) \neq 0$, and consequently the right-hand sides of (7.3) and (7.4) vanish. Since $M_0^{(0)} = M_1^{(0)} = 0$, the right-hand sides of (7.1) and (7.2) vanish as well. The determinant of the left-hand sides of these relations is $-(\hat{\lambda} - \hat{\lambda}^*) \neq 0$, and the corollary is proved. □

In a degenerate situation, one can select a contour C_I passing through the point $\hat{\lambda}$, and one might consider solving the scalar boundary value problem for *both* genera, $G = 0$ and $G = 2$, as described in §5.1.2, defining two different functions analytic in $\mathbb{C} \setminus (C_I \cup C_I^*)$. Under the conditions of Corollary 7.1.1, it follows from Lemma 5.1.2 that the scalar boundary value problems for $G = 0$ and $G = 2$ both have unique solutions. Let $F^{(0)}(\lambda)$ denote the unique solution of the scalar boundary value problem for $G = 0$ corresponding to the endpoint $\lambda_0^{\rm crit}$ and the contour C_I, and let $F_{\rm deg}^{(2)}(\lambda)$ denote the unique solution of the scalar boundary value problem for $G = 2$ corresponding to the endpoints $\lambda_0^{\rm crit}$, $\lambda_1 = \hat{\lambda}$, and $\lambda_2 = \hat{\lambda}$ and the same contour C_I. These two functions are given by apparently different explicit formulae:

$$F^{(0)}(\lambda) = \frac{R^{(0)}(\lambda)}{\pi i}\left[\int_{I_0^+ \cup I_0^-} \frac{J(2ix + 4i\eta t)}{(\lambda - \eta)R_+^{(0)}(\eta)}\,d\eta + \int_{\Gamma_I \cap C_I} \frac{i\pi\rho^0(\eta)}{(\lambda - \eta)R^{(0)}(\eta)}\,d\eta + \int_{\Gamma_I \cap C_I^*} \frac{i\pi\rho^0(\eta^*)^*}{(\lambda - \eta)R^{(0)}(\eta)}\,d\eta\right], \tag{7.9}$$

$$\begin{aligned} F_{\rm deg}^{(2)}(\lambda) = {} & \frac{R^{(0)}(\lambda)}{\pi i}(\lambda - \hat{\lambda})(\lambda - \hat{\lambda}^*) \\ & \times \left[\int_{I_0^+ \cup I_0^-} \frac{J(2ix + 4i\eta t)}{(\lambda - \eta)(\eta - \hat{\lambda})(\eta - \hat{\lambda}^*)R_+^{(0)}(\eta)}\,d\eta \right. \\ & + \text{P.V.}\int_{\Gamma_I \cap C_I} \frac{i\pi\rho^0(\eta)}{(\lambda - \eta)(\eta - \hat{\lambda})(\eta - \hat{\lambda}^*)R^{(0)}(\eta)}\,d\eta \\ & \left. + \text{P.V.}\int_{\Gamma_I \cap C_I^*} \frac{i\pi\rho^0(\eta^*)^*}{(\lambda - \eta)(\eta - \hat{\lambda})(\eta - \hat{\lambda}^*)R^{(0)}(\eta)}\,d\eta\right]. \end{aligned} \tag{7.10}$$

In both of these formulae, the paths $\Gamma_I \cap C_I$ and $\Gamma_I \cap C_I^*$ pass, respectively, through $\eta = \hat{\lambda}$ and $\eta = \hat{\lambda}^*$. To verify (7.10), one uses the degeneration of the square root function $R(\eta) = (\eta - \hat{\lambda})(\eta - \hat{\lambda}^*)R^{(0)}(\eta)$ and notes that in the general genus-$G = 2$ expression for the function $F(\lambda)$ given by $F(\lambda) = H(\lambda)R(\lambda)$ with $H(\lambda)$ given by (5.29) for $G = 2$, the paths of integration $\Gamma_I \cap C_I$ and $\Gamma_I \cap C_I^*$ may be replaced, respectively, by $(C_{I+} \cup C_{I-})/2$ and its conjugate path, where $C_{I\pm}$ are taken to lie closer to C_I than λ. This version of the formula allows one to evaluate $F(\lambda)$ when $\lambda_1 = \lambda_2 = \hat{\lambda}$, after which the contours may be collapsed to C_I and C_I^* resulting in the principal value interpretation of the singular integrals via the Plemelj formula. In this sense, the formula (7.10) is the limit of the nondegenerate genus-$G = 2$ formula as the endpoints λ_1 and λ_2 coalesce at $\hat{\lambda}$. Despite appearances, the differences in these two formulae are superficial. We have the following.

LEMMA 7.1.2 *Assume the conditions of Corollary* 7.1.1. *Then,*

$$F^{(0)}(\lambda) \equiv F^{(2)}_{\text{deg}}(\lambda) \tag{7.11}$$

for all $\lambda \in \mathbb{C} \setminus (C_I \cup C_I^*)$.

Proof. This essentially follows from the uniqueness result for the scalar boundary value problem for $G = 0$ described in Lemma 5.1.1. Indeed, both functions satisfy the same decay conditions at infinity and by the explicit formulas are analytic in $\mathbb{C}\setminus(C_I \cup C_I^*)$ with boundary values that are Hölder-continuous with exponent $1/2$. The fact that both functions satisfy the same boundary conditions almost everywhere on $C_I \cup C_I^*$ follows from taking the limit $\lambda_1 \to \hat{\lambda}$ and $\lambda_2 \to \hat{\lambda}$ in the scalar boundary value problem for genus-$G = 2$. Therefore, both functions satisfy the scalar boundary value problem for $G = 0$ and are equal by Lemma 5.1.1. □

Remark. The arguments in Lemma 7.1.2 make no explicit reference to the genus. Thus the argument actually shows that whenever a band closes up, then the function $F(\lambda)$ reduces to the solution of the scalar boundary value problem for genus-$G - 2$.

With the help of this result, we may now study the functions V_0 and R_1 on a degenerate endpoint configuration for genus-$G = 2$.

LEMMA 7.1.3 *Assume the conditions of Lemma* 7.1.2. *The reality condition* $R_1 = 0$ *is automatically satisfied by any degenerate genus-*$G = 2$ *configuration with* $\lambda_1 = \lambda_2 = \hat{\lambda}$.

Proof. From Lemma 7.1.2, the candidate density function $\rho^\sigma(\eta)$ obtained for the degenerate genus-$G = 2$ configuration from $F^{(2)}_{\text{deg}}(\lambda)$ agrees with the nondegenerate genus-$G = 0$ candidate density function. Since this function is Hölder-continuous, it is bounded. This implies that in the degenerate limit, $\rho^\sigma(\eta)$ remains uniformly bounded on I_1^+, the path of integration from λ_1 to λ_2. Since λ_1 and λ_2 both converge to $\hat{\lambda}$ in the degenerate limit, the result follows from the definition of the function R_1 by the formula (5.58) as an integral along the shrinking band I_1^+. □

LEMMA 7.1.4 *Assume the conditions of Lemma* 7.1.2. *Then for any degenerate genus-*$G = 2$ *configuration with* $\lambda_1 = \lambda_2 = \hat{\lambda}$, *we have*

$$V_0 = \Re(\tilde{\phi}^{(0)\sigma}(\hat{\lambda})), \tag{7.12}$$

where $\tilde{\phi}^{(0)\sigma}(\lambda)$ *corresponds to the* $G = 0$ *candidate density function with the single complex endpoint* λ_0^{crit}.

Proof. By the definition (5.52), V_0 is the real part of the integral of $d\tilde{\phi}^\sigma/d\lambda$ from λ_0 to λ_1, where the analytic function $\tilde{\phi}^\sigma(\lambda)$ corresponds to the endpoints λ_0, λ_1, and λ_2. By construction, $\Re(\tilde{\phi}^\sigma(\lambda_0)) = 0$, so equivalently we have $V_0 = \Re(\tilde{\phi}^\sigma(\lambda_1))$. According to Lemma 7.1.2, the $G = 2$ function $\tilde{\phi}^\sigma(\lambda)$ agrees with $\tilde{\phi}^{(0)\sigma}(\lambda)$ when $\lambda_1 = \lambda_2 = \hat{\lambda}$ since they are both derived from the same unique solution of the scalar boundary value for $G = 0$ with $\lambda_0 = \lambda_0^{\text{crit}}$. Evaluating for $\lambda_1 = \hat{\lambda}$ completes the proof. □

Remark. Both of the results contained in Lemma 7.1.3 and Lemma 7.1.4 are strengthened when we provide more detailed asymptotics near the degenerate configuration.

Combining the results of this section, we have proved the following.

THEOREM 7.1.1 *Suppose that* $x = x_{\mathrm{crit}}$, $t = t_{\mathrm{crit}}$, *and* $\lambda_0^{(0)}$ *and* $\hat{\lambda}$ *are such that the two* $G = 0$ *moment conditions* $M_0^{(0)} = 0$ *and* $M_1^{(0)} = 0$ *hold and the three real conditions contained in* (5.97) *hold true at* $\lambda = \hat{\lambda}$. *Then, holding* x_{crit} *and* t_{crit} *fixed, the complex endpoint configuration* $\lambda_0 = \lambda_0^{(0)}$, $\lambda_1 = \hat{\lambda}$, *and* $\lambda_2 = \hat{\lambda}$ *represents a degenerate solution of the equations* $M_p = 0$ *for* $p = 0, \ldots, 3$ *along with* $R_1 = 0$ *and* $V_0 = 0$. *Moreover, if we suppose that the genus-*$G = 0$ *ansatz is successful in the sense that the band* I_0^+ *exists connecting the origin to* $\lambda_0^{\mathrm{crit}}$ *on which the differential* $\rho^\sigma(\eta)\, d\eta$ *is negative real and the corresponding gap contour exists passing through* $\hat{\lambda}$ *such that* $\Re(\tilde{\phi}^{(0)\sigma}) \leq 0$ *with inequality being strict except at the endpoints and at* $\hat{\lambda}$, *then the degenerate genus-*$G = 2$ *ansatz is successful in the same sense, with exactly the same contour.*

Proof. The fact that the degenerate triple of endpoints satisfies the $G = 2$ endpoint equations follows from the chain of results already presented in this section. It remains to verify the final claim: that the inequalities persist under reinterpretation of the $G = 0$ configuration subject to the additional conditions (5.97) at $\lambda = \hat{\lambda}$ as a degenerate $G = 2$ configuration. But this follows from Lemma 7.1.2, which implies that the functions ρ^σ and $\tilde{\phi}^\sigma$ are exactly the same in both cases. □

7.2 PERTURBING THE DEGENERATE $G = 2$ ANSATZ. OPENING THE BAND I_1^+ BY VARYING x NEAR x_{crit}

Let $t = t_{\mathrm{crit}}$ be fixed. We now want to consider the possibility that the $G = 2$ ansatz exists (i.e., the endpoint equations can be solved) for x near x_{crit} but in the region of the (x, t)-plane beyond the primary caustic where the inequalities fail for the $G = 0$ ansatz due to the pinching-off of the region where $\Re(\tilde{\phi}^{(0)\sigma}) < 0$ at the point $\hat{\lambda}$. Unfortunately, a direct application of the implicit function theorem fails to establish existence, because it can be shown using the explicit formula given in §5.3 that the corresponding Jacobian determinant vanishes when evaluated on the degenerate $G = 2$ solution. In a sense, this is not surprising since we know from §5.4 that the endpoints are only determined by the constraint equations up to permutation and consequently the double point $\lambda_1 = \lambda_2 = \hat{\lambda}$ cannot be unfolded uniquely. However, the difficulties also run deeper, with the appearance of logarithms in the perturbation expansion arising from the multivaluedness (monodromy) of the function V_0 described in §5.4.

In this section, we begin with the assumption of the existence of the degenerate $G = 2$ ansatz. Therefore, we assume that for $x = x_{\mathrm{crit}}$ and $t = t_{\mathrm{crit}}$ the single complex endpoint $\lambda_0^{\mathrm{crit}}$ satisfies the two real equations $M_0^{(0)} = M_1^{(0)} = 0$, and for some nonreal $\hat{\lambda} \neq \lambda_0^{\rm crit}$ in $\mathbb{H}$, we have

$$\frac{d\tilde{\phi}^{(0)\sigma}}{d\lambda}(\hat{\lambda}) = 0, \qquad \Re(\tilde{\phi}^{(0)\sigma}(\hat{\lambda})) = 0. \tag{7.13}$$

We further assume the following conditions to rule out higher-order degeneracy:

1. The $G = 0$ endpoint equations $M_0^{(0)} = 0$ and $M_1^{(0)} = 0$ can be solved for the single complex endpoint as a function of (x, t) in a neighborhood of $(x_{\rm crit}, t_{\rm crit})$.
2. The critical point is simple, so that

$$\frac{d^2\tilde{\phi}^{(0)\sigma}}{d\lambda^2}(\hat{\lambda}) \neq 0. \tag{7.14}$$

Under these conditions, we can reduce the size of the problem somewhat.

Lemma 7.2.1 *The four $G = 2$ moment conditions $M_0 = M_1 = M_2 = M_3 = 0$ can be solved for λ_0, λ_0^*, λ_2, and λ_2^* as analytic functions of x, λ_1, and λ_1^* in a complex neighborhood of the degenerate solution. The linear terms in the implicitly defined functions are*

$$\begin{aligned}
\lambda_0(x,\lambda_1,\lambda_1^*) &= \lambda_0^{\rm crit} + J\pi\left(\left.\frac{\partial M_0^{(0)}}{\partial\lambda_0}\right|_{\rm crit}\right)^{-1}(x - x_{\rm crit}) + \cdots,\\
\lambda_0^*(x,\lambda_1,\lambda_1^*) &= \lambda_0^{{\rm crit},*} + J\pi\left(\left.\frac{\partial M_0^{(0)}}{\partial\lambda_0^*}\right|_{\rm crit}\right)^{-1}(x - x_{\rm crit}) + \cdots,\\
\lambda_2(x,\lambda_1,\lambda_1^*) &= \hat{\lambda} - 2iJ\left(\frac{d^2\tilde{\phi}^{(0)\sigma}}{d\lambda^2}(\hat{\lambda})\right)^{-1}\\
&\quad\times\frac{\lambda_0^{\rm crit} + \lambda_0^{{\rm crit},*} - 2\hat{\lambda}}{R^{(0)}(\hat{\lambda})}(x - x_{\rm crit}) - (\lambda_1 - \hat{\lambda}) + \cdots,\\
\lambda_2^*(x,\lambda_1,\lambda_1^*) &= \hat{\lambda}^* + 2iJ\left(\frac{d^2\tilde{\phi}^{(0)\sigma}}{d\lambda^2}(\hat{\lambda})^*\right)^{-1}\\
&\quad\times\frac{\lambda_0^{\rm crit} + \lambda_0^{{\rm crit},*} - 2\hat{\lambda}^*}{R^{(0)}(\hat{\lambda})^*}(x - x_{\rm crit}) - (\lambda_1^* - \hat{\lambda}^*) + \cdots,
\end{aligned} \tag{7.15}$$

where the derivatives of the $G = 0$ moment $M_0^{(0)}$ are evaluated on the critical $G = 0$ ansatz. The above coefficients that do not vanish identically are finite and strictly nonzero by our assumptions.

Remark. Note that in (7.15) we have written down *all* of the linear terms. So, in particular, the dependence of λ_0 and λ_0^* on $\lambda_1 - \hat{\lambda}$ and $\lambda_1^* - \hat{\lambda}^*$ is higher order.

Proof. We begin the proof by computing the partial derivatives of the $G = 2$ moment M_0 with respect to the endpoints, as well as the partial derivatives of the first four moments with respect to x, and evaluating them on the degenerate solution.

First, from the formula (5.162), one finds that for the moment M_p corresponding to a general genus-G ansatz

$$\begin{aligned}\frac{\partial M_p}{\partial \lambda_k} &= -\frac{J}{4}\oint_L \frac{2ix+4i\eta t}{(\eta-\lambda_k)R(\eta)}\eta^p\,d\eta + \frac{1}{4}\int_{C_{I+}\cup C_{I-}} \frac{\pi i\rho^0(\eta)\eta^p\,d\eta}{(\eta-\lambda_k)R(\eta)}\\ &\quad + \frac{1}{4}\int_{C_{I+}^*\cup C_{I-}^*} \frac{\pi i\rho^0(\eta^*)^*\eta^p\,d\eta}{(\eta-\lambda_k)R(\eta)},\\ \frac{\partial M_p}{\partial \lambda_k^*} &= -\frac{J}{4}\oint_L \frac{2ix+4i\eta t}{(\eta-\lambda_k^*)R(\eta)}\eta^p\,d\eta + \frac{1}{4}\int_{C_{I+}\cup C_{I-}} \frac{\pi i\rho^0(\eta)\eta^p\,d\eta}{(\eta-\lambda_k^*)R(\eta)}\\ &\quad + \frac{1}{4}\int_{C_{I+}^*\cup C_{I-}^*} \frac{\pi i\rho^0(\eta^*)^*\eta^p\,d\eta}{(\eta-\lambda_k^*)R(\eta)}.\end{aligned} \tag{7.16}$$

Recall that the contours C_{I+} and C_{I-} are bounded away from all of the endpoints λ_k (see figure 5.2). From this and the relation $R(\eta) = (\eta-\hat{\lambda})(\eta-\hat{\lambda}^*)R^{(0)}(\eta)$ holding for the degenerate $G = 2$ configuration, by taking linear combinations we find

$$\begin{aligned}\left.\frac{\partial}{\partial\lambda_0}(M_2 - 2\Re(\hat{\lambda})M_1 + |\hat{\lambda}|^2 M_0)\right|_{\text{deg}} &= \left.\frac{\partial M_0^{(0)}}{\partial\lambda_0}\right|_{\text{crit}},\\ \left.\frac{\partial}{\partial\lambda_0^*}(M_2 - 2\Re(\hat{\lambda})M_1 + |\hat{\lambda}|^2 M_0)\right|_{\text{deg}} &= \left.\frac{\partial M_0^{(0)}}{\partial\lambda_0^*}\right|_{\text{crit}},\end{aligned} \tag{7.17}$$

where on the left-hand side the derivatives of genus-$G = 2$ moments are evaluated on the degenerate configuration and on the right-hand side the derivatives of $M_0^{(0)}$ are evaluated on the corresponding critical $G = 0$ configuration. Using the relations developed in §5.3 that allow derivatives of higher moments to be expressed in terms of derivatives of M_0, these relations imply

$$\begin{aligned}\left.\frac{\partial M_0}{\partial\lambda_0}\right|_{\text{deg}} &= \frac{1}{(\lambda_0^{\text{crit}}-\hat{\lambda})(\lambda_0^{\text{crit}}-\hat{\lambda}^*)}\left.\frac{\partial M_0^{(0)}}{\partial\lambda_0}\right|_{\text{crit}},\\ \left.\frac{\partial M_0}{\partial\lambda_0^*}\right|_{\text{deg}} &= \frac{1}{(\lambda_0^{\text{crit},*}-\hat{\lambda})(\lambda_0^{\text{crit},*}-\hat{\lambda}^*)}\left.\frac{\partial M_0^{(0)}}{\partial\lambda_0^*}\right|_{\text{crit}}.\end{aligned} \tag{7.18}$$

Second, from the permutation invariance of the moments (cf. §5.4), a chain rule calculation shows that

$$\left.\frac{\partial M_p}{\partial\lambda_1}\right|_{\text{deg}} = \left.\frac{\partial M_p}{\partial\lambda_2}\right|_{\text{deg}} = \frac{1}{2}\frac{\partial}{\partial\hat{\lambda}}\left(M_p\big|_{\text{deg}}\right), \tag{7.19}$$

where on the right-hand side M_p is first evaluated on the degenerate endpoint configuration and then differentiated with respect to $\hat{\lambda}$. Therefore, using Lemma 7.1.1, we find

$$\left.\left(\frac{\partial}{\partial\lambda_{1,2}}(M_1 - \hat{\lambda}^* M_0)\right)\right|_{\text{deg}} = \frac{\pi^2}{2}\frac{dY^{(0)}(\hat{\lambda})}{d\hat{\lambda}} = \frac{-i\pi}{2R^{(0)}(\hat{\lambda})}\frac{d^2\tilde{\phi}^{(0)\sigma}}{d\lambda^2}(\hat{\lambda}), \tag{7.20}$$

where we have simplified the result with the help of (7.8) and (7.13). At the same time, the left-hand side can be expressed in terms of derivatives of M_0 by the reasoning of §5.3, yielding

$$\left(\frac{\partial}{\partial \lambda_{1,2}}(M_1 - \hat{\lambda}^* M_0)\right)\bigg|_{\text{deg}} = (\hat{\lambda} - \hat{\lambda}^*)\frac{\partial M_0}{\partial \lambda_{1,2}}\bigg|_{\text{deg}}. \tag{7.21}$$

Putting these results together along with a similar calculation involving derivatives with respect to $\lambda_{1,2}^*$ and the linear combination $M_1 - \hat{\lambda} M_0$, one finds

$$\frac{\partial M_0}{\partial \lambda_{1,2}}\bigg|_{\text{deg}} = \frac{-i\pi}{2(\hat{\lambda} - \hat{\lambda}^*)R^{(0)}(\hat{\lambda})}\frac{d^2\tilde{\phi}^{(0)\sigma}}{d\lambda^2}(\hat{\lambda}) = \left(\frac{\partial M_0}{\partial \lambda_{1,2}^*}\bigg|_{\text{deg}}\right)^*. \tag{7.22}$$

Note that this identity, together with condition (7.14), implies that

$$\frac{\partial M_0}{\partial \lambda_{1,2}}\bigg|_{\text{deg}} \neq 0. \tag{7.23}$$

Third, to calculate the partial derivatives of the moments M_p with respect to x, we use (5.162) and the expansion of $R(\eta) = (\eta - \hat{\lambda})(\eta - \hat{\lambda}^*)R^{(0)}(\eta)$ for large η to find

$$\begin{aligned}\frac{\partial M_p}{\partial x}\bigg|_{\text{deg}} &= iJ\oint_L \eta^{p-3}\left(1 + \frac{2\hat{\lambda} + 2\hat{\lambda}^* + \lambda_0^{\text{crit}} + \lambda_0^{\text{crit},*}}{2\eta} + \cdots\right)d\eta \\ &= \begin{cases} 0, & p = 0, 1, \\ -2\pi J, & p = 2, \\ -\pi J(2\hat{\lambda} + 2\hat{\lambda}^* + \lambda_0^{\text{crit}} + \lambda_0^{\text{crit},*}), & p = 3. \end{cases}\end{aligned} \tag{7.24}$$

Now, according to the calculations presented in §5.3, the relevant Jacobian matrix for eliminating λ_0, λ_0^*, λ_2, and λ_2^* is a product of a Vandermonde matrix and a diagonal matrix:

$$\frac{\partial(M_0, M_1, M_2, M_3)}{\partial(\lambda_0, \lambda_0^*, \lambda_2, \lambda_2^*)} = \begin{bmatrix} 1 & 1 & 1 & 1 \\ \lambda_0 & \lambda_0^* & \lambda_2 & \lambda_2^* \\ \lambda_0^2 & \lambda_0^{*2} & \lambda_2^2 & \lambda_2^{*2} \\ \lambda_0^3 & \lambda_0^{*3} & \lambda_2^3 & \lambda_2^{*3} \end{bmatrix} \cdot \text{diag}\left(\frac{\partial M_0}{\partial \lambda_0}, \frac{\partial M_0}{\partial \lambda_0^*}, \frac{\partial M_0}{\partial \lambda_2}, \frac{\partial M_0}{\partial \lambda_2^*}\right). \tag{7.25}$$

Evaluating the Jacobian determinant on the degenerate configuration, we find that the Vandermonde determinant factor is nonzero because λ_0^{crit}, $\lambda_0^{\text{crit},*}$, $\hat{\lambda}$, and $\hat{\lambda}^*$ are all distinct, and the diagonal determinant is nonzero according to the above calculations and the assumption (7.14). It follows from the implicit function theorem that we can solve for λ_0, λ_0^*, λ_2, and λ_2^*. The partial derivatives of these implicitly defined functions with respect to the remaining independent variables x, λ_1, and λ_1^* are then obtained by Cramer's rule, using the expressions for the derivatives of the moments obtained above. This yields the expressions (7.15) and completes the proof. □

We have therefore reduced our problem to the study of the two equations $R_1 = 0$ and $V_0 = 0$ involving x, λ_1, and λ_1^*. As we no longer use analyticity properties in any essential way, we work from now on in the real subspace where λ_1^* is the complex conjugate of λ_1, and we seek a solution near $\lambda_1 = \hat{\lambda}$ for x near x_{crit}. Note that it is clear that the functions R_1 and V_0 can be defined for λ_1 near $\hat{\lambda}$; the question is only about their local behavior. Let $\epsilon \geq 0$ be a small parameter. Dominant balance considerations ultimately justified by the proof of Lemma 7.2.2 to follow suggest the following scalings of the remaining variables:

$$\lambda_1 = \hat{\lambda} + \epsilon r e^{i\theta}, \qquad \lambda_1^* = \hat{\lambda}^* + \epsilon r e^{-i\theta}, \qquad x = x_{\text{crit}} + \epsilon^2 \operatorname{Log}(\epsilon^{-1}) \cdot \chi. \quad (7.26)$$

Our new variables are therefore $r > 0$, θ, and χ, all real. Define real parameters $P > 0$, α, and c by

$$Pe^{i\alpha} := \frac{i}{2}\frac{d^2\tilde{\phi}^{(0)\sigma}}{d\lambda^2}(\hat{\lambda}), \qquad c := 2J\Im(R^{(0)}(\hat{\lambda})), \quad (7.27)$$

and consider the functions

$$\begin{aligned} R_1^{\text{model}}(r,\theta,\chi) &:= Pr^2[\sin(\alpha)\cdot\sin(2\theta) - \cos(\alpha)\cdot\cos(2\theta)], \\ V_0^{\text{model}}(r,\theta,\chi) &:= c\chi + Pr^2[\cos(\alpha)\cdot\sin(2\theta) + \sin(\alpha)\cdot\cos(2\theta)]. \end{aligned} \quad (7.28)$$

Lemma 7.2.2 *Let r, θ, and χ be fixed. Then, as $\epsilon \downarrow 0$,*

$$\epsilon^{-2}R_1 \to R_1^{\text{model}}(r,\theta,\chi), \qquad [\epsilon^2\operatorname{Log}(\epsilon^{-1})]^{-1}V_0 \to V_0^{\text{model}}(r,\theta,\chi). \quad (7.29)$$

The partial derivatives with respect to r and θ also converge pointwise:

$$\begin{aligned} &\frac{\partial}{\partial r}(\epsilon^{-2}R_1) \to \frac{\partial R_1^{\text{model}}}{\partial r}(r,\theta,\chi), \quad \frac{\partial}{\partial r}([\epsilon^2\operatorname{Log}(\epsilon^{-1})]^{-1}V_0) \to \frac{\partial V_0^{\text{model}}}{\partial r}(r,\theta,\chi), \\ &\frac{\partial}{\partial \theta}(\epsilon^{-2}R_1) \to \frac{\partial R_1^{\text{model}}}{\partial \theta}(r,\theta,\chi), \quad \frac{\partial}{\partial \theta}([\epsilon^2\operatorname{Log}(\epsilon^{-1})]^{-1}V_0) \to \frac{\partial V_0^{\text{model}}}{\partial \theta}(r,\theta,\chi). \end{aligned} \quad (7.30)$$

For fixed χ, the convergence is uniform in any finite annulus $0 < r_{\min} \leq r \leq r_{\max} < \infty$.

Proof. Using the relation (5.35) and rewriting the integral over $I_1^{\pm}$ as half of a loop integral around the band, we find

$$\epsilon^{-2}R_1 = -\frac{1}{4i\epsilon^2}\oint_{L(\hat{\lambda})} R(\eta)Y(\eta)\,d\eta - \frac{1}{4i\epsilon^2}\oint_{L(\hat{\lambda}^*)} R(\eta)Y(\eta)\,d\eta, \quad (7.31)$$

where $L(\lambda)$ denotes a sufficiently small positively oriented contour surrounding λ that is held fixed as ϵ tends to zero. On both paths of integration, the following approximation for $R(\eta)$ holds uniformly:

$$\begin{aligned} R(\eta) = {} & \frac{1}{2}R^{(0)}(\eta)\cdot(\eta-\hat{\lambda})(\eta-\hat{\lambda}^*) \\ & \times\Bigg[2 - \epsilon^2\operatorname{Log}(\epsilon^{-1})\Bigg(\frac{\partial\lambda_0}{\partial x}\bigg|_{\text{deg}}\cdot\frac{\chi}{\eta-\lambda_0^{\text{crit}}} + \frac{\partial\lambda_0^*}{\partial x}\bigg|_{\text{deg}}\cdot\frac{\chi}{\eta-\lambda_0^{\text{crit},*}} \\ & + \frac{\partial\lambda_2}{\partial x}\bigg|_{\text{deg}}\cdot\frac{\chi}{\eta-\hat{\lambda}} + \frac{\partial\lambda_2^*}{\partial x}\bigg|_{\text{deg}}\cdot\frac{\chi}{\eta-\hat{\lambda}^*}\Bigg) \\ & - \epsilon^2\left(\frac{r^2e^{2i\theta}}{(\eta-\hat{\lambda})^2} + \frac{r^2e^{-2i\theta}}{(\eta-\hat{\lambda}^*)^2}\right) + O(\epsilon^3\operatorname{Log}(\epsilon^{-1}))\Bigg], \end{aligned} \quad (7.32)$$

where the partial derivatives with respect to x may be obtained explicitly from (7.15) if desired. But in fact, they are not necessary for the present calculation; since $R^{(0)}(\eta)$ and $Y(\eta)$ are analytic and uniformly bounded inside each loop as ϵ tends to zero ($L(\hat{\lambda})$ and $L(\hat{\lambda}^*)$ taken small enough to exclude λ_0^{crit} and $\lambda_0^{\text{crit},*}$ for all sufficiently small ϵ), we find from the residue theorem that

$$\epsilon^{-2}R_1 = \frac{\pi r^2 e^{2i\theta}}{4}(\hat{\lambda} - \hat{\lambda}^*)R^{(0)}(\hat{\lambda})Y(\hat{\lambda}) + \frac{\pi r^2 e^{-2i\theta}}{4}(\hat{\lambda}^* - \hat{\lambda})R^{(0)}(\hat{\lambda}^*)Y(\hat{\lambda}^*) + O(\epsilon\,\mathrm{Log}(\epsilon^{-1})). \tag{7.33}$$

Now, as $\epsilon \downarrow 0$, by analytic dependence on the endpoints, $Y(\hat{\lambda})$ and $Y(\hat{\lambda}^*)$ converge to the corresponding quantities evaluated on the degenerate $G = 2$ endpoint configuration. That is, from formula (5.41) with $\lambda = \hat{\lambda}$ and using the degenerate configuration,

$$\begin{aligned} Y(\hat{\lambda}) &\to \frac{-1}{2\pi i}\int_{C_{I+}\cup C_{I-}} \frac{\rho^0(\eta)\,d\eta}{(\eta-\hat{\lambda})^2(\eta-\hat{\lambda}^*)R^{(0)}(\eta)} + \frac{-1}{2\pi i}\int_{C_{I+}^*\cup C_{I-}^*} \frac{\rho^0(\eta^*)^*\,d\eta}{(\eta-\hat{\lambda})^2(\eta-\hat{\lambda}^*)R^{(0)}(\eta)}, \\ Y(\hat{\lambda}^*) &\to \frac{-1}{2\pi i}\int_{C_{I+}\cup C_{I-}} \frac{\rho^0(\eta)\,d\eta}{(\eta-\hat{\lambda})(\eta-\hat{\lambda}^*)^2R^{(0)}(\eta)} + \frac{-1}{2\pi i}\int_{C_{I+}^*\cup C_{I-}^*} \frac{\rho^0(\eta^*)^*\,d\eta}{(\eta-\hat{\lambda})(\eta-\hat{\lambda}^*)^2R^{(0)}(\eta)}. \end{aligned} \tag{7.34}$$

Comparing with (7.16) for $k = 1$ or $k = 2$ evaluated on the degenerate configuration, we find

$$Y(\hat{\lambda}) \to \left.\frac{2}{\pi^2}\frac{\partial M_0}{\partial \lambda_{1,2}}\right|_{\text{deg}}, \qquad Y(\hat{\lambda}^*) \to \left.\frac{2}{\pi^2}\frac{\partial M_0}{\partial \lambda_{1,2}^*}\right|_{\text{deg}}. \tag{7.35}$$

Finally, using the explicit representation of these derivatives of M_0 given in (7.22), we find

$$\begin{aligned} \epsilon^{-2}R_1 &= -\frac{i}{4}\frac{d^2\tilde{\phi}^{(0)\sigma}}{d\lambda^2}(\hat{\lambda})\cdot r^2 e^{2i\theta} + \text{complex conjugate} + o(1) \\ &= R_1^{\text{model}}(r,\theta,\chi) + o(1), \end{aligned} \tag{7.36}$$

where we have used the definition of the parameters P and α. Passing to the limit $\epsilon \downarrow 0$ then completes the first part of the proof.

To establish the convergence of the partial derivatives with respect to r and θ, we note that since λ_0, λ_0^*, λ_2, and λ_2^* depend differentially on λ_1, λ_1^*, and x and since the functions R and Y depend analytically on λ_0, λ_0^*, λ_2, λ_2^*, and x, it follows from formula (7.31) that $\epsilon^{-2}R_1$ is differentiable uniformly in ϵ. This yields the convergence of derivatives of $\epsilon^{-2}R_1$ expressed in (7.30).

Now we carry out a similar analysis of the function V_0, which is a bit more complicated due to a logarithmic divergence. First, we show that V_0 is differentiable with respect to x at $x = x_{\text{crit}}$, $\lambda_1 = \hat{\lambda}$, and $\lambda_1^* = \hat{\lambda}^*$. Using the chain rule and the expressions for the partial derivatives of V_0 with respect to endpoints obtained in §5.3, we find

$$\begin{aligned}\frac{\partial V_0}{\partial x} = \frac{1}{2\pi i}\int_{\Gamma_1^+ \cup \Gamma_1^-} \frac{(\eta - \lambda_1)(\eta - \lambda_1^*)}{R(\eta)} \Bigg(&\frac{\partial M_0}{\partial \lambda_0}\frac{\partial \lambda_0}{\partial x}(\eta - \lambda_0^*)(\eta - \lambda_2)(\eta - \lambda_2^*) \\ &+ \frac{\partial M_0}{\partial \lambda_0^*}\frac{\partial \lambda_0^*}{\partial x}(\eta - \lambda_0)(\eta - \lambda_2)(\eta - \lambda_2^*) + \frac{\partial M_0}{\partial \lambda_2}\frac{\partial \lambda_2}{\partial x}(\eta - \lambda_0)(\eta - \lambda_0^*)(\eta - \lambda_2^*) \\ &+ \frac{\partial M_0}{\partial \lambda_2^*}\frac{\partial \lambda_2^*}{\partial x}(\eta - \lambda_0)(\eta - \lambda_0^*)(\eta - \lambda_2)\Bigg)\, d\eta. \end{aligned} \tag{7.37}$$

One can solve for the products $(\partial M_0/\partial \lambda_k)(\partial \lambda_k/\partial x)$ explicitly (it is an exact Vandermonde system) and evaluate the result on the degenerate configuration, yielding

$$\left.\frac{\partial V_0}{\partial x}\right|_{\text{deg}} = \frac{1}{2\pi i}\left(\int_{\lambda_0^{\text{crit}}}^{\hat{\lambda}} + \int_{\hat{\lambda}^*}^{\lambda_0^{\text{crit},*}}\right)\frac{J\pi(2\eta - \lambda_0^{\text{crit}} - \lambda_0^{\text{crit},*})}{R^{(0)}(\eta)}\, d\eta = 2J\Im(R^{(0)}(\hat{\lambda})). \tag{7.38}$$

It follows that as $\epsilon \downarrow 0$,

$$\frac{\partial}{\partial \chi}([\epsilon^2 \operatorname{Log}(\epsilon^{-1})]^{-1} V_0) = 2J\Im(R^{(0)}(\hat{\lambda})) + o(1), \tag{7.39}$$

with the error being uniformly small for χ in any compact neighborhood of $\chi = 0$. Therefore, we have

$$\begin{aligned}[\epsilon^2 \operatorname{Log}(\epsilon^{-1})]^{-1} V_0 &= [\epsilon^2 \operatorname{Log}(\epsilon^{-1})]^{-1} V_0|_{\chi=0} + \int_0^{\chi}(2J\Im(R^{(0)}(\hat{\lambda})) + o(1))\, d\chi' \\ &= [\epsilon^2 \operatorname{Log}(\epsilon^{-1})]^{-1} V_0|_{\chi=0} + 2J\Im(R^{(0)}(\hat{\lambda}))\cdot \chi + o(1). \end{aligned} \tag{7.40}$$

Thus, it remains to analyze $[\epsilon^2 \operatorname{Log}(\epsilon^{-1})]^{-1} V_0$ with $\chi = 0$. Recall from §5.4 the formula for V_0 in terms of the functions R and Y:

$$V_0 = \frac{i\pi}{2}\int_{\lambda_0(\epsilon)}^{\lambda_1(\epsilon)} R(\eta) Y(\eta)\, d\eta + \text{complex conjugate}. \tag{7.41}$$

We note several asymptotic properties of R and Y that follow from the form of the linear terms in the series expansions of the endpoints in positive powers of ϵ (for $\chi = 0$). For fixed λ, we have

$$R(\eta) = (\eta - \hat{\lambda})(\eta - \hat{\lambda}^*) R^{(0)}(\hat{\lambda}) + O(\epsilon^2), \tag{7.42}$$

where the order-ϵ terms cancel because the symmetric contributions from $\lambda_1(\epsilon)$ and $\lambda_2(\epsilon)$ at this order cancel exactly (cf. (7.15)). The error is uniformly small for η in any compact set not containing the points λ_0^{crit} or $\hat{\lambda}$ or their conjugates. Using this result in the formula (5.41) for Y, we see from the fact that the contours of integration lie a fixed distance from these points that for *all* η in between the contours C_{I+} and C_{I-},

$$Y(\eta) = Y_{\text{deg}}(\eta) + O(\epsilon^2), \tag{7.43}$$

where $Y_{\text{deg}}(\eta)$ means the $G = 2$ function Y constructed for $\epsilon = 0$, that is, on the degenerate configuration. While the approximation (7.42) of $R(\eta)$ holds for intermediate points on the contour of integration in the formula (7.41) for V_0, it fails near both limits. To give an approximation uniformly valid near the lower limit of integration, let $T_\epsilon(\eta)$ be the function defined by the relation $T_\epsilon(\eta)^2 = \eta - \lambda_0(\epsilon)$, cut along the band I_0^+ and the negative imaginary axis, and normalized so that for $\eta - \lambda_0(\epsilon)$ sufficiently large and positive real, $T_\epsilon(\eta)$ is positive real. Then we have

$$R(\eta) = T_\epsilon(\eta)[-(\eta - \hat{\lambda})(\eta - \hat{\lambda}^*)T_0(\eta^*)^* + O(\epsilon^2)] \tag{7.44}$$

holding uniformly in a sufficiently small (but fixed as $\epsilon \downarrow 0$) neighborhood of $\eta = \lambda_0^{\text{crit}}$. To approximate the square root near the upper limit of integration, let $S_\epsilon(\eta)$ be the function defined by the relation $S_\epsilon(\eta)^2 = (\eta - \lambda_1(\epsilon))(\eta - \lambda_2(\epsilon))$, cut along the shrinking band I_1^+ and normalized so that for large η, $S_\epsilon(\eta) \sim \eta$. Then we have

$$R(\eta) = S_\epsilon(\eta)[(\eta - \hat{\lambda}^*)R^{(0)}(\eta) + O(\epsilon^2)] \tag{7.45}$$

holding uniformly in a sufficiently small fixed neighborhood of $\eta = \hat{\lambda}$. We stress that these expansions are only valid when $\chi = 0$. For $\chi \neq 0$, larger terms come into play, which we have already taken into account by computing the derivative with respect to χ.

We take the path of integration in (7.41) to pass through two points q_0 and q_1 that are fixed as $\epsilon \downarrow 0$ and lie, respectively, in the regions of validity of (7.44) and (7.45). Then we have

$$\begin{aligned} V_0|_{\chi=0} = &-\frac{i\pi}{2}\int_{\lambda_0(\epsilon)}^{q_0} T_\epsilon(\eta)[(\eta - \hat{\lambda})(\eta - \hat{\lambda}^*)Y_{\text{deg}}(\eta)T_0(\eta^*)^*]\,d\eta \\ &+ \frac{i\pi}{2}\int_{q_0}^{q_1} (\eta - \hat{\lambda})(\eta - \hat{\lambda}^*)R^{(0)}(\eta)Y_{\text{deg}}(\eta)\,d\eta \\ &+ \frac{i\pi}{2}\int_{q_1}^{\lambda_1(\epsilon)} S_\epsilon(\eta)[(\eta - \hat{\lambda}^*)R^{(0)}(\eta)Y_{\text{deg}}(\eta)]\,d\eta \\ &+ \text{complex conjugate} + O(\epsilon^2). \end{aligned} \tag{7.46}$$

To handle the first term, make the change of variables $\tau = T_0(\eta)$:

$$\begin{aligned} W_0 :=& -\frac{i\pi}{2}\int_{\lambda_0(\epsilon)}^{q_0} T_\epsilon(\eta)[(\eta - \hat{\lambda})(\eta - \hat{\lambda}^*)Y_{\text{deg}}(\eta)T_0(\eta^*)^*]\,d\eta \\ =& -i\pi\int_{T_0(\lambda_0(\epsilon))}^{T_0(q_0)} T_\epsilon(\tau^2 + \lambda_0^{\text{crit}})[\tau(\tau^2 - (\hat{\lambda} - \lambda_0^{\text{crit}}))(\tau^2 - (\hat{\lambda}^* - \lambda_0^{\text{crit}})) \\ &\times Y_{\text{deg}}(\tau^2 + \lambda_0^{\text{crit}})T_0(\tau^{*2} + \lambda_0^{\text{crit},*})^*]\,d\tau. \end{aligned} \tag{7.47}$$

The quantity in square brackets has a convergent expansion in odd powers of τ with coefficients c_n that are independent of ϵ. Similarly, $T_\epsilon(\tau^2+\lambda_0^{\rm crit})$ has the convergent expansion

$$T_\epsilon(\tau^2+\lambda_0^{\rm crit}) = \tau \sum_{m=0}^{\infty} s_m \left(\frac{\lambda_0^{\rm crit}-\lambda_0(\epsilon)}{\tau^2}\right)^m, \tag{7.48}$$

where s_m are the Taylor coefficients of $\sqrt{1+x}$. Since the convergence is uniform on the path of integration, the order of integration and summation may be exchanged:

$$\begin{aligned} W_0 &= -i\pi \sum_{m=0}^{\infty}\sum_{n=0}^{\infty} c_{2n+1} s_m (\lambda_0^{\rm crit}-\lambda_0(\epsilon))^m \int_{T_0(\lambda_0(\epsilon))}^{T_0(q_0)} \tau^{2(n-m+1)}\,d\tau \\ &= -i\pi \sum_{m=0}^{\infty}\sum_{n=0}^{\infty} \frac{c_{2n+1}s_m}{2n-2m+3}\Big[(\lambda_0^{\rm crit}-\lambda_0(\epsilon))^m T_0(q_0)^{2n-2m+3} \\ &\qquad\qquad -(-1)^m T_0(\lambda_0(\epsilon))^{2n+3}\Big] \\ &= -\pi i \sum_{n=0}^{\infty} \frac{c_{2n+1}}{2n+3} T_0(q_0)^{2n+3} + O(\epsilon^2), \end{aligned} \tag{7.49}$$

since for $\chi = 0$, $\lambda_0(\epsilon) - \lambda_0^{\rm crit} = O(\epsilon^2)$. Upon dividing by $\epsilon^2 \operatorname{Log}(\epsilon^{-1})$, the main contribution must necessarily come from the third term,

$$W_1 := \frac{i\pi}{2}\int_{q_1}^{\lambda_1(\epsilon)} S_\epsilon(\eta)\big[(\eta-\hat{\lambda}^*)R^{(0)}(\eta)Y_{\rm deg}(\eta)\big]\,d\eta. \tag{7.50}$$

Here, the quantity in square brackets has a uniformly convergent expansion in positive powers of $\eta - \hat{\lambda}$, with coefficients d_n that are independent of ϵ, while $S_\epsilon(\eta)$ has the uniformly convergent Laurent expansion:

$$S_\epsilon(\eta) = (\eta-\hat{\lambda})\sum_{m=0}^{\infty} s_m \sum_{p=0}^{m} \binom{m}{p} \frac{B_{m,p}(\epsilon)}{(\eta-\hat{\lambda})^{2m-p}}, \tag{7.51}$$

where

$$B_{m,p}(\epsilon) := [(\hat{\lambda}-\lambda_1(\epsilon)) + (\hat{\lambda}-\lambda_2(\epsilon))]^p (\hat{\lambda}-\lambda_1(\epsilon))^{m-p}(\hat{\lambda}-\lambda_2(\epsilon))^{m-p}. \tag{7.52}$$

Note that for $\chi = 0$, the terms proportional to ϵ in the square brackets cancel (cf. (7.15)), and therefore $B_{m,p}(\epsilon)$ is a quantity of order $O(\epsilon^{2m})$. Exchanging the order of summation and integration by uniform convergence, we have

$$W_1 = \frac{i\pi}{2}\sum_{n=0}^{\infty}\sum_{m=0}^{\infty}\sum_{p=0}^{m}\binom{m}{p} d_n s_m B_{m,p}(\epsilon)\int_{q_1}^{\lambda_1(\epsilon)} (\eta-\hat{\lambda})^{1+n+p-2m}\,d\eta. \tag{7.53}$$

As long as $1+n+p-2m \neq -1$,

$$\begin{aligned} B_{m,p}(\epsilon)\int_{q_1}^{\lambda_1(\epsilon)}(\eta-\hat{\lambda})^{1-n+p-2m}\,d\eta &= \frac{B_{m,p}(\epsilon)}{(\lambda_1(\epsilon)-\hat{\lambda})^{2m}}\cdot\frac{(\lambda_1(\epsilon)-\hat{\lambda})^{2+n+p}}{2+n+p-2m} \\ &\quad - \frac{B_{m,p}(\epsilon)(q_1-\hat{\lambda})^{2+n+p-2m}}{2+n+p-2m} \\ &= O(\epsilon^{n+p+2}) - O(\epsilon^{2m}). \end{aligned} \tag{7.54}$$

These terms give constant contributions only for $m = 0$, with all other terms being of order at least $O(\epsilon^2)$. On the other hand, if $1 + n + p - 2m = -1$, then there are logarithmic contributions. Thus, using $s_1 = 1/2$, we have

$$\begin{aligned} W_1 = &-\frac{i\pi}{2}\sum_{n=0}^{\infty} d_n \frac{(q_1 - \hat{\lambda})^{n+2}}{n+2} \\ &+ \frac{i\pi}{2}\left[\frac{d_0}{2}B_{1,0}(\epsilon) + \sum_{m=2}^{\infty}\sum_{n=m-2}^{2m-2}\binom{m}{2m-2-n} d_n s_m B_{m,2m-2-n}(\epsilon)\right] \\ &\times \int_{q_1}^{\lambda_1(\epsilon)} \frac{d\eta}{\eta - \hat{\lambda}} + O(\epsilon^2). \end{aligned} \tag{7.55}$$

Regardless of the path of integration, as $\epsilon \downarrow 0$,

$$\int_{q_1}^{\lambda_1(\epsilon)} \frac{d\eta}{\eta - \hat{\lambda}} = -\mathrm{Log}(\epsilon^{-1}) + O(1) \tag{7.56}$$

and consequently

$$\begin{aligned} W_1 = &-\frac{i\pi}{2}\sum_{n=0}^{\infty} d_n \frac{(q_1 - \hat{\lambda})^{n+2}}{n+2} + \frac{i\pi}{2}\left[\frac{1}{2}(\hat{\lambda} - \hat{\lambda}^*)R^{(0)}(\hat{\lambda})Y_{\mathrm{deg}}(\hat{\lambda})r^2 e^{2i\theta} + O(\epsilon)\right] \\ &\times \epsilon^2 \mathrm{Log}(\epsilon^{-1}) + O(\epsilon^2). \end{aligned} \tag{7.57}$$

When we combine this expression with the other components of V_0 at $\chi = 0$, we recall that the sum of the constant terms in V_0 vanishes because $\Re(\tilde{\phi}^{(0)\sigma}(\hat{\lambda})) = 0$, and thus we find

$$\begin{aligned} [\epsilon^2 \mathrm{Log}(\epsilon^{-1})]^{-1} V_0 &= c\chi + \left[-\frac{i}{2}Pr^2 e^{i\alpha} e^{2i\theta} + \text{complex conjugate}\right] + o(1) \\ &= V_0^{\mathrm{model}}(r, \theta, \chi) + o(1), \end{aligned} \tag{7.58}$$

as $\epsilon \downarrow 0$. This establishes the desired convergence of V_0.

To verify the convergence of the corresponding partial derivatives, we use the following exact formula, which can be obtained by direct application of the chain rule and substitution from the Vandermonde-type system used to eliminate λ_0, λ_0^*, λ_2, and λ_2^*:

$$\frac{dV_0}{d\lambda_1} = \frac{1}{2\pi i}\cdot\frac{\partial M_0}{\partial \lambda_1}\cdot\frac{\det \mathbf{V}(\lambda_0, \lambda_0^*, \lambda_1, \lambda_2, \lambda_2^*)}{\det \mathbf{V}(\lambda_0, \lambda_0^*, \lambda_2, \lambda_2^*)}\int_{\Gamma_1^+ \cup \Gamma_1^-} \frac{\eta - \lambda_1^*}{R(\eta)}\, d\eta, \tag{7.59}$$

where $\mathbf{V}(a_1, \ldots, a_N)$ denotes the $N \times N$ Vandermonde matrix. In this formula, the notation $dV_0/d\lambda_1$ refers to the derivative after λ_0, λ_0^*, λ_2, and λ_2^* have been eliminated in favor of λ_1, λ_1^*, and x. Evaluating the determinants explicitly gives

$$\frac{\det \mathbf{V}(\lambda_0, \lambda_0^*, \lambda_1, \lambda_2, \lambda_2^*)}{\det \mathbf{V}(\lambda_0, \lambda_0^*, \lambda_2, \lambda_2^*)} = (\lambda_0 - \lambda_1)(\lambda_0^* - \lambda_1)(\lambda_1 - \lambda_2)(\lambda_1 - \lambda_2^*). \tag{7.60}$$

Substituting the expansions of the endpoints in terms of ϵ, one finds that as $\epsilon \downarrow 0$,

$$[\epsilon \mathrm{Log}(\epsilon^{-1})]^{-1}\frac{dV_0}{d\lambda_1} = -iPre^{i\alpha}e^{i\theta} + o(1). \tag{7.61}$$

Combining this relation with its complex conjugate and using the chain rule relations

$$\frac{\partial}{\partial r} = \epsilon e^{i\theta}\frac{\partial}{\partial \lambda_1} + \epsilon e^{-i\theta}\frac{\partial}{\partial \lambda_1^*}, \qquad \frac{\partial}{\partial \theta} = i\epsilon r e^{i\theta}\frac{\partial}{\partial \lambda_1} - i\epsilon r e^{-i\theta}\frac{\partial}{\partial \lambda_1^*}, \tag{7.62}$$

one obtains the desired convergence of the partial derivatives of $[\epsilon^2 \operatorname{Log}(\epsilon^{-1})]^{-1} V_0$ with respect to r and θ.

We complete the proof with a simple remark about the uniformity of these limits. The statement that for fixed χ the region of uniform validity is an arbitrary fixed annulus in the (r, θ)–polar plane is mirrored in the expressions for the functions R_1^{model} and V_0^{model}, which become meaningless if r tends to infinity or zero. □

The model equations $R_1^{\text{model}}(r, \theta, \chi) = 0$ and $V_0^{\text{model}}(r, \theta, \chi) = 0$ are easily solved. The graph of $R_1^{\text{model}}(r, \theta, \chi) = 0$ in the (r, θ) polar plane is independent of χ and is simply the union of two perpendicular lines through the origin:

$$\theta = \theta_n := \frac{\pi}{4} - \frac{\alpha}{2} + \frac{n\pi}{2}, \qquad n \in \mathbb{Z}. \tag{7.63}$$

We then have

$$V_0^{\text{model}}(r, \theta_n, \chi) = c\chi + (-1)^n P r^2. \tag{7.64}$$

When $c\chi > 0$, this equation is consistent only for $n \in 2\mathbb{Z} - 1$, while for $c\chi < 0$ we must have $n \in 2\mathbb{Z}$. Under this condition, the radial coordinate is uniquely determined:

$$r(\chi) = \sqrt{\frac{|c\chi|}{P}}. \tag{7.65}$$

So, for each χ there are two opposite solutions (r, θ) with $-\pi < \theta \le \pi$. As χ moves through zero, these two solutions coalesce at the origin and reemerge moving in the perpendicular direction.

Theorem 7.2.1 *Let $t = t_{\text{crit}}$ be fixed, and suppose that for $x = x_{\text{crit}}$ there is a simply degenerate $G = 2$ ansatz (in the sense of the two additional assumptions given at the beginning of this section). Then for each x with $|x - x_{\text{crit}}|$ sufficiently small, there exists a nondegenerate solution of the $G = 2$ endpoint equations that is unique up to permutation of the endpoints and is continuous in x.*

Proof. By Lemma 7.1.1, it is sufficient to prove that the equations $R_1 = 0$ and $V_0 = 0$ can be solved for λ_1 and λ_1^* when x is near x_{crit}. Fix an arbitrary $\chi \neq 0$, and for all $\epsilon > 0$ define

$$\begin{aligned} R_1^{\text{family}}(r, \theta; \epsilon) &:= \epsilon^{-2} R_1(x_{\text{crit}} + \epsilon^2 \operatorname{Log}(\epsilon^{-1}) \cdot \chi, \hat{\lambda} + \epsilon r e^{i\theta}, \hat{\lambda}^* + \epsilon r e^{-i\theta}), \\ V_1^{\text{family}}(r, \theta; \epsilon) &:= [\epsilon^2 \operatorname{Log}(\epsilon^{-1})]^{-1} \\ &\quad \times V_0(x_{\text{crit}} + \epsilon^2 \operatorname{Log}(\epsilon^{-1}) \cdot \chi, \hat{\lambda} + \epsilon r e^{i\theta}, \hat{\lambda}^* + \epsilon r e^{-i\theta}). \end{aligned} \tag{7.66}$$

By Lemma 7.2.2, these functions are differentiable with respect to r and θ, and the partial derivatives are continuous down to $\epsilon = 0$. Of course when $\epsilon = 0$, we have

$$R_1^{\text{family}}(r, \theta; 0) = R_1^{\text{model}}(r, \theta, \chi), \qquad V_0^{\text{family}}(r, \theta; 0) = V_0^{\text{model}}(r, \theta, \chi). \tag{7.67}$$

When $\epsilon = 0$, the Jacobian determinant of these relations is

$$\begin{vmatrix} \partial R_1^{\text{model}}/\partial r & \partial R_1^{\text{model}}/\partial \theta \\ \partial V_0^{\text{model}}/\partial r & \partial V_0^{\text{model}}/\partial \theta \end{vmatrix} = -4P^2 r^3, \tag{7.68}$$

which is not zero when evaluated on either of the two explicit solutions of the model equations for $\chi \neq 0$. It follows from the implicit function theorem that for each of the two solutions of the model problem and for sufficiently small positive ϵ there is a solution $r(\epsilon)$ and $\theta(\epsilon)$, continuous in ϵ, of the equations $R_1^{\text{family}}(r, \theta; \epsilon) = V_0^{\text{family}}(r, \theta; \epsilon) = 0$. The corresponding solution of the endpoint equations for genus-$G = 2$ is given in terms of these functions for $x = x_{\text{crit}} + \epsilon^2 \operatorname{Log}(\epsilon^{-1}) \cdot \chi$ by

$$\begin{aligned} \lambda_1 &= \hat{\lambda} + \epsilon r(\epsilon) e^{i\theta(\epsilon)}, \\ \lambda_0 &= \lambda_0(x_{\text{crit}} + \epsilon^2 \operatorname{Log}(\epsilon^{-1}) \cdot \chi, \hat{\lambda} + \epsilon r(\epsilon) e^{i\theta(\epsilon)}, \hat{\lambda}^* + \epsilon r(\epsilon) e^{-i\theta(\epsilon)}), \\ \lambda_2 &= \lambda_2(x_{\text{crit}} + \epsilon^2 \operatorname{Log}(\epsilon^{-1}) \cdot \chi, \hat{\lambda} + \epsilon r(\epsilon) e^{i\theta(\epsilon)}, \hat{\lambda}^* + \epsilon r(\epsilon) e^{-i\theta(\epsilon)}), \end{aligned} \tag{7.69}$$

with similar formulas for the complex conjugates. □

Having established the existence of a nondegenerate genus-$G = 2$ endpoint configuration for all x in a sufficiently small deleted neighborhood of x_{crit}, we now consider whether the necessary inequalities can be satisfied by the $G = 2$ ansatz. For $x_{\text{crit}} \neq 0$, the local unfolding will take place for x values totally of one sign or the other. We suppose from now on that the critical $G = 0$ ansatz and the degenerate $G = 2$ ansatz that agrees with it both correspond to the choice $J = \operatorname{sign}(x_{\text{crit}})$. The unfolding of the degenerate ansatz corresponds to the same value of J for all x under consideration. Of course, we found that this choice of J was necessary for the small-time existence theory of the $G = 0$ ansatz (cf. §6.2.2), and even in the global analysis carried out with the help of the computer, we found this choice to lead to a workable ansatz right up to the primary caustic.

In order to proceed, we make one further assumption about the degenerate ansatz at $x = x_{\text{crit}}$ and $t = t_{\text{crit}}$:

$$\textbf{Shadow condition: } \Im(R^{(0)}(\hat{\lambda})) < 0. \tag{7.70}$$

By virtue of the normalization condition $R^{(0)}(\lambda) \sim -\lambda$ for λ near infinity, this condition holds for all $\hat{\lambda} \in \mathbb{H}$ outside a bounded region that is the "shadow" of the band I_0^+, the region enclosed by I_0^+ and the vertical segment descending from the endpoint λ_0^{crit} to the real axis. See figure 7.1. We note that the computer plots presented in chapter 6 indicate that for the Satsuma-Yajima initial data the critical point $\hat{\lambda}$ indeed always lies outside the shadow of the band I_0^+, and therefore (7.70) is always satisfied at the primary caustic.

Under these assumptions, we see that the constant c appearing in the model equation $V_0^{\text{model}}(r, \theta, \chi) = 0$ always has the opposite sign of x_{crit}. From the exact solution of the model problem and the fact that it is a good approximation for small ϵ (equivalently for x sufficiently close to x_{crit}) to the true dynamics of the endpoint $\lambda_1(\epsilon) = \hat{\lambda} + \epsilon r(\epsilon) e^{i\theta(\epsilon)}$, we can easily deduce that when x is tuned away from x_{crit} toward $x = 0$, the two endpoints move apart in the direction of steepest descent of the function $\Re(\tilde{\phi}^{(0)\sigma}(\lambda))$ at the saddle point $\hat{\lambda}$. On the other hand, when x is

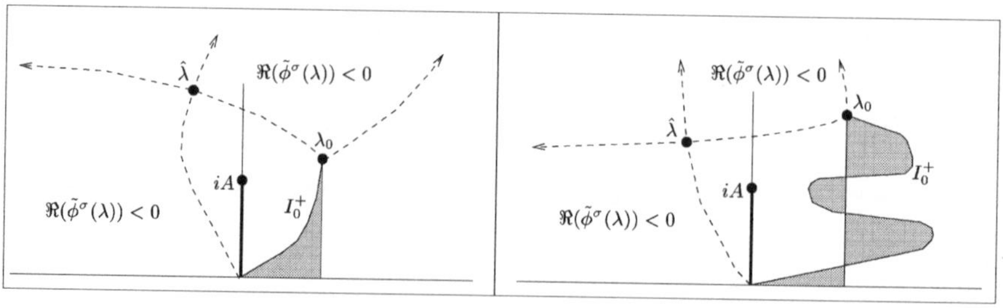

Figure 7.1 A schematic diagram showing the shadow S of the band I_0^+. A simple case similar to the numerical results described in chapter 6 is shown on the left, where the shadow is shaded. If I_0^+ is more complicated, the shadow may consist of several components as shown on the right. In all cases we assume that the critical point $\hat\lambda$ lies outside the shadow.

tuned away from x_{crit} in the direction of increasing $|x|$, the endpoints separate in the direction of the sharpest increase of this function.

Once separation has occurred, the assumption (7.14) ensures that for $|x - x_{\text{crit}}|$ sufficiently small the function $R(\lambda)Y(\lambda)$ vanishes *exactly* like a square root at both endpoints emerging from the critical point $\hat\lambda$. For small $|x - x_{\text{crit}}|$, the function $Y(\lambda)$ can be approximated locally by the constant value $Y_{\text{deg}}(\hat\lambda)$, and in a rescaled ϵ-neighborhood of $\hat\lambda$ the function $R(\lambda)$ takes on a canonical form. These facts allow fixed-point arguments similar to those used in the proofs of the local continuation theorems in §5.2 to be used to prove that as x passes through x_{crit} the local orbit structure of the differential equation (5.56) switches between the two cases illustrated in figure 7.2.

Furthermore, the continuation arguments show that for x just inside the primary caustic (i.e., for $|x| < |x_{\text{crit}}|$), exactly one of the two new endpoints born from $\hat\lambda$ lies on a trajectory of (5.56) connecting to $\lambda_0(x, t_{\text{crit}})$. We break symmetry by calling this endpoint $\lambda_1(x, t_{\text{crit}})$, which makes the other endpoint $\lambda_2(x, t_{\text{crit}})$. Clearly, the

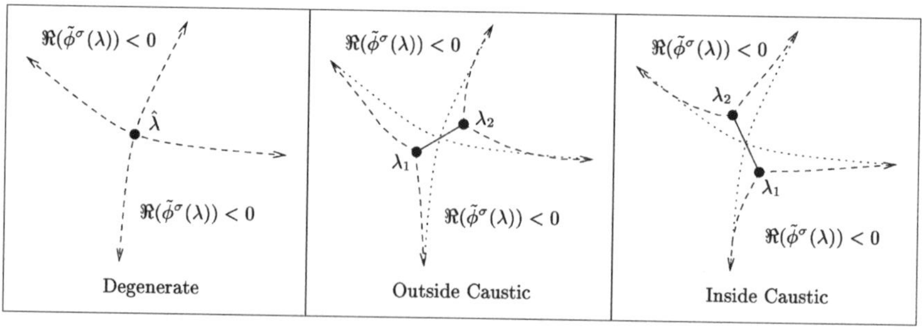

Figure 7.2 The two ways the degenerate $G = 2$ endpoint configuration in the neighborhood of $\lambda = \hat\lambda$ (left) unfolds for x near x_{crit}. The dashed curves are the zero-level sets of $\Re(\tilde\phi^\sigma(\lambda))$. Outside the caustic (center), the newly born band I_1^+, shown with a solid curve connecting the endpoints λ_1 and λ_2, does not form a bridge between regions where $\Re(\tilde\phi^\sigma(\lambda)) < 0$. Inside the caustic (right) however, the new band I_1^+ indeed forms just such a bridge.

genus-$G = 0$ gap contour Γ_1^+ that connects $\lambda_0^{\rm crit}$ to $\lambda = -\sigma 0$ and passes through the critical point $\hat{\lambda}$ at $x = x_{\rm crit}$ can be taken to split into two new gap contours, Γ_1^+ connecting $\lambda_0(x, t_{\rm crit})$ to $\lambda_1(x, t_{\rm crit})$ and Γ_2^+ connecting $\lambda_2(x, t_{\rm crit})$ to $\lambda = -\sigma 0$. Both of these contours can be chosen for $|x_{\rm crit}| - |x|$ sufficiently small and positive so that the inequality $\Re(\tilde{\phi}^\sigma(\lambda)) < 0$ holds except at the endpoints.

With the function $R(\lambda)$ taken to be cut between $\lambda_1(x, t_{\rm crit})$ and $\lambda_2(x, t_{\rm crit})$ along the zero level I_1^+ of $\Re(\tilde{\phi}^\sigma(\lambda))$, it remains to verify the inequality $\rho^\sigma(\eta)\, d\eta \in \mathbb{R}_-$ in this newly born band. It follows from the fixed-point argument for the existence of this small orbit of (5.56) for small $|x - x_{\rm crit}|$ that the band I_1^+ is smooth; this implies that there are no internal zeros of $\rho^\sigma(\eta)$ and therefore that the differential $\rho^\sigma(\eta)\, d\eta$ is necessarily real and of one sign in I_1^+. From the sign of $\Re(\tilde{\phi}^\sigma(\lambda))$ indicated for the configuration in the right-hand diagram of figure 7.2 and from the relation between the functions $\tilde{\phi}^\sigma(\lambda)$ and $\rho^\sigma(\lambda)$, we can easily compute the sign of the differential $\rho^\sigma(\eta)\, d\eta$ for all $\eta \in I_1^+$. Select the sign of the differential $d\eta$ according to the orientation of I_1^+ starting at $\lambda_1(x, t_{\rm crit})$ and ending at $\lambda_2(x, t_{\rm crit})$. Just beyond the endpoint $\lambda_2(x, t_{\rm crit})$ in the direction tangent to I_1^+, we see that the differential $[d\tilde{\phi}^\sigma/d\eta]\, d\eta$ is negative real because $d\eta$ is oriented in the direction of steepest decrease of $\Re(\tilde{\phi}^\sigma(\eta))$ and $\Im(\tilde{\phi}^\sigma(\eta)) = 0$ along this same trajectory. Since $d\tilde{\phi}^\sigma/d\eta = \pi i R(\eta) Y(\eta)$ with $Y(\eta)$ analytic and $R(\eta)$ vanishing like a simple square root at $\lambda_2(x, t_{\rm crit})$, this formula continues analytically around the endpoint in the counterclockwise direction to the cut I_1^+ as $\pi i R_+(\eta) Y(\eta)$, which we identify as $\pi i \rho^\sigma(\eta)$. In the process of continuing around the square root branch point, a factor of i is contributed:

$$R_+(\lambda_2(x, t_{\rm crit}) - d\eta) \approx i R(\lambda_2(x, t_{\rm crit}) + d\eta). \tag{7.71}$$

Therefore, we have

$$\begin{aligned}\rho^\sigma(\lambda_2(x, t_{\rm crit}) - d\eta)\, d\eta &= R_+(\lambda_2(x, t_{\rm crit}) - d\eta) Y(\lambda_2(x, t_{\rm crit}) - d\eta)\, d\eta \\ &\approx i R(\lambda_2(x, t_{\rm crit}) + d\eta) Y(\lambda_2(x, t_{\rm crit}) + d\eta)\, d\eta \\ &= \frac{1}{\pi} \frac{d\tilde{\phi}^\sigma}{d\eta}(\lambda_2(x, t_{\rm crit}) + d\eta)\, d\eta,\end{aligned} \tag{7.72}$$

which is negative real, as desired. We have therefore proved the following.

THEOREM 7.2.2 *If the genus-$G = 0$ ansatz undergoes a failure at $(x_{\rm crit}, t_{\rm crit})$ characterized by the (simple) pinching-off of the gap contour at a point $\hat{\lambda}$ not in the shadow of the band I_0^+, then there exists a genus-$G = 2$ ansatz for $|x_{\rm crit}| - |x|$ small and positive that satisfies all inequalities and becomes degenerate at $x = x_{\rm crit}$ with the closing of the band I_1^+, where it matches onto the critical $G = 0$ solution.*

Remark. The shadow condition (7.70) may seem somewhat artificial. However, it is equivalent to the statement that the region where $|x| > |x_{\rm crit}|$ corresponds to a genus-zero unfolding. If it is known a priori that this region "outside" the primary caustic corresponds to a genus-zero ansatz that first becomes critical when $|x|$ is decreased to $|x_{\rm crit}|$, then the shadow condition (7.70) must *automatically* be satisfied in order for the unfolding that occurs to be consistent. Under these conditions, the shadow condition need not be checked at all.

Remark. The scalings (7.26) of the variables $\lambda_1 - \hat{\lambda}$, $\lambda_1^* - \hat{\lambda}^*$, and $x - x_{\text{crit}}$ are significant in that they determine the size of the new band that opens up in the complex λ-plane as x is tuned into the genus-two region. In particular, to obtain a band of length $|\lambda_2 - \lambda_1| \sim \epsilon$, one must have $|x - x_{\text{crit}}| \sim \epsilon^2 \operatorname{Log}(\epsilon^{-1})$.

Chapter Eight

Variational Theory of the Complex Phase

Apart from relying heavily on the analyticity of the function $\rho^0(\eta)$ characterizing the asymptotic density of eigenvalues on the imaginary interval $[0, iA]$ by the WKB formula (3.2), in the direct construction of the complex phase function $g^\sigma(\lambda)$ presented in chapter 5 there was no way to determine a priori the value of the genus G for which a successful ansatz could be constructed for given values of x and t, nor indeed whether such a finite G exists at all. To begin to address these issues, we need to reformulate the conditions for an admissible density function $\rho^\sigma(\eta)$ for generating a complex phase function $g^\sigma(\lambda)$ given in Definition 4.2.5 in a more abstract form.

The Green's function for Laplace's equation in the upper half-plane $\mathbb{C}_+$ with Dirichlet boundary conditions on the real axis is

$$G(\lambda;\eta) = \log\left|\frac{\lambda-\eta^*}{\lambda-\eta}\right| \tag{8.1}$$

for λ and η in $\mathbb{C}_+$. For $\lambda \in \mathbb{C}_+$ and also in the domain of analyticity of $\rho^0(\lambda)$ (which of course is the whole upper half-plane for the special case of the Satsuma-Yajima ensemble, when $\rho^0(\lambda) \equiv \rho^0_{\text{SY}}(\lambda) \equiv i$), define the "external field"

$$\begin{aligned}\varphi^\sigma(\lambda) = -\Re\bigg(\int_0^{iA} L^0_\eta(\lambda)\rho^0(\eta)\,d\eta + \int_{-iA}^0 L^0_\eta(\lambda)\rho^0(\eta^*)^*\,d\eta \\ + i\pi\sigma\int_\lambda^{iA}\rho^0(\eta)\,d\eta + 2iJ(\lambda x+\lambda^2 t)\bigg).\end{aligned} \tag{8.2}$$

Note that this field is a sum of a harmonic part and a subharmonic potential part. Let $d\mu^0(\eta)$ be the nonnegative measure $-\rho^0(\eta)d\eta$ on the segment $[0, iA]$ oriented from 0 to iA. Then we can write

$$\varphi^\sigma(\lambda) = -\int G(\lambda;\eta)\,d\mu^0(\eta) - \Re\left(i\pi\sigma\int_\lambda^{iA}\rho^0(\eta)\,d\eta + 2iJ(\lambda x+\lambda^2 t)\right), \tag{8.3}$$

which displays the field $\varphi^\sigma(\lambda)$ as a sum of a Green's potential of a system of fixed *negative* charges distributed on the segment $[0, iA]$ and an "ambient" harmonic contribution. This is the explicit Riesz decomposition of the superharmonic function $-\varphi^\sigma(\lambda)$ in the upper half-plane [ST97].

Let $d\mu$ be a nonnegative Borel measure with support contained in the closure of $\mathbb{C}_+$, and consider the weighted energy functional

$$E[d\mu] := \frac{1}{2}\int d\mu(\lambda)\int G(\lambda;\eta)\,d\mu(\eta) + \int \varphi^\sigma(\lambda)\,d\mu(\lambda). \tag{8.4}$$

This can be interpreted physically as the potential energy of a given system of *positive* charges with distribution $d\mu$ in the upper half-plane with the real axis as a conducting boundary, in the presence of the external potential field $\varphi^\sigma(\lambda)$. The first term in $E[d\mu]$ is the self-energy of the charge distribution $d\mu$, and the second term is the interaction energy with the field $\varphi^\sigma(\lambda)$. From the remarks above, one term in this interaction energy is the Green's energy of interaction between the positive charges in $d\mu$ and the fixed negative charges with distribution $-d\mu^0$.

THEOREM 8.0.1 *Let $\rho^\sigma(\eta)$ be an admissible density function on the oriented loop contour C_σ lying in $\mathbb{H}$ as in Definition* 4.2.5. *Then*

$$E[-\rho^\sigma(\eta)d\eta] = \inf_{d\mu\in\mathcal{B}_+(C)} E[d\mu], \tag{8.5}$$

where the infimum is taken over $\mathcal{B}_+(C)$, the set of all nonnegative Borel measures with support in the closure of C having finite total mass and finite Green's energy, that is, measures for which

$$\int d\mu(\lambda) < \infty \qquad \textit{and} \qquad \int d\mu(\lambda)\int G(\lambda;\eta)d\mu(\eta) < \infty. \tag{8.6}$$

Proof. With the orientation σ of the contour C, the admissible differential $-\rho^\sigma(\eta)d\eta$ is a real nonnegative Borel measure on C with finite mass. Let $d\mu \in \mathcal{B}_+(C)$. Then

$$\begin{aligned} E[d\mu] - E[-\rho^\sigma(\eta)d\eta] =& \frac{1}{2}\int d\Delta(\lambda)\int d\Delta(\eta)G(\lambda;\eta) \\ &+ \int d\Delta(\lambda)\left[\varphi^\sigma(\lambda) + \int_{C_\sigma} G(\lambda;\eta)\rho^\sigma(\eta)d\eta\right], \end{aligned} \tag{8.7}$$

where $d\Delta(\eta) := d\mu(\eta)+\rho^\sigma(\eta)d\eta$ with $d\eta$ defined on C_σ by the orientation σ. First, note that the term that is quadratic in $d\Delta$ is always nonnegative, being the Green's energy of a signed measure with finite positive and negative parts, each of which has finite Green's energy. Indeed, the nonnegativity of the Green's energy for such measures is, for example, the content of Theorem II.5.6 in [ST97]. Next, observe that for $\lambda \in C$ and with the value of the interpolant index K chosen according to (5.10), we have

$$\Re(\tilde{\phi}^\sigma(\lambda)) = -\left[\varphi^\sigma(\lambda) + \int_{C_\sigma} G(\lambda;\eta)\rho^\sigma(\eta)d\eta\right]. \tag{8.8}$$

Thus we have

$$E[d\mu] - E[-\rho^\sigma(\eta)d\eta] \geq -\int \Re(\tilde{\phi}^\sigma(\lambda))d\Delta(\lambda). \tag{8.9}$$

Since according to Definition 4.2.5 we have $\Re(\tilde{\phi}^\sigma(\lambda)) \equiv 0$ for λ in the support of $\rho^\sigma(\eta)d\eta$, the integral on the right-hand side may be taken over the gaps of C. Therefore,

$$E[d\mu] - E[-\rho^\sigma(\eta)d\eta] \geq -\int_{\cup_k\Gamma_k^+} \Re(\tilde{\phi}^\sigma)d\mu \geq 0, \tag{8.10}$$

because $d\mu$ is a nonnegative measure and according to Definition 4.2.5 we have $\Re(\tilde{\phi}^\sigma(\lambda)) \leq 0$ for λ in the gaps of C. □

Remark. Note that the weaker condition that $\Re(\tilde{\phi}^\sigma(\lambda)) \leq 0$ in the gaps suffices in the proof of the theorem. Therefore, $-\rho^\sigma(\eta)d\eta$ is a minimizer even if the inequality is not strict in the gaps.

Therefore, the measure $-\rho^\sigma(\eta)d\eta$ on the oriented contour C_σ solves the energy minimization problem for positive charge distributions on the contour C. It is an *equilibrium measure* corresponding to the contour C, and the corresponding value of E is the *equilibrium energy* $E_{\min}[C]$ of C. Although we have so far only considered contours C that support admissible density functions $\rho^\sigma(\eta)$, the equilibrium energy $E_{\min}[C]$ of an arbitrary loop contour C can be defined by the infimum on the right-hand side of (8.5). Note that the equilibrium measure is by no means unique due to the requirement that the curve C meet the origin, which lies on the boundary of the domain for the Green's function. Thus, the support of a measure $d\mu$ can contain the origin, and two measures differing only by a Dirac mass at the origin always have the same energy because $\varphi^\sigma(0) = 0$. This is a nontrivial issue because the support of $-\rho^\sigma(\eta)d\eta$ on C always includes the origin according to Definition 4.2.5.

Next, we consider the variations of the energy as the contour C undergoes small deformations.

Theorem 8.0.2 *Let $\rho^\sigma(\eta)$ be an admissible density function on an oriented contour C_σ in the sense of Definition 4.2.5. For each function $\kappa(\eta)$ analytic in a neighborhood of the the support of $-\rho^\sigma(\eta)d\eta$ in C and satisfying $\kappa(0) = 0$ and for each real ϵ with $|\epsilon|$ sufficiently small, define a measure $d\mu_\epsilon^\kappa$ as follows: the support of $d\mu_\epsilon^\kappa$ is the image of that of $-\rho^\sigma(\eta)d\eta$ under the near-identity map*

$$\nu_\epsilon^\kappa : \eta \mapsto \eta + \epsilon\kappa(\eta), \tag{8.11}$$

and the measure $\mu_\epsilon^\kappa(M)$ of each measurable subset M of its support is defined to be the integral of $-\rho^\sigma(\eta)d\eta$ over the inverse image of M under the map ν_ϵ^κ. Then, with the function $\kappa(\eta)$ held fixed,

$$\left.\frac{d}{d\epsilon}E[d\mu_\epsilon^\kappa]\right|_{\epsilon=0} = 0. \tag{8.12}$$

Proof. For each $\kappa(\eta)$, we have $d\mu_0^\kappa(\eta) = -\rho^\sigma(\eta)d\eta$. By definition of the deformed measure $d\mu_\epsilon^\kappa$, we have

$$\begin{aligned} E[d\mu_\epsilon^\kappa] = & \frac{1}{2}\int d\mu_0^\kappa(\lambda)\int d\mu_0^\kappa(\eta)G(\nu_\epsilon^\kappa(\lambda);\, \nu_\epsilon^\kappa(\eta)) \\ & + \int d\mu_0^\kappa(\lambda)\varphi^\sigma(\nu_\epsilon^\kappa(\lambda)). \end{aligned} \tag{8.13}$$

First, we expand the quadratic term for ϵ small, using the fact that for any branch of the logarithm, $G(\lambda;\eta) = \Re(\log(\lambda-\eta^*)) - \Re(\log(\lambda-\eta))$,

$$\begin{aligned}
&\frac{1}{2}\int d\mu_0^\kappa(\lambda)\int d\mu_0^\kappa(\eta) G(\nu_\epsilon^\kappa(\lambda);\nu_\epsilon^\kappa(\eta)) \\
&= \frac{1}{2}\int d\mu_0^\kappa(\lambda)\int d\mu_0^\kappa(\eta) G(\lambda;\eta) \\
&\quad + \frac{\epsilon}{2}\Re\left(\int d\mu_0^\kappa(\lambda)\int d\mu_0^\kappa(\eta)\frac{\kappa(\lambda)-\kappa(\eta)^*}{\lambda-\eta^*}\right) \\
&\quad - \frac{\epsilon}{2}\Re\left(\int d\mu_0^\kappa(\lambda)\int d\mu_0^\kappa(\eta)\frac{\kappa(\lambda)-\kappa(\eta)}{\lambda-\eta}\right) + O(\epsilon^2).
\end{aligned} \tag{8.14}$$

The second integral proportional to ϵ above is nonsingular because κ is analytic on the support of $d\mu_0^\kappa$. Upon regularization by interpreting one or the other of the iterated integrals in the sense of the Cauchy principal value, the terms in the numerator can be separated. Thus,

$$\begin{aligned}
\int d\mu_0^\kappa(\lambda)\int d\mu_0^\kappa(\nu)\frac{\kappa(\lambda)-\kappa(\eta)}{\lambda-\eta} &= \int d\mu_0^\kappa(\lambda)\kappa(\lambda)\text{P.V.}\int\frac{d\mu_0^\kappa(\eta)}{\lambda-\eta} \\
&\quad - \int d\mu_0^\kappa(\eta)\kappa(\eta)\text{P.V.}\int\frac{d\mu_0^\kappa(\lambda)}{\lambda-\eta} \\
&= 2\int d\mu_0^\kappa(\lambda)\kappa(\lambda)\text{P.V.}\int\frac{d\mu_0^\kappa(\eta)}{\lambda-\eta} \\
&= -2\int d\mu_0^\kappa(\lambda)\kappa(\lambda)\text{P.V.}\int_{C_\sigma}\frac{\rho^\sigma(\eta)d\eta}{\lambda-\eta}.
\end{aligned} \tag{8.15}$$

The first integral proportional to ϵ can be handled without regularization. Here, for the real part we find

$$\begin{aligned}
\Re\left(\int d\mu_0^\kappa(\lambda)\int d\mu_0^\kappa(\eta)\frac{\kappa(\lambda)-\kappa(\eta)^*}{\lambda-\eta^*}\right) &= 2\Re\left(\int d\mu_0^\kappa(\lambda)\kappa(\lambda)\int\frac{d\mu_0^\kappa(\eta)}{\lambda-\eta^*}\right) \\
&= 2\Re\left(\int d\mu_0^\kappa(\lambda)\kappa(\lambda)\int_{[C^*]_\sigma}\frac{\rho^\sigma(\eta^*)^*d\eta}{\lambda-\eta}\right).
\end{aligned} \tag{8.16}$$

Next, we expand the linear term in the energy for small ϵ. We find

$$\begin{aligned}
\frac{d}{d\epsilon}\varphi^\sigma(\nu_\epsilon^\kappa(\lambda))\bigg|_{\epsilon=0} &= -\Re\bigg[\kappa(\lambda)\bigg(\int_0^{iA}\frac{\rho^0(\eta)d\eta}{\lambda-\eta} + \int_{-iA}^0\frac{\rho^0(\eta^*)^*d\eta}{\lambda-\eta} - i\pi\sigma\rho^0(\lambda) \\
&\quad + 2iJ(x+2\lambda t)\bigg)\bigg].
\end{aligned} \tag{8.17}$$

In a now-familiar step (cf. chapter 5), we introduce a path of integration $C_I : 0 \to iA$ that agrees with C_σ in the support of $d\mu_0^\kappa$ and then connects the final point of support to iA. Then, using analyticity of $\rho^0(\eta)$, this expression becomes

$$\frac{d}{d\epsilon}\varphi^\sigma(\nu_\epsilon^\kappa(\lambda))\bigg|_{\epsilon=0} = -\Re\left[\kappa(\lambda)\left(\text{P.V.}\int_{C_I}\frac{\rho^0(\eta)d\eta}{\lambda-\eta} + \int_{C_I^*}\frac{\rho^0(\eta^*)^*d\eta}{\lambda-\eta}\right.\right.$$
$$\left.\left. + 2iJ(x+2\lambda t)\right)\right]. \tag{8.18}$$

Combining these calculations and making an identification with the derivative of $\tilde{\phi}^\sigma(\lambda)$ along the contour C, we find that

$$\frac{d}{d\epsilon}E[d\mu_\epsilon^\kappa]\bigg|_{\epsilon=0} = -\int d\mu_0^\kappa(\lambda)\Re\left[\kappa(\lambda)\frac{d}{d\lambda}\tilde{\phi}^\sigma(\lambda)\right], \tag{8.19}$$

where the derivative along the contour is meant. It is sufficient to integrate over the support of $d\mu_0^\kappa = -\rho^\sigma(\eta)d\eta$. By Definition 4.2.5, the function $\tilde{\phi}^\sigma(\lambda)$ is constant along each component of the support of $d\mu_0^\kappa$, which proves the theorem. □

Remark. A contour C for which the variations described in the statement of Theorem 8.0.2 all vanish is said to have the *S-property* [GR87]. This terminology appears in the approximation theory literature, where S stands for "symmetry." Clearly, it might just as well stand for "stationary" or, in the context of applications to steepest-descent-type asymptotic analysis of Riemann-Hilbert problems, "steepest." Further information about the S-property and its variational interpretation can be found in the preprint [PR94].

Remark. In some applications, it may be enough that Theorem 8.0.2 holds for a dense subset of analytic functions $\kappa(\eta)$. For example, one often restricts attention to *Schiffer variations* [S50] in which $\kappa(\eta)$ has the form of a simple rational function

$$\kappa(\eta) = \frac{\alpha\eta}{\eta-\eta_0} \tag{8.20}$$

for $\alpha \in \mathbb{C}$ and η_0 not lying on the contour C. The condition that $\kappa(0) = 0$ simply fixes the contour to the origin under deformation.

The results described in Theorem 8.0.1 and Theorem 8.0.2 indicate that the conditions that characterize the complex phase function $g^\sigma(\lambda)$ (cf. Definition 4.2.5) are equivalent to the existence of a certain kind of critical point for the energy functional, where variations with respect to both the measure *and the contour of support* are permitted. *Thus, we have obtained a generalization of the method of Lax and Levermore* [LL83], *who considered the restricted problem of minimizing the energy of measures supported on a fixed and given contour.* In this connection, it is attractive to consider whether an appropriate variational problem can be well-posed whose solution is exactly a critical point of the desired type. This would effectively complement the ansatz-based construction of $g^\sigma(\lambda)$ given in chapter 5 by allowing techniques of functional analysis and logarithmic potential theory to be applied to determine properties of the complex phase function, such as existence, uniqueness, and genus.

If we suppose that for each given analytic function $\kappa(\lambda)$ as in the statement of Theorem 8.0.2 the equilibrium energy $E_{\min}[\nu_\epsilon^\kappa(C)]$ is differentiable with respect to ϵ, then the existence of a loop contour C for which

$$\left.\frac{d}{d\epsilon}E_{\min}[\nu_\epsilon^\kappa(C)]\right|_{\epsilon=0} = 0 \tag{8.21}$$

implies that C has the S-property. To see this, let $d\mu$ be an equilibrium measure for C, and consider the corresponding family of measures $d\mu_\epsilon^\kappa$ supported on the curve $\nu_\epsilon^\kappa(C)$ as in the statement of Theorem 8.0.2. Clearly, $E[d\mu_0^\kappa] = E_{\min}[C] = E_{\min}[\nu_0^\kappa(C)]$, and since $E[d\mu_\epsilon^\kappa]$ is not generally an equilibrium measure for $\nu_\epsilon^\kappa(C)$, we have $E[d\mu_\epsilon^\kappa] \geq E_{\min}[\nu_\epsilon^\kappa(C)]$. Now as a function of ϵ, $E[d\mu_\epsilon^\kappa]$ is clearly differentiable at $\epsilon = 0$, and it follows from its domination of the equilibrium energy that (8.21) implies that the derivative of $E[d\mu_\epsilon^\kappa]$ with respect to ϵ vanishes at $\epsilon = 0$.

Thus, differentiability of the equilibrium energy implies that the object that may be taken to be stationary at a curve with the S-property is the equilibrium energy $E_{\min}[C]$ itself as a functional of the loop contour C. This suggests posing a "stationary-min" problem for the energy functional E, with possible special case variants "min-min" and "max-min." It is not difficult to argue that the "min-min" problem, that is, finding a contour C for which the equilibrium energy $E_{\min}[C]$ is minimal, has no solution. This is because the external field $\varphi^\sigma(\lambda)$ goes to $-\infty$ as $\lambda \to \infty$ in a sector of the upper half-plane (depending on x and t), and consequently, the equilibrium energy can be made arbitrarily negative by considering a sequence of contours expanding into this sector. On the other hand, the "max-min" problem, that is, finding a contour C for which the equilibrium energy $E_{\min}[C]$ is as large as possible, is a version of the well-studied problem of finding sets of *minimal weighted logarithmic capacity* satisfying certain geometrical constraints. (Here, the geometrical constraint is that the set must be a contour surrounding the imaginary interval $[0, iA]$ and connecting $0-$ to $0+$.) This sort of problem is sometimes referred to in the literature as a Chebotarev problem [GR87]. In circumstances significantly simpler than those of our problem, the minimal capacity problem is known to have a solution that is unique in the support of the equilibrium measure.

We pose the "max-min" problem in the following conjecture.

CONJECTURE 8.0.1 *Suppose for simplicity that $A(x)$ is such that $\rho^0(\eta)$ defined by (3.1) is entire. Let $\mathcal{C}$ be a family of loop contours C in the cut upper half-plane $\mathbb{H}$ that begin and end at the origin. For each contour $C \in \mathcal{C}$, let $d\mu_C^*$ be a measure minimizing the weighted energy E:*

$$E[d\mu_C^*] = \inf_{d\mu\geq 0, \mathrm{supp}(d\mu)\subset C} E[d\mu]. \tag{8.22}$$

Suppose $C^* \in \mathcal{C}$ can be found such that

$$E[d\mu_{C^*}^*] = \sup_{C\in\mathcal{C}} E[d\mu_C^*]. \tag{8.23}$$

Then, the extremal measure $d\mu_{C^*}^*$ is unique modulo point masses at the origin, and its support consists of a finite number of analytic arcs, one of which meets the origin. Writing $d\mu_{C^*}^* = -\rho^\sigma(\eta)d\eta$ defines a density function $\rho^\sigma(\eta)$ that is admissible in *the sense of Definition 4.2.5 and thus generates a complex phase function permitting the asymptotic analysis of the semiclassical soliton ensemble corresponding to the initial data* $A(x)$.

Posing the semiclassical limit for the focusing nonlinear Schrödinger equation as a constrained minimum capacity problem thus closes the circle. We began our analysis of the inverse problem in §4.1 with the observation that what we were essentially dealing with was a problem of rational interpolation of entire functions; indeed, the set of minimal weighted capacity has played a central role in the theory of rational approximation for several years. We plan to address these issues more carefully in the future.

Chapter Nine

Conclusion and Outlook

The generalized steepest-descent scheme we have described in detail for analyzing the semiclassical limit of the initial-value problem (1.1) for the focusing nonlinear Schrödinger equation provides what we believe to be the first rigorous result of its kind: that solutions of a sequence of well-posed problems (i.e., the initial-value problem (1.1) with WKB-modified initial data corresponding to true data of the form $\psi_0(x) = A(x)$ for the sequence $\hbar = \hbar_N$) converge to an object whose macroscopic properties (weak limits of conserved local densities) are described by a system of elliptic modulation equations, whose initial-value problem is significantly less well behaved. Our methods allow the convergence to be effectively analyzed for times that are not necessarily small and in particular for times beyond which the solutions become wild and oscillatory.

From the point of view of the semiclassical limit for the initial-value problem (1.1), the tools we have developed in the preceding pages provide an avenue toward the analysis of many open problems. For example, questions of the way that limits of solutions depend on the analyticity properties of the initial data can be systematically addressed (see [CM02] for some recent considerations in this direction).

But also from the point of view of other problems that can be attacked by Riemann-Hilbert methods (e.g., long-time asymptotics for integrable partial differential equations, problems in approximation theory and statistical analysis of random matrix ensembles, and some related combinatorial problems), the generalization of the steepest-descent method of Deift and Zhou that we have presented here is likely to be useful as a general technique. For example, certain problems in the theory of orthogonal polynomials involving exotic orthogonality conditions can be treated by our methods.

In this final chapter, we would like to outline several ways that we would like to consider extending what we have presented.

9.1 GENERALIZATION FOR NONQUANTUM VALUES OF $\hbar$

An essential role was played in our work by the assumption that the semiclassical parameter $\hbar$ should be restricted to a particular discrete sequence of values as it goes to zero. Thus, technically speaking we have only established that the explicit model we have presented in terms of Riemann theta functions is a strong limit point for the semiclassical asymptotics. However unlikely it may seem, we cannot a priori rule out the possibility that there could be other limit points as well, which one might find by considering values of $\hbar$ intermediate to those in the sequence $\hbar = \hbar_N$.

Therefore, it appears that some advantage would be gained by addressing the asymptotic behavior of soliton ensembles without the quantization restriction on $\hbar$. For the special case of the Satsuma-Yajima initial data $\psi_0(x) = A(x)$, the *exact* spectral data becomes somewhat more complicated when general values of $\hbar$ are considered, since there is a nonzero reflection coefficient when $\hbar \neq \hbar_N$ for any N. In fact, the reflection coefficient is not even uniformly small in any neighborhood of $\lambda = 0$ as $\hbar \to 0$.

In this special case, however (and also in the recent cases described in [TV00]), at least one has an exact formula for the reflection coefficient, and since it is small except near $\lambda = 0$, it could be taken into account at the level of the local model Riemann-Hilbert Problem 4.4.5 for the matrix $\hat{\mathbf{F}}^{\sigma}(\zeta)$. In fact, the reader will observe that without the incorporation of the reflection coefficient into this Riemann-Hilbert problem, the jump matrices will not even satisfy the compatibility condition (A.2) for general values of $\hbar$ (we needed to assume that $\hbar = \hbar_N$ to obtain the desired compatibility), and consequently the problem will be unsolvable.

As interesting as it will be to see how to handle the intermediate values of $\hbar$ for the Satsuma-Yajima special case, it is of more interest to have a scheme that works for general soliton ensembles. Here, the difficulty is that an accurate WKB approximation is required for the reflection coefficient in a small enough (shrinking) neighborhood of the origin. Formal WKB theory simply predicts the pointwise (fixed-λ) convergence of the reflection coefficient to zero and fails to capture the asymptotic structure of the coefficient near the origin. This structure would be needed to generalize our techniques to general values of $\hbar$ for arbitrary soliton ensembles.

9.2 EFFECT OF COMPLEX SINGULARITIES IN $\rho^0(\eta)$

For the Satsuma-Yajima ensemble, the function $\rho^0(\eta)$ defined by (3.1) is entire, and there is no obstruction whatsoever to the placement of the contour C anywhere in the cut upper half-plane $\mathbb{H}$. However, for other real-analytic, bell-shaped initial data, even data with sufficient decay and curvature at its peak to admit analytic continuation of $\rho^0(\eta)$ to a complex neighborhood of the interval $[0, iA]$, there could be singularities some distance from $[0, iA]$ in the complex plane that could ultimately constrain the free motion of the contour C according to the variational conditions.

Since the presence of complex singularities will be the rule rather than the exception, it is of some interest to determine the effect of these on the dynamics of the contour motion. For example, it may be the case that the contour C is typically repelled by any singularities. On the other hand, if it is possible for the contour to collide with a singularity of $\rho^0(\eta)$ for finite x and t, what can one expect to happen to the ansatz-based construction of the complex phase function $g^{\sigma}(\lambda)$ at such a moment? Is it somehow still possible for the phase function to exist, possibly by passing to a higher-genus ansatz? In other words, is the collision of the contour with a complex singularity of $\rho^0(\eta)$ a possible mechanism for phase transitions? Or might it even be the case that upon meeting a singularity, the support of the equilibrium measure becomes irregular, with an infinite number of bands and gaps?

9.3 UNIFORMITY OF THE ERROR NEAR $t = 0$

We have noted that for general soliton ensembles, we cannot control the error of our approximation exactly at $t = 0$ with the methods described in this book because here the variational conditions select a contour loop C, part of which coincides with a subset of the imaginary interval $[0, iA]$. Since this is the locus of accumulation of the poles in the meromorphic Riemann-Hilbert Problem 2.0.1, the specified contour C does not do the job of surrounding the poles, and thus our error analysis fails.

At the same time, however, we know that our approximation remains valid when $t = 0$, at least in the L^2 sense, because this calculation can paradoxically be done directly by Lax-Levermore methods. Indeed this was shown explicitly in [EJLM93]. From this point of view, one might suppose that the reason we cannot control the error is that we are trying to bound the error in a uniform approximation of the eigenfunction in the complex plane, rather than just worrying about the error in the potential, $|\psi - \tilde{\psi}|$. The eigenfunction is simply more complicated when $t = 0$ than for nonzero t because there is an endpoint of the support of the equilibrium measure (near which the local behavior of the eigenfunction should be described in terms of Airy functions) superimposed on the locus of accumulation of eigenvalues. So approximation of the eigenfunction would appear to be just a different problem at $t = 0$ than for nonzero t.

Nonetheless, there is considerable advantage in a unified Riemann-Hilbert-based approach to semiclassical asymptotics for (1.1) that works for all t. Exactly such an approach has recently been described in [M02], and strong convergence of arbitrary semiclassical soliton ensembles to the function $A(x)$ at $t = 0$ in the limit $\hbar \downarrow 0$ has thus been established. The central idea in [M02] is to take advantage of the fact that a genus-zero ansatz may be constructed at $t = 0$ in two distinct ways, since the sign parameter σ is arbitrary. The arbitrariness of σ can be traced back to a freedom of analytic interpolant of the proportionality constants $\{\gamma_k\}$ at the eigenvalues $\{\lambda_k\}$. One may therefore choose some curve C_M connecting $\lambda = 0$ to $\lambda = iA$ and lying in the domain D containing the eigenvalues and enclosed by the loop contour C containing iA that divides D into a left half and a right half. One then uses different interpolants of the $\{\gamma_k\}$ in either side of C_M in D. The contour C_M then plays the role played by C_I in chapter 5 of this book, although it has the advantage that it need not avoid the imaginary interval $[0, iA]$ where the eigenvalues are accumulating since the exact jump matrix on C_M is analytic there. This latter fact allows the error in the continuum limit approximation to be controlled even if C_M meets $[0, iA]$. In particular for $t = 0$, the genus-zero ansatz requires that C_M *coincide* with the imaginary interval $[0, iA]$; with the error controlled the ansatz thus leads to a convergence proof.

9.4 ERRORS INCURRED BY MODIFYING THE INITIAL DATA

If we desire to interpret our completely rigorous results regarding asymptotics for semiclassical soliton ensembles in the context of the semiclassical limit for the initial-value problem (1.1), then there is a step missing in our analysis. Namely, we would need to provide an estimate for arbitrary x and t of the errors we make by replacing the $\hbar$-parametrized family of initial-value problems (1.1) by another family of problems in which the initial data has been modified so that its spectrum is replaced with its purely discrete WKB approximation. The latter is what we have been referring to throughout as a soliton ensemble and for which we compute accurate asymptotics.

This modification of the initial data was also an essential ingredient in Lax and Levermore's original analysis of the zero-dispersion limit for the Korteweg–de Vries equation. It has also been built into each analogous study of an integrable system since that time. As pointed out above, the Lax-Levermore method proves L^2 convergence of the modified initial data to the true initial data. And in the light of the local well-posedness of the hyperbolic Whitham equations that they prove governs the limit, it is a reasonable claim that this convergence also holds for finite nonzero times.

But in our study of the focusing nonlinear Schrödinger equation, we find that the limit is governed by Whitham equations that are elliptic. Without any local well-posedness for this asymptotic dynamical model, we must look elsewhere if we wish to control the errors introduced by modifying the initial data.

The best bet may be to study more carefully the forward-scattering problem for the nonselfadjoint Zakharov-Shabat eigenvalue problem in the $\hbar \downarrow 0$ limit. If it would be possible to directly estimate the error of the WKB approximation at the level of the scattering data, then this error could be built into the Riemann-Hilbert analysis (cf. §4.5) as another layer of approximation to be expanded in a Neumann series and consequently controlled.

9.5 ANALYSIS OF THE MAX-MIN VARIATIONAL PROBLEM

At the philosophical heart of our work is Conjecture 8.0.1, that the "max-min" problem for the weighted Green's energy functional described in chapter 8 is a natural and appropriate generalization of the celebrated variational principle of Lax and Levermore's zero-dispersion analysis and that it characterizes the semiclassical limit of the initial-value problem (1.1) for the focusing nonlinear Schrödinger equation. We may even speculate that a variational principle of this type characterizes the limit for a class of initial data that is much more general than what we have considered here.

In the Lax-Levermore method, the variational principle plays a central role. It is the equilibrium measure that determines the zero-dispersion limit for the Korteweg–de Vries equation in general, and if the initial data is smooth enough, then it can be shown [DKM98] that the support of the equilibrium measure is sufficiently regular (and in particular consists of a finite number of bands and gaps) that it may be constructed by an ansatz-based method of which what we presented in chapter 5 is a generalization.

The reader will observe that our approach to the semiclassical limit for the focusing nonlinear Schrödinger equation has been quite different. We began with analyticity of the initial data and constructed the complex phase function $g^{\sigma}(\lambda)$ directly, by ansatz. Then, after the fact, we observed in chapter 8 that the complex phase function could be given a variational interpretation. Indeed, if it turned out that the max-min problem had a solution for which the support of the equilibrium measure were severely irregular, we would not know how to use it in the Riemann-Hilbert analysis to asymptotically reduce the phase-conjugated Riemann-Hilbert Problem 4.1.1 to a simple form. In short, the Riemann-Hilbert approach would appear to manifestly depend on analyticity (e.g., the deformation of "opening the lenses" is only possible if $\rho^{\sigma}(\eta)\,d\eta$ is an analytic measure).

At the same time, we would need to develop existence and regularity arguments for the max-min problem. Some results of this type do indeed exist in the literature for problems that are perhaps not too different from ours. Even if it turns out that regularity properties are somehow built in from the start, it would be useful to have a sufficiently developed theory of the max-min problem that we could estimate the genus for given values of x and t in terms of elementary properties of the initial data.

9.6 INITIAL DATA WITH $S(x) \not\equiv 0$

Throughout, we have assumed $S(x) \equiv 0$ and considered the initial-value problem (1.1) with purely real initial data. Dropping the assumption $S(x) \equiv 0$ is expected to require new alterations in our method. In this case, the existing numerical evidence [B96] and formal calculations [M01] suggest that for analytic potentials the eigenvalues accumulate as $\hbar$ tends to zero on a union of curves in the complex plane. In the examples presented in the literature, the asymptotic spectrum is distributed on a "Y-shaped" curve consisting of a "neck" on the imaginary axis connected to the origin, out of which are born two "branches" on which the eigenvalues have nonzero real parts. While there exist analogs of the Bohr-Sommerfeld quantization rule for these Y-shaped spectra [M01], there is not an adequate WKB description of the proportionality constant γ_k connected with each discrete
eigenvalue.

Even with the forward-scattering analysis in this primitive state of affairs, it may be possible to apply our methods to semiclassical soliton ensembles appropriately defined from the formal WKB approximations for $S(x) \not\equiv 0$. The basic setup should be the same, with a holomorphic Riemann-Hilbert problem being posed relative to a loop contour that surrounds the whole (possibly Y-shaped) locus of accumulation of the eigenvalues in the upper half-plane. The conditions imposed on the complex phase function $g^{\sigma}(\lambda)$ in §4.2 should be unchanged; similarly the error analysis should require only technical modifications. The part of the method that will need to be rethought almost completely is the ansatz-based construction of $g^{\sigma}(\lambda)$ (cf. chapter 5), since for an asymptotic eigenvalue measure $\rho^{0}(\eta)\,d\eta$ with complicated support properties it is not clear how to "push" the measure by analyticity onto the loop contour C. We plan to generalize our method to handle these more general semiclassical soliton ensembles in the near future.

9.7 FINAL REMARKS

Thus, the story of the semiclassical limit for the focusing nonlinear Schrödinger equation is far from complete. Even for analytic initial data, much work remains on the front of rigorous WKB theory for the nonselfadjoint Zakharov-Shabat operator, as well as on the development of the variational-theoretic aspect of the inverse problem described in chapter 8, which has the potential to be a very powerful analytical tool. And once the problem is rigorously understood for given analytic initial data, the next task is to determine the sensitivity of the resulting semiclassical limit within this analytic class. How (un)stable are the caustic curves (phase transitions) separating one kind of local behavior from another with respect to, say, L^2-small analytic perturbations? Is there sense in putting a probability measure on some class of initial data and determining the statistics of the local genus $G(x, t)$ considered as a random variable? Sometimes, instability can be mollified by statistical averaging, and this could be one avenue toward giving useful physical meaning to the semiclassical limit of the focusing nonlinear Schrödinger equation.

Appendix A

Hölder Theory of Local Riemann-Hilbert Problems

In this appendix we collect a number of results, some quite classical, with the aim of rigorously establishing in some generality the existence, uniqueness, and decay properties of solutions of "local" Riemann-Hilbert problems of the type that often arise in steepest-descent-type calculations. Throughout this appendix, a fixed norm $\|\cdot\|$ on matrices is assumed.

A.1 LOCAL RIEMANN-HILBERT PROBLEMS. STATEMENT OF RESULTS

Let us define what we mean by a *local Riemann-Hilbert problem*. Let $\Sigma_{\mathbf{L}}$ be a union of an even number of straight-line rays emanating from the origin in the ζ-plane, and a circle of radius $R \neq 1$ centered at the origin. Note that $R \neq 1$ can always be arranged by a simple rescaling of ζ. The contour $\Sigma_{\mathbf{L}}$ divides the ζ-plane into two disjoint regions, $\Omega_{\mathbf{L}}^{+}$ and $\Omega_{\mathbf{L}}^{-}$, each of which may comprise several simply connected regions, such that each ray or circular arc of $\Sigma_{\mathbf{L}}$ forms part of the boundary of both $\Omega_{\mathbf{L}}^{+}$ and $\Omega_{\mathbf{L}}^{-}$. Choose some labeling of the regions consistent with this description; now each is a component of either $\Omega_{\mathbf{L}}^{+}$ or $\Omega_{\mathbf{L}}^{-}$. Consider $\Sigma_{\mathbf{L}}$ oriented such that it forms the positively oriented boundary of the region $\Omega_{\mathbf{L}}^{+}$. See figure A.1. The jump matrices we consider are defined as follows.

Definition A.1.1 (Admissible jump matrices for local problems). *Let $\zeta_1, \ldots, \zeta_N$ denote the points of intersection of the rays of $\Sigma_{\mathbf{L}}$ with the circle. An admissible jump matrix for a local Riemann-Hilbert problem is a matrix-valued function $\mathbf{v}_{\mathbf{L}}(\zeta)$ defined for $\zeta \in \Sigma_{\mathbf{L}} \backslash \{0, \zeta_1, \ldots, \zeta_N\}$ satisfying for some $0 < \nu \leq 1$ and some $K > 0$ the following conditions:*

1. **Unimodularity:** *For all $\zeta \in \Sigma_{\mathbf{L}} \backslash \{0, \zeta_1, \ldots, \zeta_N\}$, $\det(\mathbf{v}_{\mathbf{L}}(z)) = 1$.*
2. **Interior smoothness:** *Whenever ζ_1 and ζ_2 belong to the same smooth component of $\Sigma_{\mathbf{L}}$ (either ray segment or circular arc), the Hölder condition $\|\mathbf{v}_{\mathbf{L}}(\zeta_2) - \mathbf{v}_{\mathbf{L}}(\zeta_1)\| \leq K|\zeta_2 - \zeta_1|^{\nu}$ holds.*
3. **Compatibility at self-intersection points:** *Let ζ_0 denote any of the points $0, \zeta_1, \ldots, \zeta_N$. By interior smoothness, it follows that on each smooth component $\Sigma_{\mathbf{L}}^{(k)}$ of $\Sigma_{\mathbf{L}}$ meeting ζ_0, the limit*

$$\mathbf{v}_{\mathbf{L}}^{(k)} := \lim_{\substack{\zeta \to \zeta_0 \\ \zeta \in \Sigma_{\mathbf{L}}^{(k)}}} \mathbf{v}_{\mathbf{L}}(\zeta) \tag{A.1}$$

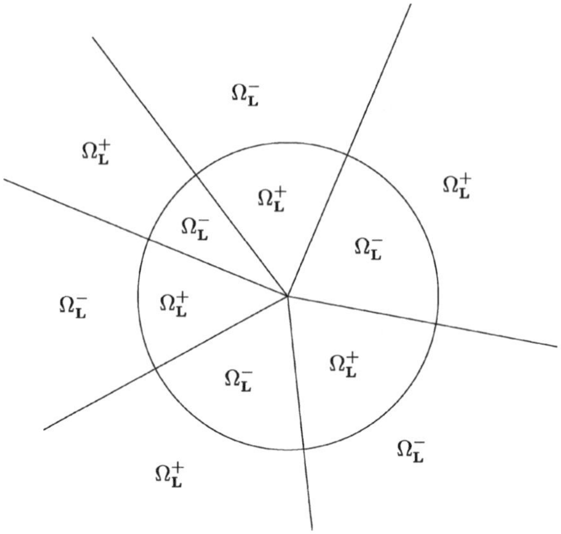

Figure A.1 A contour $\Sigma_{\mathbf{L}}$ for a local Riemann-Hilbert problem. The contour is oriented to be simultaneously the positively oriented boundary of $\Omega_{\mathbf{L}}^{+}$ and the negatively oriented boundary of the complementary domain $\Omega_{\mathbf{L}}^{-}$.

exists. Let the components $\Sigma_{\mathbf{L}}^{(k)}$ be ordered with increasing k in the counterclockwise direction about ζ_0, starting with any given contour component. Then the condition

$$\mathbf{v}_{\mathbf{L}}^{(1)}\mathbf{v}_{\mathbf{L}}^{(2)-1}\mathbf{v}_{\mathbf{L}}^{(3)}\mathbf{v}_{\mathbf{L}}^{(4)-1}\cdots\mathbf{v}_{\mathbf{L}}^{(n-1)}\mathbf{v}_{\mathbf{L}}^{(n)-1} = \mathbb{I} \tag{A.2}$$

is to be satisfied, where n is the even number of contour components meeting at ζ_0 (this number is exactly 4 for $\zeta_0 = \zeta_1, \ldots, \zeta_N$ and N for $\zeta_0 = 0$).

4. **Decay:** *As $\zeta \to \infty$ on any ray of $\Sigma_{\mathbf{L}}$, $\|\mathbf{v}_{\mathbf{L}}(\zeta) - \mathbb{I}\| = O(|\zeta|^{-\nu})$.*

Let $\mathbf{N}_{\mathbf{L}}$ be any constant matrix. Given the data $(\Sigma_{\mathbf{L}}, \mathbf{v}_{\mathbf{L}}(\zeta), \mathbf{N}_{\mathbf{L}})$ we pose the following problem.

Riemann-Hilbert Problem A.1.1 (Local problem in the Hölder sense). *Let $\mathbf{v}_{\mathbf{L}}(\zeta)$ be an admissible jump matrix defined on a contour $\Sigma_{\mathbf{L}}$ as in Figure* A.1. *Let ν be the Hölder exponent of the jump matrix. Find a matrix function $\mathbf{L}(\zeta)$ with the following properties:*

1. **Analyticity:** *The matrix function $\mathbf{L}(\zeta)$ is holomorphic in $\mathbb{C}\setminus\Sigma_{\mathbf{L}}$.*
2. **Boundary behavior:** *For each $\mu < \nu$, $\mathbf{L}(\zeta)$ assumes Hölder-continuous boundary values from each connected component of its domain of analyticity, including corner points, with Hölder exponent μ. More precisely, for each $\zeta \in \Sigma_{\mathbf{L}}\setminus\{0, \zeta_1, \ldots, \zeta_N\}$, the boundary values*

$$\mathbf{L}_{\pm}(\zeta) := \lim_{\substack{\lambda\to\zeta \\ \lambda\in\Omega_{\mathbf{L}}^{\pm}}} \mathbf{L}(\lambda) \tag{A.3}$$

exist independently of the path of approach. For all positive $\mu < \nu$, *there exists a constant* K' *such that for all* $\zeta \in \Sigma_{\mathbf{L}}\backslash\{0, \zeta_1, \ldots, \zeta_N\}$ *and all* $\lambda \in \Omega_{\mathbf{L}}^{\pm}$, $\|\mathbf{L}(\lambda) - \mathbf{L}_{\pm}(\zeta)\| \leq K'|\lambda - \zeta|^{\mu}$. *Also, whenever* ζ_1 *and* ζ_2 *belong to the same smooth component of* $\Sigma_{\mathbf{L}}$, $\|\mathbf{L}_{\pm}(\zeta_2) - \mathbf{L}_{\pm}(\zeta_1)\| \leq K'|\zeta_2 - \zeta_1|^{\mu}$. *Also, whenever* $\Sigma_{\mathbf{L}}^{(j)}$ *and* $\Sigma_{\mathbf{L}}^{(k)}$ *are two smooth components of the contour bounding a connected component of* $\Omega_{\mathbf{L}}^{+}$ *and meeting at* $\zeta = \zeta_0$, *we have*

$$\lim_{\substack{\zeta \to \zeta_0 \\ \zeta \in \Sigma_{\mathbf{L}}^{(j)}}} \mathbf{L}_{+}(\zeta) = \lim_{\substack{\zeta \to \zeta_0 \\ \zeta \in \Sigma_{\mathbf{L}}^{(k)}}} \mathbf{L}_{+}(\zeta), \tag{A.4}$$

and both limits are finite. Similarly, if $\Sigma_{\mathbf{L}}^{(j)}$ *and* $\Sigma_{\mathbf{L}}^{(k)}$ *meet at* ζ_0 *and bound a connected component of* $\Omega_{\mathbf{L}}^{-}$, *then*

$$\lim_{\substack{\zeta \to \zeta_0 \\ \zeta \in \Sigma_{\mathbf{L}}^{(j)}}} \mathbf{L}_{-}(\zeta) = \lim_{\substack{\zeta \to \zeta_0 \\ \zeta \in \Sigma_{\mathbf{L}}^{(k)}}} \mathbf{L}_{-}(\zeta), \tag{A.5}$$

with both limits being finite.

3. **Jump conditions:** *For each* $\zeta \in \Sigma_{\mathbf{L}}\backslash\{0, \zeta_1, \ldots, \zeta_N\}$, *the boundary values are related by* $\mathbf{L}_{+}(\zeta) = \mathbf{L}_{-}(\zeta)\mathbf{v}_{\mathbf{L}}(\zeta)$.
4. **Normalization:** *As* $\zeta \to \infty$ *in each ray of* $\Sigma_{\mathbf{L}}$, $\|\mathbf{L}_{\pm}(\zeta) - \mathbf{N}_{\mathbf{L}}\| = O(|\zeta|^{-\mu})$ *for all positive* $\mu < \nu$. *The same estimate with* $\mathbf{L}_{\pm}(\zeta)$ *replaced by* $\mathbf{L}(\zeta)$ *holds uniformly in all other directions.*

Note that in formulating any local Riemann-Hilbert problem, the condition that $\Sigma_{\mathbf{L}}$ should consist of an even number of rays can always be achieved by adjoining a new ray on which the jump matrix is taken to be the identity matrix. Also, we emphasize that in many applications the circle will be absent altogether, which can also be accomplished by taking the jump to be the identity there. There is no cost for adding identity jumps to the contour for Riemann-Hilbert problems posed in the class of Hölder-continuous boundary values. This follows from an elementary Cauchy integral argument establishing that whenever the jump matrix is equal to the identity on a smooth contour segment and the solution takes on boundary values as above (and in particular in the sense of uniform continuity), then the solution matrix is in fact analytic *on* the contour segment.

We establish here the following results.

THEOREM A.1.1 (Local Fredholm alternative). *Suppose that a set of data* ($\Sigma_{\mathbf{L}}$, $\mathbf{v}_{\mathbf{L}}(\zeta), \mathbf{N}_{\mathbf{L}}$) *is such that the jump matrix* $\mathbf{v}_{\mathbf{L}}(\zeta)$ *satisfies the conditions of Definition* A.1.1. *Then the associated local Riemann-Hilbert Problem* A.1.1 *has a unique solution if and only if the corresponding homogeneous problem with data* ($\Sigma_{\mathbf{L}}, \mathbf{v}_{\mathbf{L}}(\zeta), \mathbf{0}$) *has only the trivial solution* $\mathbf{L}(\zeta) \equiv \mathbf{0}$.

We note that if a solution $\mathbf{L}(\zeta)$ exists for some particular Hölder exponent $\mu_0 < \nu$, then $\mathbf{L}(\zeta)$ also serves as a solution for all positive Hölder exponents $\mu < \mu_0$. Therefore, to establish solvability with Hölder exponent μ for *all* $\mu < \nu$, it is sufficient to find a number $\mu_0 < \nu$ such that all homogeneous solutions with the Hölder exponent $\mu > \mu_0$ are trivial. As not every problem is solvable, there is no completely general method for ruling out nontrivial vanishing solutions, and a

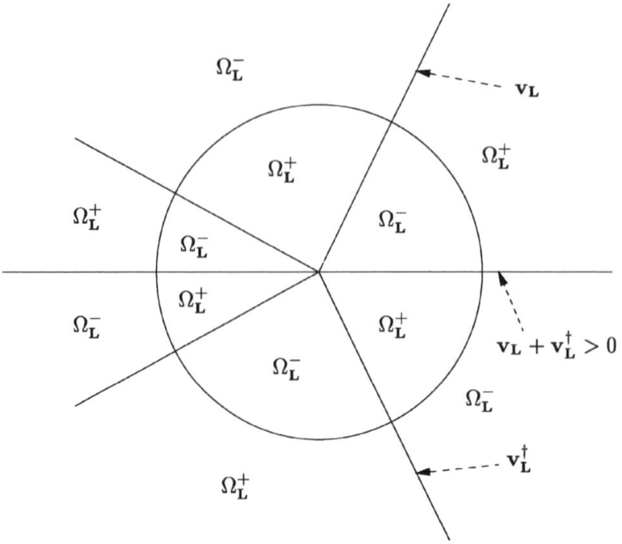

Figure A.2 A contour $\Sigma_{\mathbf{L}}$ with Schwarz reflection symmetry.

variety of techniques and results from complex analysis are useful. What appears to be the most general result holds for contours and jump matrices that satisfy the following symmetry criterion:

Schwarz reflection symmetry: The contour $\Sigma_{\mathbf{L}}$ contains the real axis, is invariant under complex conjugation, and for all $\zeta \in \Sigma_{\mathbf{L}}$ with $\Im(\zeta) \neq 0$, $\mathbf{v_L}(\zeta) = \mathbf{v_L}(\zeta^*)^{\dagger}$, while for all real ζ, the matrix $\mathbf{v_L}(\zeta) + \mathbf{v_L}(\zeta)^{\dagger}$ is strictly positive definite. (A.6)

The condition that the real axis be part of a contour satisfying (A.6) may always be satisfied without loss of generality by taking the jump matrix to be the identity there (this obviously satisfies the positive definiteness condition). A contour with the symmetry of (A.6) is illustrated in figure A.2.

Then, we have the following theorem.

THEOREM A.1.2 (Local unique solvability). *Suppose that the data set* $(\Sigma_{\mathbf{L}}, \mathbf{v_L}(\zeta), \mathbf{N_L})$ *satisfies condition* (A.6) *in addition to the requirements of Theorem* A.1.1. *Then the associated local Riemann-Hilbert Problem* A.1.1 *has a unique solution.*

Remark. The main difficulty lies in proving the existence of solutions of the required type. Indeed, uniqueness holds even without the assumption (A.6). One simply considers the matrix quotient of any two solutions $\mathbf{Q} := \mathbf{L}(\zeta)\mathbf{L}'(\zeta)^{-1}$, which by the Banach algebra property of the Hölder spaces (see §A.3 for precise definitions of these spaces) again takes on boundary values in the Hölder (and in particular uniformly continuous) sense. The boundary values $\mathbf{Q}_+(\zeta)$ and $\mathbf{Q}_-(\zeta)$ are easily seen to be the same for all points $\zeta \in \Sigma_{\mathbf{L}} \backslash \{0\}$, and $\mathbf{Q}(\zeta) \to \mathbb{I}$ as $\zeta \to \infty$. Given the behavior of the boundary values, an elementary Cauchy integral argument shows that $\mathbf{Q}(\zeta)$ is in fact analytic in the whole plane, and then it follows from

Liouville's theorem that $\mathbf{Q}(\zeta) \equiv \mathbb{I}$. It is possible to push through the same argument with less control on the boundary values; in [D99] this is explained in the 2×2 case when the Hölder-smoothness property of the boundary values is replaced with L^2 convergence, so that in particular one admits solutions for which the boundary values become unbounded. Therefore, if there exists a solution with Hölder-class boundary values, it is unique in much larger spaces, including at least L^2.

Finally, some refinement of the decay of the solution of the local Riemann-Hilbert Problem A.1.1 is possible under certain additional conditions.

THEOREM A.1.3 (Enhanced decay). *Suppose that on the interior of each smooth arc of $\Sigma_{\mathbf{L}}$ the jump matrix $\mathbf{v}_{\mathbf{L}}(\zeta)$ is analytic and that from any ray component $\Sigma_{\mathbf{L}}^{(k)}$ of $\Sigma_{\mathbf{L}}$ the jump matrix may be continued to either side within a strip S_k bounded by two parallel rays. Suppose that for $\zeta \in \Sigma_{\mathbf{L}}^{(k)}$ we can write*

$$\zeta(\mathbf{v}_{\mathbf{L}}(\zeta) - \mathbb{I}) := \Phi_k(\zeta) = \langle \Phi_k \rangle + \delta\Phi_k(\zeta), \tag{A.7}$$

where $\langle \Phi_k \rangle$ is a constant matrix and $\delta\Phi_k(\zeta)$ is a bounded matrix function that satisfies

$$\int_{\zeta_0^{(k)}}^{\zeta} \delta\Phi_k(s)\, ds = O(\log|\zeta|) \quad \text{as } \zeta \to \infty \text{ uniformly for } \zeta \in S_k, \tag{A.8}$$

where $\zeta_0^{(k)}$ is an arbitrary fixed basepoint on $\Sigma_{\mathbf{L}}^{(k)}$. If there exists a solution $\mathbf{L}(\zeta)$ of Riemann-Hilbert Problem A.1.1 *and if the condition*

$$\sum_{\text{all rays } k} \langle \Phi_k \rangle = \mathbf{0} \tag{A.9}$$

holds, then there is a constant $M > 0$ such that uniformly for all ζ sufficiently large,

$$\|\mathbf{L}(\zeta) - \mathbb{I}\| \leq M|\zeta|^{-1}. \tag{A.10}$$

Thus we obtain $O(\zeta^{-1})$ decay to the identity if the moments, defined by the asymptotic behavior of $\zeta(\mathbf{v}_{\mathbf{L}}(\zeta) - \mathbb{I})$ as ζ tends to infinity along each ray, have average values in an appropriate sense that sum to zero.

Remark. The refinement afforded by Theorem A.1.3 is significant, since without the condition (A.9) the typical decay is only $O(|\zeta|^{-1}\log|\zeta|)$. Note also that the condition (A.9) may be viewed as a "higher-order" version of the compatibility condition (A.2), which is necessary for solvability.

A.2 UMBILICAL RIEMANN-HILBERT PROBLEMS

We now consider posing a certain type of auxiliary Riemann-Hilbert problem in the z-plane on a *compact contour*. Let us now describe what we mean by an *umbilical Riemann-Hilbert problem*. Let Σ_{U} be any compact contour consisting of a union of a finite number of smooth closed arcs terminating at self-intersection points

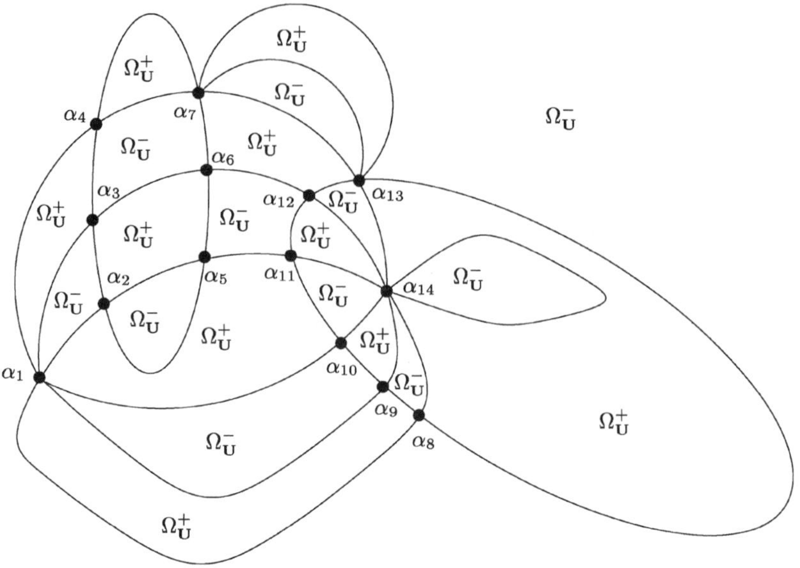

Figure A.3 An example of an umbilical contour Σ_U. The contour is oriented so as to form the positively oriented boundary of a compact domain Ω_U^+ and the negatively oriented boundary of the noncompact complementary domain Ω_U^-.

$z = \alpha_k$ such that Σ_U divides the z-plane into two disjoint regions, Ω_U^+ and Ω_U^-, while serving as the positively oriented boundary of Ω_U^+ and at the same time the negatively oriented boundary of Ω_U^-. This means in particular that an even number of arcs meet at each intersection point α_k. The geometry of a typical umbilical contour is shown in figure A.3.

Definition A.2.1 (Admissible jump matrices for umbilical problems). *An admissible jump matrix for an umbilical Riemann-Hilbert problem is a matrix-valued function* $\mathbf{v}_U(z)$ *defined for* $z \in \Sigma_U \setminus \{\alpha_k\}$ *with the following properties for some* $K'' > 0$ *and* $0 < \nu \leq 1$*:*

1. **Unimodularity:** *For all* $z \in \Sigma_U \setminus \{\alpha_k\}$, $\det(\mathbf{v}_U(z)) = 1$.
2. **Interior smoothness:** *Whenever* z_1 *and* z_2 *belong to the same smooth arc of* Σ_U*, the Hölder condition* $\|\mathbf{v}_U(z_2) - \mathbf{v}_U(z_1)\| \leq K''|z_2 - z_1|^\nu$ *holds.*
3. **Compatibility at self-intersection points:** *Let* z_0 *be any of the points* α_i *of self-intersection of* Σ_U*, let* $\Sigma_U^{(1)}, \dots, \Sigma_U^{(N)}$ *be the open arcs meeting at* $z = \alpha_i$ *enumerated in counterclockwise order, and define*

$$\mathbf{v}_U^{(k)} := \lim_{\substack{z \to z_0 \\ z \in \Sigma_U^{(k)}}} \mathbf{v}_U(z). \tag{A.11}$$

Then for each such point z_0*, the condition*

$$\mathbf{v}_U^{(1)} \mathbf{v}_U^{(2)-1} \mathbf{v}_U^{(3)} \mathbf{v}_U^{(4)-1} \dots \mathbf{v}_U^{(N-1)} \mathbf{v}_U^{(N)-1} = \mathbb{I} \tag{A.12}$$

holds.

Let $\mathbf{N}_{\mathbf{U}}$ be any constant matrix. The umbilical Riemann-Hilbert problem associated with the data $(\Sigma_{\mathbf{U}}, \mathbf{v}_{\mathbf{U}}(z), \mathbf{N}_{\mathbf{U}})$ is posed as follows.

Riemann-Hilbert Problem A.2.1 (Umbilical problem in the Hölder sense). *Find a matrix function $\mathbf{U}(z)$ with the following properties:*

1. **Analyticity:** *The matrix function $\mathbf{U}(z)$ is holomorphic in $\mathbb{C}\setminus\Sigma_{\mathbf{U}}$ and also at $z = \infty$.*
2. **Boundary behavior:** *$\mathbf{U}(z)$ assumes Hölder-continuous boundary values from each connected component of its domain of analyticity, including self-intersection points, with all exponents $\mu < \nu$. More precisely, for each $z \in \Sigma_{\mathbf{U}}\setminus\{\alpha_k\}$, the boundary values*
$$\mathbf{U}_{\pm}(z) := \lim_{\substack{w\to z \\ w\in\Omega_{\mathbf{U}}^{\pm}}} \mathbf{U}(w) \tag{A.13}$$
exist independently of the path of approach. For all positive $\mu < \nu$, there exists a constant K' such that for all $z \in \Sigma_{\mathbf{U}}\setminus\{\alpha_k\}$ and all $w \in \Omega_{\mathbf{U}}^{\pm}$, $\|\mathbf{U}(w) - \mathbf{U}_{\pm}(z)\| \leq K'|w-z|^{\mu}$. Also, whenever z_1 and z_2 belong to the same arc of $\Sigma_{\mathbf{U}}$, $\|\mathbf{U}_{\pm}(z_2) - \mathbf{U}_{\pm}(z_1)\| \leq K'|z_2 - z_1|^{\mu}$. Also, whenever $\Sigma_{\mathbf{U}}^{(j)}$ and $\Sigma_{\mathbf{U}}^{(k)}$ are two arcs meeting at a self-intersection point $z = \alpha_i$ and bounding a connected component of $\Omega_{\mathbf{U}}^{+}$, we have
$$\lim_{\substack{z\to\alpha_i \\ z\in\Sigma_{\mathbf{U}}^{(j)}}} \mathbf{U}_{+}(z) = \lim_{\substack{z\to\alpha_i \\ z\in\Sigma_{\mathbf{U}}^{(k)}}} \mathbf{U}_{+}(z), \tag{A.14}$$
and both limits are finite. Similarly, if $\Sigma_{\mathbf{U}}^{(j)}$ and $\Sigma_{\mathbf{U}}^{(k)}$ meet at $z = \alpha_i$ and bound a connected component of $\Omega_{\mathbf{U}}^{-}$, then
$$\lim_{\substack{z\to\alpha_i \\ z\in\Sigma_{\mathbf{U}}^{(j)}}} \mathbf{U}_{-}(z) = \lim_{\substack{z\to\alpha_i \\ z\in\Sigma_{\mathbf{U}}^{(k)}}} \mathbf{U}_{-}(z), \tag{A.15}$$
with both limits being finite.
3. **Jump condition:** *For each $z \in \Sigma_{\mathbf{U}}\setminus\{\alpha_i\}$, the boundary values are related by $\mathbf{U}_{+}(z) = \mathbf{U}_{-}(z)\mathbf{v}_{\mathbf{U}}(z)$.*
4. **Normalization:** $\mathbf{U}(\infty) = \mathbf{N}_{\mathbf{U}}$.

Each local Riemann-Hilbert problem is equivalent to an umbilical Riemann-Hilbert problem.

Lemma A.2.1 *Consider a local Riemann-Hilbert Problem* A.1.1 *for an unknown matrix $\mathbf{L}(\zeta)$ corresponding to the data $(\Sigma_{\mathbf{L}}, \mathbf{v}_{\mathbf{L}}(\zeta), \mathbf{N}_{\mathbf{L}})$, with $\mathbf{N}_{\mathbf{L}}$ invertible and with the jump matrix satisfying Definition* A.1.1. *Let θ be any angle different from those of all the rays of $\Sigma_{\mathbf{L}}$, and introduce the automorphism of the Riemann sphere given by*
$$z(\zeta) = \frac{\zeta e^{-i\theta} - 1}{\zeta e^{-i\theta} + 1} \qquad \textit{with inverse} \qquad \zeta(z) = e^{i\theta}\frac{1+z}{1-z}. \tag{A.16}$$
This transformation defines an umbilical contour in the z-plane by $\Sigma_{\mathbf{U}} := z(\Sigma_{\mathbf{L}})$, preserving orientation of each smooth component. On $\Sigma_{\mathbf{U}}$, define a jump matrix

by $\mathbf{v}_{\mathrm{U}}(z) := \mathbf{v}_{\mathrm{L}}(\zeta(z))$. *Then the jump matrix so defined satisfies the conditions of Definition* A.2.1, *and the solution(s) of the umbilical Riemann-Hilbert Problem* A.2.1 *with data* $(\Sigma_{\mathrm{U}}, \mathbf{v}_{\mathrm{U}}(z), \mathbf{N}_{\mathrm{U}})$ *and* $\mathbf{N}_{\mathrm{U}}$ *invertible are in one-to-one correspondence with those of the local Riemann-Hilbert Problem* A.1.1 *with data* $(\Sigma_{\mathrm{L}}, \mathbf{v}_{\mathrm{L}}(\zeta), \mathbf{N}_{\mathrm{L}})$.

Proof. Under the transformation $\zeta \rightarrow z(\zeta)$, the image of the straight line making an angle ϕ with the positive real axis in the ζ-plane is the graph of

$$|z + i\cot(\phi - \theta)|^2 = \csc^2(\phi - \theta), \tag{A.17}$$

a circle for all $\phi \neq \theta$, which always contains the two points $z = \pm 1$. This transformation fixes the straight line in the ζ-plane making an angle $\phi = \theta$ with the positive real axis and takes the origin to $z = -1$ and the point at infinity to $z = +1$. Furthermore, the real-ζ line is mapped to a complete circle (for $\theta \neq 0$), and all other ray components of Σ_{L} are mapped to various circular arcs connecting $z = -1$ to $z = +1$. The circle of radius $R \neq 1$ centered at the origin in the ζ-plane is mapped to a circle of radius $2R/(R^2-1)$ centered at $z = (R^2+1)/(R^2-1)$. Since θ differs from the angles of all components of Σ_{L} and since we are assuming that $R \neq 1$, the image $\Sigma_{\mathrm{U}} := z(\Sigma_{\mathrm{L}})$ of the contour is compact. For contours Σ_{L} with the additional symmetry (A.6), which by definition contain the real axis in the ζ-plane, the condition $\theta \neq 0$ is necessary for compactness of Σ_{U}. The image of the contour Σ_{L} under this transformation is illustrated in figure A.4.

It is immediate from the definition and the unimodularity of $\mathbf{v}_{\mathrm{L}}(\zeta)$ that the jump matrix $\mathbf{v}_{\mathrm{U}}(z) := \mathbf{v}_{\mathrm{L}}(\zeta(z))$ is also unimodular. The fact that $\mathbf{v}_{\mathrm{U}}(z)$ satisfies the interior smoothness condition follows from the corresponding local Hölder smoothness of $\mathbf{v}_{\mathrm{L}}(\zeta)$ on finite parts of Σ_{L}, which is preserved under composition with the smooth map $\zeta(z)$, and the local smoothness near $z = 1$ follows from the decay condition

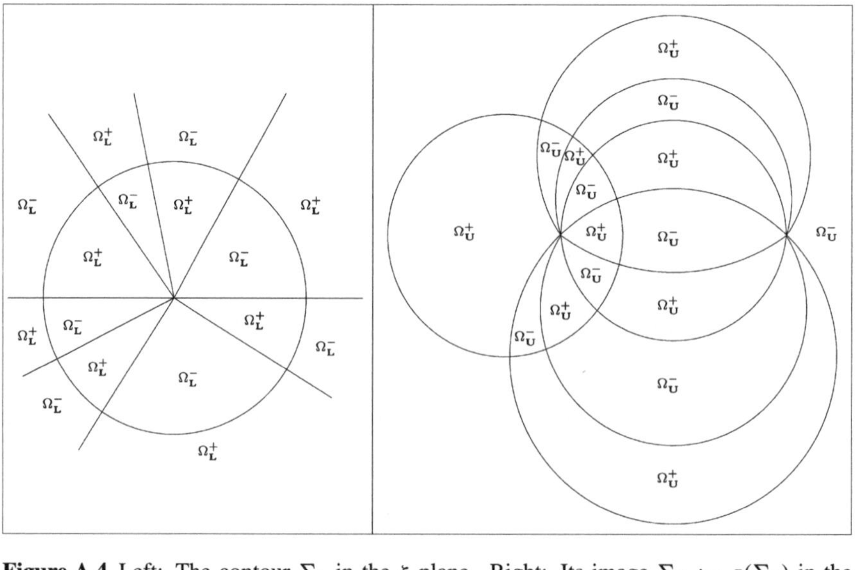

Figure A.4 Left: The contour Σ_{L} in the ζ-plane. Right: Its image $\Sigma_{\mathrm{U}} := z(\Sigma_{\mathrm{L}})$ in the z-plane for some $\theta > 0$. Also shown are the regions $\Omega_{\mathrm{U}}^{+} := z(\Omega_{\mathrm{L}}^{+})$ and $\Omega_{\mathrm{U}}^{-} := z(\Omega_{\mathrm{L}}^{-})$.

satisfied by $\mathbf{v_L}(\zeta)$. The jump matrix $\mathbf{v_U}(z)$ is compatible at the self-intersection points $\alpha_1 = -1$ and $\alpha_{k+1} = z(\zeta_k)$ by the corresponding property of $\mathbf{v_L}(\zeta)$ and the continuity of the map $z(\zeta)$ at $\zeta = 0$ and $\zeta = \zeta_k$. At the other self-intersection point $z = +1$, the compatibility condition follows from the decay of $\mathbf{v_L}(\zeta)$ at infinity.

The correspondence between solutions of the two Riemann-Hilbert problems is set up as follows. Given a matrix $\mathbf{L}(\zeta)$ solving the local Riemann-Hilbert Problem A.1.1, the matrix $\mathbf{U}(z) := \mathbf{N_U}\mathbf{L}(-e^{i\theta})^{-1}\mathbf{L}(\zeta(z))$ is a solution of the umbilical Riemann-Hilbert Problem A.2.1. Conversely, given a matrix $\mathbf{U}(z)$ satisfying the umbilical Riemann-Hilbert Problem A.2.1, the matrix $\mathbf{L}(\zeta) := \mathbf{N_L}\mathbf{U}(1)^{-1}\mathbf{U}(z(\zeta))$ satisfies the local Riemann-Hilbert Problem A.1.1. These formulae make sense because by taking determinants of the jump conditions for the two problems and using the unimodularity of the jump matrices in conjunction with the Hölder smoothness of the boundary values and Liouville's theorem it follows that $\det(\mathbf{L}(\zeta)) \equiv \det(\mathbf{N_L}) \neq 0$ and $\det(\mathbf{U}(z)) \equiv \det(\mathbf{N_U}) \neq 0$. A similar argument (taking ratios of matrices and using the jump relations, Hölder boundary conditions, and Liouville's theorem) shows that these formulae are injective transformations. □

We now restrict our study to the umbilical Riemann-Hilbert Problem A.2.1 subject to the jump matrix satisfying the conditions of Definition A.2.1. Before we begin, we need to review some elementary results.

A.3 REVIEW OF HÖLDER RESULTS FOR SIMPLE CONTOURS

Here, we recall some classical facts about Hölder spaces of functions on contours and the associated Cauchy integral operators. The basic references are Muskhelishvili [M53] and Prössdorf [P78]. Let C be a piecewise-smooth closed compact contour, where the points at which C is not smooth are at worst corner points with finite angles. The sort of contour we have in mind for future applications is shown in figure A.5. Let $\mathrm{Lip}^\nu(C)$ denote the space of all Hölder-continuous complex matrix-valued functions $\mathbf{f}(z)$ on C with exponent ν, equipped with the norm

$$\|\mathbf{f}\|_{\mathrm{Lip}^\nu(C)} := \sup_{z\in C}\|\mathbf{f}(z)\| + \sup_{z_1,z_2\in C}\frac{\|\mathbf{f}(z_2)-\mathbf{f}(z_1)\|}{|z_2-z_1|^\nu}, \qquad \text{(A.18)}$$

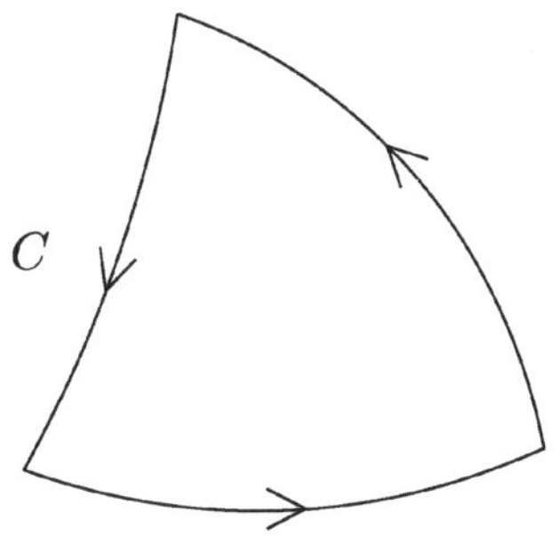

Figure A.5 The piecewise-smooth closed contour C. It is given an arbitrary orientation.

where $\|\cdot\|$ denotes some matrix norm. This norm makes $\mathrm{Lip}^{\nu}(C)$ a Banach space. If $\mathbf{f} \in \mathrm{Lip}^{\nu}(C)$ for any $\nu > 0$, then $\mathbf{f}(z)$ is uniformly continuous on C. Also, it is easy to see that if $\mathbf{f} \in \mathrm{Lip}^{\nu}(C)$, then $\mathbf{f} \in \mathrm{Lip}^{\mu}(C)$ whenever $0 < \mu < \nu$. This fact defines an inclusion operator $\mathcal{I}_{\nu\to\mu} : \mathrm{Lip}^{\nu}(C) \to \mathrm{Lip}^{\mu}(C)$. The remarkable and useful fact about this inclusion map is the following.

Lemma A.3.1 *Whenever $0 < \mu < \nu$, the inclusion operator $\mathcal{I}_{\nu\to\mu} : \mathrm{Lip}^{\nu}(C) \to \mathrm{Lip}^{\mu}(C)$ is compact.*

A proof of this statement is given by Prössdorf (see pages 102 and 103 of [P78]) in the scalar case when C is a smooth compact contour. The proof relies heavily on the Arzelá-Ascoli theorem. There are only cosmetic differences in extending the result to the matrix case, but Prössdorf's proof needs to be extended to admit corner points in the contour C. The estimates required to carry out this extension involve bounding arc-length distance above and below by the shortest distance for pairs of points near a corner point of C and can be found in the appendixes of [M53].

The space $\mathrm{Lip}^{\nu}(C)$ is also a Banach algebra in that it is closed under multiplication and satisfies the estimate

$$\|\mathbf{fg}\|_{\mathrm{Lip}^{\nu}(C)} \leq \|\mathbf{f}\|_{\mathrm{Lip}^{\nu}(C)} \|\mathbf{g}\|_{\mathrm{Lip}^{\nu}(C)}. \tag{A.19}$$

This follows from simply writing $\mathbf{f}(z_2)\mathbf{g}(z_2) - \mathbf{f}(z_1)\mathbf{g}(z_1)$ as $(\mathbf{f}(z_2) - \mathbf{f}(z_1))\mathbf{g}(z_2) + \mathbf{f}(z_1)(\mathbf{g}(z_2) - \mathbf{g}(z_1))$, the domination of the L^{∞} norm by the Hölder norm, and the fact that the matrices with the norm $\|\cdot\|$ are themselves a Banach algebra satisfying an estimate of the same form as (A.19). If $\mathbf{f}(z)$ is in $\mathrm{Lip}^{\nu}(C)$ and is invertible for each $z \in C$, then $\mathbf{g}(z) = \mathbf{f}(z)^{-1}$ is also in $\mathrm{Lip}^{\nu}(C)$. The estimate (A.19) immediately gives the following result.

Lemma A.3.2 *Let $\mathbf{g} \in \mathrm{Lip}^{\nu}(C)$, and let $\mathcal{R}_{\mathbf{g}}$ and $\mathcal{L}_{\mathbf{g}}$ denote the operators of pointwise right and left multiplication by $\mathbf{g}(z)$. Then $\mathcal{R}_{\mathbf{g}} : \mathrm{Lip}^{\nu}(C) \to \mathrm{Lip}^{\nu}(C)$ and $\mathcal{L}_{\mathbf{g}} : \mathrm{Lip}^{\nu}(C) \to \mathrm{Lip}^{\nu}(C)$ are bounded linear operators.*

Consider C to be oriented. For $\mathbf{f} \in \mathrm{Lip}^{\nu}(C)$, the Cauchy contour integral

$$(\mathcal{C}^{C}\mathbf{f})(z) := \frac{1}{2\pi i} \oint_{C} (s - z)^{-1}\mathbf{f}(s)\, ds \tag{A.20}$$

defines a holomorphic matrix-valued function in the multiply connected domain $\mathbb{C}\backslash C$. For $z \in C$, denote the boundary values of $(\mathcal{C}^{C}\mathbf{f})(w)$ as w tends to z from the left (respectively, right) of C by $(\mathcal{C}^{C}_{+}\mathbf{f})(z)$ (respectively, $(\mathcal{C}^{C}_{-}\mathbf{f})(z)$). Then, we have the following lemma.

Lemma A.3.3 (Plemelj-Privalov). *Let $\mathbf{f} \in \mathrm{Lip}^{\nu}(C)$. The boundary values $(\mathcal{C}^{C}_{+}\mathbf{f})(z)$ and $(\mathcal{C}^{C}_{-}\mathbf{f})(z)$ exist independently of the path of approach to the boundary and are in $\mathrm{Lip}^{\nu}(C)$ as functions of $z \in C$. Moreover, the linear operators $\mathcal{C}^{C}_{\pm} : \mathrm{Lip}^{\nu}(C) \to \mathrm{Lip}^{\nu}(C)$ taking $\mathbf{f}(z)$ to the boundary values of the Cauchy integral are bounded with respect to the Hölder norm.*

The proof of this statement is in §§19 and 22 of Muskhelishvili [M53], with adjustments for the corners as described in his appendixes 1 and 2. It is a scalar result, but the matrix generalization requires only cosmetic alterations. The simple nature of the contour C guarantees that the boundary value operators are complementary projections so that the Plemelj formula holds: $(\mathcal{C}_+^C\mathbf{f})(z) - (\mathcal{C}_-^C\mathbf{f})(z) = \mathbf{f}(z)$. Also we have the operator identity $\mathcal{C}_+^C \circ \mathcal{C}_-^C = \mathcal{C}_-^C \circ \mathcal{C}_+^C = 0$.

The next statement of interest concerns certain commutators. Let $\mathbf{g} \in \mathrm{Lip}^\beta(C)$ for $0 < \beta < 1$, and consider the commutators $[\mathcal{C}_+^C, \mathcal{L}_\mathbf{g}]$, $[\mathcal{C}_-^C, \mathcal{L}_\mathbf{g}]$, $[\mathcal{C}_+^C, \mathcal{R}_\mathbf{g}]$, and $[\mathcal{C}_-^C, \mathcal{R}_\mathbf{g}]$. These operators can be interpreted as nonsingular integral operators by the formulas

$$\begin{aligned}([\mathcal{C}_+^C, \mathcal{L}_\mathbf{g}]\mathbf{f})(z) = ([\mathcal{C}_-^C, \mathcal{L}_\mathbf{g}]\mathbf{f})(z) &= \frac{1}{2\pi i}\oint_C (s-z)^{-1}(\mathbf{g}(s) - \mathbf{g}(z))\mathbf{f}(s)\,ds,\\ ([\mathcal{C}_+^C, \mathcal{R}_\mathbf{g}]\mathbf{f})(z) = ([\mathcal{C}_-^C, \mathcal{R}_\mathbf{g}]\mathbf{f})(z) &= \frac{1}{2\pi i}\oint_C (s-z)^{-1}\mathbf{f}(s)(\mathbf{g}(s) - \mathbf{g}(z))\,ds.\end{aligned} \tag{A.21}$$

These operators are essentially as nice as the function $\mathbf{g}(z)$ is. We have the following result.

Lemma A.3.4 *Let $\mathbf{g} \in \mathrm{Lip}^\beta(C)$. The operator $[\mathcal{C}_+^C, \mathcal{L}_\mathbf{g}] = [\mathcal{C}_-^C, \mathcal{L}_\mathbf{g}]$ and the operator $[\mathcal{C}_+^C, \mathcal{R}_\mathbf{g}] = [\mathcal{C}_-^C, \mathcal{R}_\mathbf{g}]$ are bounded operators from $\mathrm{Lip}^\alpha(C)$ to $\mathrm{Lip}^\beta(C)$ as long as $0 < \alpha \le 1$ and $0 \le \beta < 1$.*

This statement is proved for scalar functions (trivially extended to the matrix case, however) by Prössdorf in his Lemma 4.1 on page 100 and Corollary 4.2 on page 102 of [P78]. Again, some technical adjustment as described by Muskhelishvili [M53] will need to be used in order to admit corner points in the closed compact contour C. Note that the statement holds even if $\beta > \alpha$. In this sense, the commutator can *improve* the smoothness of the functions on which it acts.

A.4 GENERALIZATION FOR UMBILICAL CONTOURS

For an umbilical Riemann-Hilbert problem, let the simply connected components of Ω_U^+ be denoted $\Omega_\mathrm{U}^{+(k)}$ and the components of Ω_U^- be denoted $\Omega_\mathrm{U}^{-(k)}$. All but one of these components are bounded domains. For any simply connected domain D (possibly unbounded) with piecewise-smooth boundary, let ∂D denote the boundary oriented with D on the left. Then the oriented contour for the umbilical Riemann-Hilbert problem can be written either as

$$\Sigma_\mathrm{U} = \sum_k \partial\Omega_\mathrm{U}^{+(k)} \quad \text{or} \quad \Sigma_\mathrm{U} = -\sum_k \partial\Omega_\mathrm{U}^{-(k)}. \tag{A.22}$$

Note that while these formulas hold when we regard the contours as paths of integration of Hölder class functions, they do not hold in the set-theoretic sense of disjoint union since the self-intersection points $z = \alpha_i$ are each counted several times on the right-hand side and only once on the left; this is reflected in the use of the sum notation and the use of signs to denote orientation.

We begin by introducing some spaces. Let $A_\pm^\nu$ denote the set of matrix-valued functions $\mathbf{f}(z)$ on $\Sigma_\mathrm{U}\backslash\{\alpha_i\}$, which for each k may be assigned values at $z = \alpha_i$ so that the restriction of $\mathbf{f}(z)$ to $\partial\Omega_\mathrm{U}^{\pm(k)}$ is in $\mathrm{Lip}^\nu(\partial\Omega_\mathrm{U}^{\pm(k)})$. Note that at a fixed intersection point $z = \alpha_i$, the values given to $\mathbf{f}(z)$ to establish the required continuity will generally be different for each k. The continuity properties of functions in these

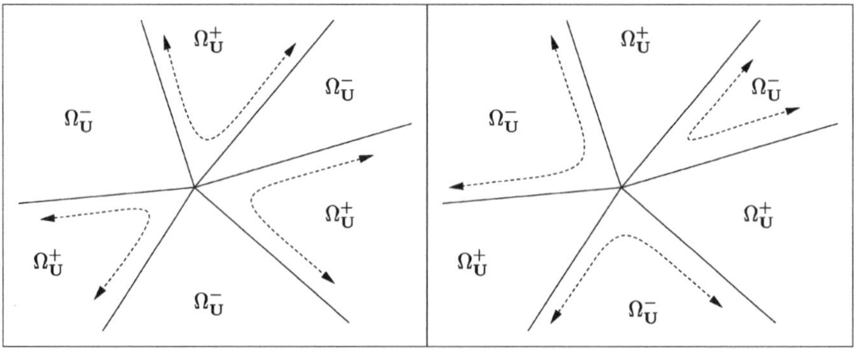

Figure A.6 Left: The dashed arrows indicate the continuity properties of a function in A_+^ν near a self-intersection point $z = \alpha_i$ of Σ_U. Right: The same for A_-^ν.

spaces are illustrated in figure A.6. These sets are Banach algebras with the norm being taken as the sum of the norms of the individual components:

$$\|\mathbf{f}\|_{A_\pm^\nu} := \sum_k \|\mathbf{f}\|_{\mathrm{Lip}^\nu(\partial\Omega_U^{\pm(k)})}, \tag{A.23}$$

where in each term in the sum on the right-hand side, the matrix function $\mathbf{f}$ is considered to be assigned definite values at all intersection points $z = \alpha_i$ (possibly different in each term) and subsequently to be restricted to the simple closed contour $\partial\Omega_U^{\pm(k)}$. Finally, note that every matrix function $\mathbf{f}$ in $A_+^\nu \cap A_-^\nu$ has a unique continuous extension to the whole closed contour Σ_U, and thus note that $A_+^\nu \cap A_-^\nu$ can be identified with the Banach algebra $\mathrm{Lip}^\nu(\Sigma_U)$ of matrix functions $\mathbf{f}(z)$ defined for $z \in \Sigma_U$ with the norm

$$\|\mathbf{f}\|_{\mathrm{Lip}^\nu(\Sigma_U)} := \sup_{z\in\Sigma_U} \|\mathbf{f}(z)\| + \sup_{z_1,\, z_2\in\Sigma_U} \frac{\|\mathbf{f}(z_2) - \mathbf{f}(z_1)\|}{|z_2 - z_1|^\nu}. \tag{A.24}$$

Convergence in $\mathrm{Lip}^\nu(\Sigma_U)$ is equivalent to simultaneous convergence in both A_+^ν and A_-^ν.

Now we use these spaces to state the appropriate generalizations of the results from §A.3 to the case of the umbilical contour Σ_U. First of all, Lemma A.3.1 can be applied to individual components of the boundary of $\Omega_U^\pm$ to prove an analogous result for these spaces.

Lemma A.4.1 *For $0 < \mu \le \nu$, the inclusion operator $\mathcal{I}_{\nu\to\mu}$ can be defined from A_+^ν to A_+^μ, from A_-^ν to A_-^μ, or from* $\mathrm{Lip}^\nu(\Sigma_U)$ *to* $\mathrm{Lip}^\mu(\Sigma_U)$. *It is a compact operator in all of these instances whenever $\mu < \nu$.*

Similarly, the obvious generalization of Lemma A.3.2 is the following.

Lemma A.4.2 *Let $\mathbf{g} \in A_+^\nu$. Then the multiplication operators $\mathcal{L}_\mathbf{g}$ and $\mathcal{R}_\mathbf{g}$ are bounded on A_+^ν as well as from* $\mathrm{Lip}^\nu(\Sigma_U)$ *to A_+^ν. Similarly, if $\mathbf{g} \in A_-^\nu$, then $\mathcal{L}_\mathbf{g}$ and $\mathcal{R}_\mathbf{g}$ are bounded on A_-^ν as well as from* $\mathrm{Lip}^\nu(\Sigma_U)$ *to A_-^ν. Finally, if $\mathbf{g} \in$* $\mathrm{Lip}^\nu(\Sigma_U)$, *then $\mathcal{L}_\mathbf{g}$ and $\mathcal{R}_\mathbf{g}$ are bounded on* $\mathrm{Lip}^\nu(\Sigma_U)$, *$A_+^\nu$, and A_-^ν.*

For a function $\mathbf{f} \in A^\nu_+$, we can write the corresponding Cauchy integral over the whole contour Σ_{U} in terms of Cauchy integrals over simple closed contours as defined in §A.3:

$$(\mathcal{C}^{\Sigma_{\mathrm{U}}}\mathbf{f})(z) := \frac{1}{2\pi i}\int_{\Sigma_{\mathrm{U}}}(s-z)^{-1}\mathbf{f}(s)\,ds = \sum_k \left(\mathcal{C}^{\partial\Omega_{\mathrm{U}}^{+(k)}}\mathbf{f}\right)(z), \tag{A.25}$$

where on the right-hand side we use the same symbol $\mathbf{f}$ to denote the various $\mathrm{Lip}^\nu(\partial\Omega_{\mathrm{U}}^{+(k)})$ completions and restrictions to the closed compact contours $\partial\Omega_{\mathrm{U}}^{+(k)}$. Likewise, for a function $\mathbf{f} \in A^\nu_-$, we can write

$$(\mathcal{C}^{\Sigma_{\mathrm{U}}}\mathbf{f})(z) = -\sum_k \left(\mathcal{C}^{\partial\Omega_{\mathrm{U}}^{-(k)}}\mathbf{f}\right)(z). \tag{A.26}$$

These Cauchy integrals are of course analytic in $\mathbb{C}\backslash\Sigma_{\mathrm{U}}$. These formulae allow the boundary values of the Cauchy integral over Σ to be expressed in terms of the boundary values of Cauchy integrals of Hölder class functions over simple closed contours as developed in §A.3. This allows those results to be generalized to the umbilical contour Σ_{U} with self-intersection points.

An intermediate result that we need is the following.

LEMMA A.4.3 *Let $j \neq k$ and consider the Cauchy operators $\mathcal{C}^{\partial\Omega_{\mathrm{U}}^{+(k)}}|_{\partial\Omega_{\mathrm{U}}^{+(j)}}$ and $\mathcal{C}^{\partial\Omega_{\mathrm{U}}^{-(k)}}|_{\partial\Omega_{\mathrm{U}}^{-(j)}}$, that is, the restrictions of the Cauchy operators over one simple component of $\partial\Omega_{\mathrm{U}}^{\pm}$ to another. Then the first of these is a bounded operator from $\mathrm{Lip}^\nu(\partial\Omega_{\mathrm{U}}^{+(k)})$ to $\mathrm{Lip}^\nu(\partial\Omega_{\mathrm{U}}^{+(j)})$, and the second of these is a bounded operator from $\mathrm{Lip}^\nu(\partial\Omega_{\mathrm{U}}^{-(k)})$ to $\mathrm{Lip}^\nu(\partial\Omega_{\mathrm{U}}^{-(j)})$.*

For a proof of (a stronger version of) this statement, see §22 of Muskhelishvili's book [M53]. To apply his result, we need only to observe that either $\partial\Omega_{\mathrm{U}}^{+(j)} \subset \Omega_{\mathrm{U}}^{+(k)} \cup \partial\Omega_{\mathrm{U}}^{+(k)}$ or $\partial\Omega_{\mathrm{U}}^{+(j)} \subset \mathbb{C}\backslash\Omega_{\mathrm{U}}^{+(k)}$, and similarly for $\partial\Omega_{\mathrm{U}}^{-(j)}$.

For $z \in \Sigma_{\mathrm{U}}\backslash\{\alpha_i\}$, we denote by $(\mathcal{C}_\pm^{\Sigma_{\mathrm{U}}}\mathbf{f})(z)$ the boundary value of the Cauchy integral $(\mathcal{C}^{\Sigma_{\mathrm{U}}}\mathbf{f})(w)$ as w tends to z from the region $\Omega_{\mathrm{U}}^{\pm}$. It is a consequence of the representations (A.25) and (A.26) of the Cauchy integral, along with Lemma A.4.3, that the following generalization of Lemma A.3.3 holds.

LEMMA A.4.4 Generalized Plemelj-Privalov). *The operator $\mathcal{C}_+^{\Sigma_{\mathrm{U}}}$ is bounded on A^ν_+ and is bounded from A^ν_- to the smaller space $\mathrm{Lip}^\nu(\Sigma_{\mathrm{U}})$. Similarly, the operator $\mathcal{C}_-^{\Sigma_{\mathrm{U}}}$ is bounded on A^ν_- and is bounded from A^ν_+ to $\mathrm{Lip}^\nu(\Sigma_{\mathrm{U}})$. Also, for $\mathbf{f} \in A^\nu_+ \cup A^\nu_-$ the Plemelj formula holds,*

$$(\mathcal{C}_+^{\Sigma_{\mathrm{U}}}\mathbf{f})(z) - (\mathcal{C}_-^{\Sigma_{\mathrm{U}}}\mathbf{f})(z) = \mathbf{f}(z), \tag{A.27}$$

as well as the relation $(\mathcal{C}_+^{\Sigma_{\mathrm{U}}} \circ \mathcal{C}_-^{\Sigma_{\mathrm{U}}}\mathbf{f})(z) = (\mathcal{C}_-^{\Sigma_{\mathrm{U}}} \circ \mathcal{C}_+^{\Sigma_{\mathrm{U}}}\mathbf{f})(z) \equiv 0$.

Proof. Consider first $\mathcal{C}_+^{\Sigma_{\mathrm{U}}}$ acting on $\mathbf{f} \in A^\nu_+$. Using the representation (A.25), the restriction of the result to a component $\partial\Omega_{\mathrm{U}}^{+(j)}$ of the boundary can be written as

$$(\mathcal{C}_+^{\Sigma_{\mathrm{U}}}\mathbf{f})|_{\partial\Omega_{\mathrm{U}}^{+(j)}} = \mathcal{C}_+^{\partial\Omega_{\mathrm{U}}^{+(j)}}\mathbf{f}|_{\partial\Omega_{\mathrm{U}}^{+(j)}} + \sum_{k\neq j}\left(\mathcal{C}^{\partial\Omega_{\mathrm{U}}^{+(k)}}\mathbf{f}\right)|_{\partial\Omega_{\mathrm{U}}^{+(j)}}. \tag{A.28}$$

That this operator is bounded to $\text{Lip}^\nu(\partial\Omega_{\mathrm{U}}^{+(j)})$ is clear from Lemma A.3.3 and Lemma A.4.3. Summing the norm estimates over the components $\partial\Omega_{\mathrm{U}}^{+(j)}$ then gives the boundedness of $\mathcal{C}_+^{\Sigma_{\mathrm{U}}} : A_+^\nu \to A_+^\nu$. Now consider $\mathcal{C}_+^{\Sigma_{\mathrm{U}}}$ acting on $\mathbf{f} \in A_-^\nu$. Using the representation (A.26), the restriction of the result to a component $\partial\Omega_{\mathrm{U}}^{-(j)}$ of the boundary can be written as

$$(\mathcal{C}_+^{\Sigma_{\mathrm{U}}}\mathbf{f})|_{\partial\Omega_{\mathrm{U}}^{-(j)}} = -\mathcal{C}_-^{\partial\Omega_{\mathrm{U}}^{-(j)}}\mathbf{f}\Big|_{\partial\Omega_{\mathrm{U}}^{-(j)}} - \sum_{k\neq j}\big(\mathcal{C}^{\partial\Omega_{\mathrm{U}}^{-(k)}}\mathbf{f}\big)\big|_{\partial\Omega_{\mathrm{U}}^{-(j)}}. \tag{A.29}$$

The boundary value on the right-hand side is "−" because the approach to the boundary is from the right of $\partial\Omega_{\mathrm{U}}^{-(j)}$ with its orientation. (Recall that by convention we are taking the boundary ∂D of a simply connected domain D to be oriented with D on the left.) By similar arguments, it is then clear that $\mathcal{C}_+^{\Sigma_{\mathrm{U}}}$ is bounded on A_-^ν. However, in this case, more is true. If $z = \alpha_i$ is one of the self-intersection points of Σ_{U}, then at the point $\alpha_i \in \partial\Omega_{\mathrm{U}}^{-(j)}$ the above formula reads

$$(\mathcal{C}_+^{\Sigma_{\mathrm{U}}}\mathbf{f})|_{\partial\Omega_{\mathrm{U}}^{-(j)}}(\alpha_i) = -\sum_k\big(\mathcal{C}_-^{\partial\Omega_{\mathrm{U}}^{-(k)}}\mathbf{f}\big)\big|_{\partial\Omega_{\mathrm{U}}^{-(j)}}(\alpha_i), \tag{A.30}$$

where the sum is taken over all components k, and it is clear that the value at the mutual corner point α_i is the same in all components $\partial\Omega_{\mathrm{U}}^{-(j)}$. This, along with the usual argument (as in [M53], page 13) that piecewise Hölder functions that are continuous are globally Hölder implies that $\mathcal{C}_+^{\Sigma_{\mathrm{U}}}$ is actually bounded from A_-^ν to $\text{Lip}^\nu(\Sigma_{\mathrm{U}})$. Similar arguments establish the analogous results for $\mathcal{C}_-^{\Sigma_{\mathrm{U}}}$. □

Similarly, the generalization of Lemma A.3.4 is the following.

Lemma A.4.5 *Let* $0 < \alpha \le 1$ *and* $0 \le \beta < 1$. *Suppose that* $\mathbf{g} \in \text{Lip}^\beta(\Sigma_{\mathrm{U}})$. *Then the commutators* $[\mathcal{C}_+^{\Sigma_{\mathrm{U}}}, \mathcal{R}_{\mathbf{g}}]$ *and* $[\mathcal{C}_+^{\Sigma_{\mathrm{U}}}, \mathcal{L}_{\mathbf{g}}]$ *are bounded from* A_-^α *to* $\text{Lip}^\beta(\Sigma_{\mathrm{U}})$, *and the commutators* $[\mathcal{C}_-^{\Sigma_{\mathrm{U}}}, \mathcal{R}_{\mathbf{g}}]$ *and* $[\mathcal{C}_-^{\Sigma_{\mathrm{U}}}, \mathcal{L}_{\mathbf{g}}]$ *are bounded from* A_+^α *to* $\text{Lip}^\beta(\Sigma_{\mathrm{U}})$. *Now suppose only that* $\mathbf{g} \in A_+^\beta$. *Then the commutators* $[\mathcal{C}_+^{\Sigma_{\mathrm{U}}}, \mathcal{R}_{\mathbf{g}}]$ *and* $[\mathcal{C}_+^{\Sigma_{\mathrm{U}}}, \mathcal{L}_{\mathbf{g}}]$ *are bounded from* A_+^α *to* A_+^β. *Similarly if* $\mathbf{g} \in A_-^\beta$, *then* $[\mathcal{C}_-^{\Sigma_{\mathrm{U}}}, \mathcal{R}_{\mathbf{g}}]$ *and* $[\mathcal{C}_-^{\Sigma_{\mathrm{U}}}, \mathcal{L}_{\mathbf{g}}]$ *are bounded from* A_-^α *to* A_-^β.

Proof. Again, one proceeds by decomposing the Cauchy operators $\mathcal{C}_\pm^{\Sigma_{\mathrm{U}}}$ according to (A.25) or (A.26) depending on the space and then applies Lemma A.3.4 componentwise. When the range of the transformation is $\text{Lip}^\beta(\Sigma_{\mathrm{U}})$, one shows continuity at the intersection points exactly as in the proof of Lemma A.4.4. □

A.5 FREDHOLD ALTERNATIVE FOR UMBILICAL RIEMANN-HILBERT PROBLEMS

In this section, we apply the Hölder theory of Cauchy integrals on the umbilical contour Σ_{U} to establish a Fredholm alternative theorem for umbilical Riemann-Hilbert Problem A.2.1. We begin by factoring the jump matrix $\mathbf{v}_{\mathrm{U}}(z)$ in a straightforward way.

LEMMA A.5.1 *Let $\Sigma_{\mathbf{U}}$ be an umbilical contour, and let there be given on $\Sigma_{\mathbf{U}}\backslash\{\alpha_i\}$ a matrix function $\mathbf{v}_{\mathbf{U}}(z)$ satisfying the conditions of Definition* A.2.1. *Then $\mathbf{v}_{\mathbf{U}}(z)$ admits a factorization $\mathbf{v}_{\mathbf{U}}(z) = \mathbf{b}^-(z)^{-1}\mathbf{b}^+(z)$ with $\mathbf{b}^\pm(z)$ invertible, where $\mathbf{b}^+ \in A_+^\nu$ and $\mathbf{b}^- \in A_-^\nu$.*

Proof. We construct such a factorization algorithmically as follows. The factorization will be carried out locally at each intersection point, so first select for each α_i a number $r_i > 0$ sufficiently small that $\Sigma_{\mathbf{U}} \cap B_{r_i}(\alpha_i)$, where $B_{r_i}(\alpha_i)$ is the open disk of radius r_i centered at $z = \alpha_i$, contains no other intersection points and that each circle of radius $r < r_i$ centered at α_i meets each arc terminating at α_i exactly once. For $z \in \Sigma_{\mathbf{U}} \cap (\bigcap_k B_{r_k}(\alpha_k)^c)$, where the superscript c denotes the complement (i.e., outside all disks), set $\mathbf{b}^+(z) \equiv \mathbf{v}_{\mathbf{U}}(z)$ and $\mathbf{b}^-(z) \equiv \mathbb{I}$. Now, letting α_i be an intersection point, we specify the factorization for $z \in \Sigma_{\mathbf{U}} \cap B_{r_i}(\alpha_i)$. Let the open arcs meeting at α_i be enumerated in counterclockwise order about α_i: $\Sigma_{\mathbf{U}}^{(1)}, \ldots, \Sigma_{\mathbf{U}}^{(N)}$ (here N is even but may depend on i). Let the unique point in $\Sigma_{\mathbf{U}}^{(j)}$ common to the boundary of $B_{r_i}(\alpha_i)$ be denoted z_j. For $z \in \Sigma_{\mathbf{U}}^{(1)} \cap B_{r_i}(\alpha_i)$ begin the factorization by setting $\mathbf{b}^-|_{\Sigma_{\mathbf{U}}^{(1)}} \equiv \mathbb{I}$ and therefore $\mathbf{b}^+|_{\Sigma_{\mathbf{U}}^{(1)}} \equiv \mathbf{v}_{\mathbf{U}}|_{\Sigma_{\mathbf{U}}^{(1)}}$. Now suppose that a factorization $\mathbf{b}^+|_{\Sigma_{\mathbf{U}}^{(j)}}$ and $\mathbf{b}^-|_{\Sigma_{\mathbf{U}}^{(j)}}$ has been constructed on $\Sigma_{\mathbf{U}}^{(j)} \cap B_{r_i}(\alpha_i)$. We now describe how to extend the factorization to $\Sigma_{\mathbf{U}}^{(j+1)} \cap B_{r_i}(\alpha_i)$. Suppose first that the region bounded by $\Sigma_{\mathbf{U}}^{(j)}$, $\Sigma_{\mathbf{U}}^{(j+1)}$, and the boundary of the disk is a component $\Omega_{\mathbf{U}}^{+(k)}$ of the "+" region $\Omega_{\mathbf{U}}^+$. Let $\mathbf{s}(\rho)$ be a C_1-map from $[0, r_i]$ into $GL(n, \mathbb{C})$ with $\mathbf{s}(0) = \mathbf{b}^+|_{\Sigma_{\mathbf{U}}^{(j)}}(\alpha_i)$ and $\mathbf{s}(r_i) = \mathbf{v}_{\mathbf{U}}|_{\Sigma_{\mathbf{U}}^{(j+1)}}(z_{j+1})$. Such a map exists because $GL(n, \mathbb{C})$ is arcwise-connected [SW86]. Then for $z \in \Sigma_{\mathbf{U}}^{(j+1)} \cap B_{r_i}(\alpha_i)$, we set $\mathbf{b}^+(z) = \mathbf{s}(|z - \alpha_i|)$ and $\mathbf{b}^-(z) = \mathbf{s}(|z - \alpha_i|)\mathbf{v}_{\mathbf{U}}(z)^{-1}$. On the other hand, if the region bounded by $\Sigma_{\mathbf{U}}^{(j)}$, $\Sigma_{\mathbf{U}}^{(j+1)}$, and the boundary of the disk is a component of $\Omega_{\mathbf{U}}^-$, then we take $\mathbf{s}(\rho)$ to be a C_1-map from $[0, r_i]$ into $GL(n, \mathbb{C})$ with $\mathbf{s}(0) = \mathbf{b}^-|_{\Sigma_{\mathbf{U}}^{(j)}}(\alpha_i)$ and $\mathbf{s}(r_i) = \mathbb{I}$, and then for $z \in \Sigma_{\mathbf{U}}^{(j+1)} \cap B_{r_i}(\alpha_i)$ we set $\mathbf{b}^-(z) = \mathbf{s}(|z - \alpha_i|)$ and then $\mathbf{b}^+(z) = \mathbf{s}(|z - \alpha_i|)\mathbf{v}_{\mathbf{U}}(z)$. Using this algorithm, we then construct factorizations on the part of each open arc $\Sigma_{\mathbf{U}}^{(j)}$ within $B_{r_i}(\alpha_i)$ starting from $\Sigma_{\mathbf{U}}^{(1)}$ and working counterclockwise about $z = \alpha_i$. This construction, when carried out under the compatibility condition (cf. Definition A.2.1) satisfied by the limiting jump matrices $\mathbf{v}_{\mathbf{U}}|_{\Sigma_{\mathbf{U}}^{(j)}}$ at each endpoint, guarantees that the restrictions $\mathbf{b}^\pm|_{\Sigma_{\mathbf{U}}^{(j)}}$ uniformly satisfy the Hölder-continuity condition with exponent ν on each open arc $\Sigma_{\mathbf{U}}^{(j)}$. This follows from the interior smoothness condition, the continuity of the factorization at the disk boundaries, the Banach algebra property of Hölder-continuous functions, and the fact that C_1-functions are Hölder-continuous with any exponent less than or equal to one. The construction also guarantees that $\mathbf{b}^+|_{\partial\Omega_{\mathbf{U}}^{+(k)}}$ may be defined at each intersection point α_i to be continuous at the corner points for each k and likewise that $\mathbf{b}^-|_{\partial\Omega_{\mathbf{U}}^{-(k)}}$ may be defined to be continuous at the corner points for each k. Then it follows (see [M53], page 13 and the appendixes), that the functions $\mathbf{b}^+(z)$ and $\mathbf{b}^-(z)$ are in A_+^ν and A_-^ν, respectively. Finally, the invertibility (and in fact the unimodularity) of $\mathbf{b}^\pm(z)$ follows directly from the above algorithm and the unimodularity of $\mathbf{v}_{\mathbf{U}}(z)$. □

Now, set $\mathbf{w}^+(z) := \mathbf{b}^+(z) - \mathbb{I} \in A^\nu_+$ and $\mathbf{w}^-(z) := \mathbb{I} - \mathbf{b}^-(z) \in A^\nu_-$. Choose any positive μ with $\mu \le \nu$, and define the operator $\mathcal{C}_\mathbf{w}$ on the space $\mathrm{Lip}^\mu(\Sigma_\mathrm{U})$ by

$$\begin{aligned}(\mathcal{C}_\mathbf{w}\mathbf{m})(z) &:= (\mathcal{C}_+^{\Sigma_\mathrm{U}} \circ \mathcal{R}_{\mathbf{w}^-}\mathbf{m})(z) + (\mathcal{C}_-^{\Sigma_\mathrm{U}} \circ \mathcal{R}_{\mathbf{w}^+}\mathbf{m})(z) \\ &= (\mathcal{C}_+^{\Sigma_\mathrm{U}}(\mathbf{m}\mathbf{w}^-))(z) + (\mathcal{C}_-^{\Sigma_\mathrm{U}}(\mathbf{m}\mathbf{w}^+))(z). \end{aligned} \tag{A.31}$$

It follows from Lemma A.4.4 that this formula indeed defines a function in the Banach space $\mathrm{Lip}^\mu(\Sigma_\mathrm{U})$. For example, in the first term, the multiplication operator is a map from $\mathrm{Lip}^\mu(\Sigma_\mathrm{U})$ to A^μ_- (with the function $\mathbf{w}^-$ reinterpreted under the inclusion map as an element of A^μ_- since $\mu \le \nu$), and then the operator $\mathcal{C}_+^{\Sigma_\mathrm{U}}$ brings us back from this space to $\mathrm{Lip}^\mu(\Sigma_\mathrm{U})$, according to Lemma A.4.4. The second term is understood similarly. Moreover, as the composition of bounded operators, $\mathcal{C}_\mathbf{w}$ is itself a bounded operator on $\mathrm{Lip}^\mu(\Sigma_\mathrm{U})$.

The umbilical Riemann-Hilbert Problem A.2.1 can now be reformulated as a singular integral equation in $\mathrm{Lip}^\mu(\Sigma_\mathrm{U})$.

Lemma A.5.2 *Consider the umbilical Riemann-Hilbert Problem* A.2.1 *with data* $(\Sigma_\mathrm{U}, \mathbf{v}_\mathrm{U}(z), \mathbf{N}_\mathrm{U})$. *Let the normalization matrix* $\mathbf{N}_\mathrm{U}$ *be identified with a constant function in* $\mathrm{Lip}^\mu(\Sigma_\mathrm{U})$, *and suppose the jump matrix* $\mathbf{v}_\mathrm{U}(z)$ *satisfying the conditions of Definition* A.2.1 *to be factored according to Lemma* A.5.1, *with* $\mathcal{C}_\mathbf{w}$ *being the corresponding singular integral operator in* $\mathrm{Lip}^\mu(\Sigma_\mathrm{U})$. *Then, the solutions* $\mathbf{U}(z)$ *of the umbilical Riemann-Hilbert Problem* A.2.1 *are in one-to-one correspondence with the solutions* $\mathbf{m} \in \mathrm{Lip}^\mu(\Sigma_\mathrm{U})$ *of the integral equation*

$$\mathbf{m}(z) - (\mathcal{C}_\mathbf{w}\mathbf{m})(z) = \mathbf{N}_\mathrm{U}. \tag{A.32}$$

Proof. First suppose that we are given a solution $\mathbf{m}(z)$ of the integral equation (A.32) in $\mathrm{Lip}^\mu(\Sigma_\mathrm{U})$. For $z \in \mathbb{C}\backslash\Sigma_\mathrm{U}$ define

$$\begin{aligned}\mathbf{U}(z;\mathbf{m}) &:= \mathbf{N}_\mathrm{U} + (\mathcal{C}^{\Sigma_\mathrm{U}} \circ \mathcal{R}_{\mathbf{w}^+}\mathbf{m})(z) + (\mathcal{C}^{\Sigma_\mathrm{U}} \circ \mathcal{R}_{\mathbf{w}^-}\mathbf{m})(z) \\ &= \mathbf{N}_\mathrm{U} + (\mathcal{C}^{\Sigma_\mathrm{U}}(\mathbf{m}\mathbf{w}^+))(z) + (\mathcal{C}^{\Sigma_\mathrm{U}}(\mathbf{m}\mathbf{w}^-))(z). \end{aligned} \tag{A.33}$$

Then $\mathbf{U}(z;\mathbf{m})$ is a solution of the umbilical Riemann-Hilbert Problem A.2.1. The analyticity of $\mathbf{U}(z;\mathbf{m})$ in $\mathbb{C}\backslash\Sigma_\mathrm{U}$ and the normalization $\mathbf{U}(\infty;\mathbf{m}) = \mathbf{N}_\mathrm{U}$ follow directly from the representation (A.33) and the properties of elements of $\mathrm{Lip}^\mu(\Sigma_\mathrm{U})$. That the $A^\mu_\pm$ boundary values of $\mathbf{U}(z;\mathbf{m})$ satisfy the jump relations follows from simply inserting (A.33) into the jump relations and using the Plemelj formula in conjunction with (A.32).

To show the injectivity of the map $\mathbf{m}(z) \to \mathbf{U}(z;\mathbf{m})$, observe that the Cauchy integral representation (A.33) implies that $\mathbf{U}(z;\mathbf{m}_2) \equiv \mathbf{U}(z;\mathbf{m}_1)$ if and only if $(\mathbf{m}_2(z) - \mathbf{m}_1(z))(\mathbf{w}^+(z) + \mathbf{w}^-(z)) \equiv \mathbf{0}$. At the same time, since $\mathbf{m}_1(z)$ and $\mathbf{m}_2(z)$ both satisfy (A.32), it follows that

$$(\mathbf{m}_2(z) - \mathbf{m}_1(z)) - (\mathcal{C}_+^{\Sigma_\mathrm{U}}((\mathbf{m}_2 - \mathbf{m}_1)\mathbf{w}^-))(z) - (\mathcal{C}_-^{\Sigma_\mathrm{U}}((\mathbf{m}_2 - \mathbf{m}_1)\mathbf{w}^+))(z) \equiv \mathbf{0}. \tag{A.34}$$

Putting these two together gives

$$\begin{aligned}\mathbf{0} &\equiv (\mathbf{m}_2(z) - \mathbf{m}_1(z)) + (\mathcal{C}_+^{\Sigma_\mathrm{U}}((\mathbf{m}_2 - \mathbf{m}_1)\mathbf{w}^+))(z) - (\mathcal{C}_-^{\Sigma_\mathrm{U}}((\mathbf{m}_2 - \mathbf{m}_1)\mathbf{w}^+))(z) \\ &= (\mathbf{m}_2(z) - \mathbf{m}_1(z)) + (\mathbf{m}_2(z) - \mathbf{m}_1(z))\mathbf{w}^+(z) \\ &= (\mathbf{m}_2(z) - \mathbf{m}_1(z))\mathbf{b}^+(z), \end{aligned} \tag{A.35}$$

where we have used the Plemelj formula. From the invertibility of $\mathbf{b}^+(z)$, it follows that $\mathbf{m}_2(z) \equiv \mathbf{m}_1(z)$.

On the other hand, suppose we are given a solution $\mathbf{U}(z)$ of the umbilical Riemann-Hilbert Problem A.2.1. For $z \in \Sigma_\mathrm{U} \backslash \{\alpha_i\}$, set

$$\mathbf{m}(z; \mathbf{U}) := \mathbf{U}_+(z)\mathbf{b}^+(z)^{-1} = \mathbf{U}_-(z)\mathbf{b}^-(z)^{-1}. \tag{A.36}$$

That these two expressions yield the same result follows from the factorization of the jump matrix established in Lemma A.5.1 and the jump conditions satisfied by $\mathbf{U}(z)$ on Σ_U. Moreover, it is clear that $\mathbf{m}(\cdot; \mathbf{U})$ is in both A_+^μ and A_-^μ; therefore, it is an element of $\mathrm{Lip}^\mu(\Sigma_\mathrm{U})$. Also, the map $\mathbf{U}(z) \to \mathbf{m}(z; \mathbf{U})$ is injective because the matrix functions $\mathbf{b}^\pm(z)$ are invertible. Now observe that

$$\begin{aligned}
&\mathbf{m}(z; \mathbf{U}) - (\mathcal{C}_\mathbf{w}\mathbf{m}(\cdot; \mathbf{U}))(z) \\
&\quad = \mathbf{m}(z; \mathbf{U}) - (\mathcal{C}_+^{\Sigma_\mathrm{U}}(\mathbf{U}_-(\mathbf{b}^-)^{-1}\mathbf{w}^-))(z) - (\mathcal{C}_-^{\Sigma_\mathrm{U}}(\mathbf{U}_+(\mathbf{b}^+)^{-1}\mathbf{w}^+))(z) \\
&\quad = \mathbf{m}(z; \mathbf{U}) - (\mathcal{C}_+^{\Sigma_\mathrm{U}}(\mathbf{U}_-(\mathbf{b}^-)^{-1} - \mathbf{U}_-))(z) - (\mathcal{C}_-^{\Sigma_\mathrm{U}}(\mathbf{U}_+ - \mathbf{U}_+(\mathbf{b}^+)^{-1}))(z) \\
&\quad = \mathbf{m}(z; \mathbf{U}) - (\mathcal{C}_+^{\Sigma_\mathrm{U}}\mathbf{m}(\cdot; \mathbf{U}))(z) + (\mathcal{C}_-^{\Sigma_\mathrm{U}}\mathbf{m}(\cdot; \mathbf{U}))(z) \\
&\qquad + (\mathcal{C}_+^{\Sigma_\mathrm{U}}\mathbf{U}_-)(z) - (\mathcal{C}_-^{\Sigma_\mathrm{U}}\mathbf{U}_+)(z) \\
&\quad = (\mathcal{C}_+^{\Sigma_\mathrm{U}}\mathbf{U}_-)(z) - (\mathcal{C}_-^{\Sigma_\mathrm{U}}\mathbf{U}_+)(z) \\
&\quad = \mathbf{N}_\mathrm{U}.
\end{aligned} \tag{A.37}$$

Here, in the next-to-last step we have used the Plemelj formula, and in the final step we have used the continuity of the boundary values and computed a residue at $z = \infty$, which necessarily occurs within a component of either Ω_U^+ or Ω_U^-.

Finally, a similar argument shows that the composition of these two correspondences is the identity mapping. Consider (A.33) evaluated for $\mathbf{m}(z; \mathbf{U})$:

$$\begin{aligned}
\mathbf{N}_\mathrm{U} &+ (\mathcal{C}^{\Sigma_\mathrm{U}} \circ \mathcal{R}_{\mathbf{w}^+}\mathbf{m}(\cdot; \mathbf{U}))(z) + (\mathcal{C}^{\Sigma_\mathrm{U}} \circ \mathcal{R}_{\mathbf{w}^-}\mathbf{m}(\cdot; \mathbf{U}))(z) \\
&= \mathbf{N}_\mathrm{U} + (\mathcal{C}^{\Sigma_\mathrm{U}}\mathbf{U}_+)(z) + (\mathcal{C}^{\Sigma_\mathrm{U}}\mathbf{U}_-)(z) \\
&= \mathbf{U}(z),
\end{aligned} \tag{A.38}$$

with the last equality following from Cauchy's theorem. □

So, solving the umbilical Riemann-Hilbert Problem A.2.1 amounts to inverting the operator $1 - \mathcal{C}_\mathbf{w}$ on the Banach space $\mathrm{Lip}^\mu(\Sigma_\mathrm{U})$ or at least defining the inverse on the subspace of constant functions $\mathbf{N}_\mathrm{U}$. We note for future use the following corollary of Lemma A.5.2.

Corollary A.5.1 *Suppose* $\mathbf{w}^\pm(z) \in A_\pm^\nu$ *as defined following* (A.22). *Let* $\mathbf{m}_0 \in \ker(1 - \mathcal{C}_\mathbf{w})$, *with* $\mathbf{m}_0 \not\equiv 0$. *Then* $\mathbf{U}_0(z) := (\mathcal{C}^{\Sigma_\mathrm{U}}(\mathbf{m}_0\mathbf{w}^+))(z) + (\mathcal{C}^{\Sigma_\mathrm{U}}(\mathbf{m}_0\mathbf{w}^-))(z)$ *is a nontrivial solution of the homogeneous umbilical Riemann-Hilbert Problem* A.2.1 *with data* $(\Sigma_\mathrm{U}, \mathbf{v}_\mathrm{U}(z), \mathbf{0})$.

We continue our analysis by studying the integral equation (A.32). Let $\tilde{\mathbf{w}}^+(z) = \mathbf{b}^+(z)^{-1} - \mathbb{I} \in A_+^\nu$ and $\tilde{\mathbf{w}}^-(z) = \mathbb{I} - \mathbf{b}^-(z)^{-1} \in A_-^\nu$. Along with the operator $\mathcal{C}_\mathbf{w}$, we consider also another bounded operator on $\mathrm{Lip}^\mu(\Sigma_\mathrm{U})$ defined by

$$(\mathcal{C}_{\tilde{\mathbf{w}}}\mathbf{m})(z) = (\mathcal{C}_+^{\Sigma_\mathrm{U}} \circ \mathcal{R}_{\tilde{\mathbf{w}}^-}\mathbf{m})(z) + (\mathcal{C}_-^{\Sigma_\mathrm{U}} \circ \mathcal{R}_{\tilde{\mathbf{w}}^+}\mathbf{m})(z). \tag{A.39}$$

On the space $\mathrm{Lip}^{\mu}(\Sigma_{\mathbf{U}})$, we have

$$(1-\mathcal{C}_{\tilde{\mathbf{w}}})\circ(1-\mathcal{C}_{\mathbf{w}})=1+\mathcal{T}_{\tilde{\mathbf{w}},\mathbf{w}} \tag{A.40}$$

and

$$(1-\mathcal{C}_{\mathbf{w}})\circ(1-\mathcal{C}_{\tilde{\mathbf{w}}})=1+\mathcal{T}_{\mathbf{w},\tilde{\mathbf{w}}}, \tag{A.41}$$

where

$$\begin{aligned}(\mathcal{T}_{\tilde{\mathbf{w}},\mathbf{w}}\mathbf{m})(z) &:= (\mathcal{C}_{+}^{\Sigma_{\mathbf{U}}}\circ\mathcal{R}_{\tilde{\mathbf{w}}^-}\circ\mathcal{C}_{-}^{\Sigma_{\mathbf{U}}}\circ\mathcal{R}_{\mathbf{w}^+}\mathbf{m})(z)\\ &\quad+(\mathcal{C}_{+}^{\Sigma_{\mathbf{U}}}\circ\mathcal{R}_{\tilde{\mathbf{w}}^-}\circ\mathcal{C}_{-}^{\Sigma_{\mathbf{U}}}\circ\mathcal{R}_{\mathbf{w}^-}\mathbf{m})(z)\\ &\quad+(\mathcal{C}_{-}^{\Sigma_{\mathbf{U}}}\circ\mathcal{R}_{\tilde{\mathbf{w}}^+}\circ\mathcal{C}_{+}^{\Sigma_{\mathbf{U}}}\circ\mathcal{R}_{\mathbf{w}^+}\mathbf{m})(z)\\ &\quad+(\mathcal{C}_{-}^{\Sigma_{\mathbf{U}}}\circ\mathcal{R}_{\tilde{\mathbf{w}}^+}\circ\mathcal{C}_{+}^{\Sigma_{\mathbf{U}}}\circ\mathcal{R}_{\mathbf{w}^-}\mathbf{m})(z)\\ &:=(\mathcal{T}^{(1)}_{\tilde{\mathbf{w}},\mathbf{w}}\mathbf{m})(z)+(\mathcal{T}^{(2)}_{\tilde{\mathbf{w}},\mathbf{w}}\mathbf{m})(z)+(\mathcal{T}^{(3)}_{\tilde{\mathbf{w}},\mathbf{w}}\mathbf{m})(z)+(\mathcal{T}^{(4)}_{\tilde{\mathbf{w}},\mathbf{w}}\mathbf{m})(z),\end{aligned} \tag{A.42}$$

and $\mathcal{T}_{\mathbf{w},\tilde{\mathbf{w}}}$ is similarly defined with the roles of $\mathbf{w}^{\pm}$ and $\tilde{\mathbf{w}}^{\pm}$ exchanged. The formulae (A.40) and (A.41) require only the Plemelj relation and the fact that $\mathbf{w}^+(z)\tilde{\mathbf{w}}^+(z)=-(\mathbf{w}^+(z)+\tilde{\mathbf{w}}^+(z))$ and $\mathbf{w}^-(z)\tilde{\mathbf{w}}^-(z)=\mathbf{w}^-(z)+\tilde{\mathbf{w}}^-(z)$. The results from §A.4 can now be used to show the following.

Lemma A.5.3 *With $\mathbf{w}^+$ and $\tilde{\mathbf{w}}^+$ in A_+^{ν} and with $\mathbf{w}^-$ and $\tilde{\mathbf{w}}^-$ in A_-^{ν}, the operators $\mathcal{T}_{\tilde{\mathbf{w}},\mathbf{w}}$ and $\mathcal{T}_{\mathbf{w},\tilde{\mathbf{w}}}$ are compact on $\mathrm{Lip}^{\mu}(\Sigma_{\mathbf{U}})$ whenever $\mu<\nu$.*

Proof. It is sufficient to prove the result for $\mathcal{T}_{\tilde{\mathbf{w}},\mathbf{w}}$. Consider the second term. Because $\mathcal{C}_{+}^{\Sigma_{\mathbf{U}}}\circ\mathcal{C}_{-}^{\Sigma_{\mathbf{U}}}=0$ on A_-^{μ}, we can write

$$\begin{aligned}(\mathcal{T}^{(2)}_{\tilde{\mathbf{w}},\mathbf{w}}\mathbf{m})(z) &:= (\mathcal{C}_{+}^{\Sigma_{\mathbf{U}}}\circ\mathcal{R}_{\tilde{\mathbf{w}}^-}\circ\mathcal{C}_{-}^{\Sigma_{\mathbf{U}}}\circ\mathcal{R}_{\mathbf{w}^-}\mathbf{m})(z)\\ &= -(\mathcal{C}_{+}^{\Sigma_{\mathbf{U}}}\circ[\mathcal{C}_{-}^{\Sigma_{\mathbf{U}}},\mathcal{R}_{\tilde{\mathbf{w}}^-}]\circ\mathcal{R}_{\mathbf{w}^-}\mathbf{m})(z).\end{aligned} \tag{A.43}$$

Similarly, because $\mathcal{C}_{-}^{\Sigma_{\mathbf{U}}}\circ\mathcal{C}_{+}^{\Sigma_{\mathbf{U}}}=0$ on A_+^{μ}, the third term can be written in the form

$$\begin{aligned}(\mathcal{T}^{(3)}_{\tilde{\mathbf{w}},\mathbf{w}}\mathbf{m})(z) &:= (\mathcal{C}_{-}^{\Sigma_{\mathbf{U}}}\circ\mathcal{R}_{\tilde{\mathbf{w}}^+}\circ\mathcal{C}_{+}^{\Sigma_{\mathbf{U}}}\circ\mathcal{R}_{\mathbf{w}^+}\mathbf{m})(z)\\ &= -(\mathcal{C}_{-}^{\Sigma_{\mathbf{U}}}\circ[\mathcal{C}_{+}^{\Sigma_{\mathbf{U}}},\mathcal{R}_{\tilde{\mathbf{w}}^+}]\circ\mathcal{R}_{\mathbf{w}^+}\mathbf{m})(z).\end{aligned} \tag{A.44}$$

To handle the first term, we first use the Plemelj formula to decompose $\tilde{\mathbf{w}}^-(z)=\tilde{\mathbf{w}}^-_+(z)-\tilde{\mathbf{w}}^-_-(z)$, where $\tilde{\mathbf{w}}^-_+(z)=(\mathcal{C}_{+}^{\Sigma_{\mathbf{U}}}\tilde{\mathbf{w}}^-)(z)$ has an analytic continuation into $\Omega^+_{\mathbf{U}}$ and $\tilde{\mathbf{w}}^-_-(z)=(\mathcal{C}_{-}^{\Sigma_{\mathbf{U}}}\tilde{\mathbf{w}}^-)(z)$ has an analytic continuation into $\Omega^-_{\mathbf{U}}$. In whichever region contains $z=\infty$, the corresponding function decays like $1/z$. Using this decomposition, we find

$$\begin{aligned}(\mathcal{T}^{(1)}_{\tilde{\mathbf{w}},\mathbf{w}}\mathbf{m})(z) &:= (\mathcal{C}_{+}^{\Sigma_{\mathbf{U}}}\circ\mathcal{R}_{\tilde{\mathbf{w}}^-}\circ\mathcal{C}_{-}^{\Sigma_{\mathbf{U}}}\circ\mathcal{R}_{\mathbf{w}^+}\mathbf{m})(z)\\ &= (\mathcal{C}_{+}^{\Sigma_{\mathbf{U}}}\circ\mathcal{R}_{\tilde{\mathbf{w}}^-_+}\circ\mathcal{C}_{-}^{\Sigma_{\mathbf{U}}}\circ\mathcal{R}_{\mathbf{w}^+}\mathbf{m})(z)-(\mathcal{C}_{+}^{\Sigma_{\mathbf{U}}}\circ\mathcal{R}_{\tilde{\mathbf{w}}^-_-}\circ\mathcal{C}_{-}^{\Sigma_{\mathbf{U}}}\circ\mathcal{R}_{\mathbf{w}^+}\mathbf{m})(z)\\ &= (\mathcal{C}_{+}^{\Sigma_{\mathbf{U}}}\circ\mathcal{R}_{\tilde{\mathbf{w}}^-_+}\circ\mathcal{C}_{-}^{\Sigma_{\mathbf{U}}}\circ\mathcal{R}_{\mathbf{w}^+}\mathbf{m})(z).\end{aligned} \tag{A.45}$$

The term that vanishes above does so because it is of the form $\mathcal{C}_{+}^{\Sigma_{\mathbf{U}}}$ acting on a product of functions each having an analytic extension to $\Omega^-_{\mathbf{U}}$ and decaying appropriately if $\infty\in\Omega^-_{\mathbf{U}}$. Now, since by Lemma A.4.4, $\tilde{\mathbf{w}}^-_+\in A^{\mu}_+$, we can again use $\mathcal{C}_{+}^{\Sigma_{\mathbf{U}}}\circ\mathcal{C}_{-}^{\Sigma_{\mathbf{U}}}=0$ on this Banach algebra to finally write the first term of $\mathcal{T}_{\tilde{\mathbf{w}},\mathbf{w}}$ in the form

$$(\mathcal{T}^{(1)}_{\tilde{\mathbf{w}},\mathbf{w}}\mathbf{m})(z)=-(\mathcal{C}_{+}^{\Sigma_{\mathbf{U}}}\circ[\mathcal{C}_{-}^{\Sigma_{\mathbf{U}}},\mathcal{R}_{\tilde{\mathbf{w}}^-_+}]\circ\mathcal{R}_{\mathbf{w}^+}\mathbf{m})(z). \tag{A.46}$$

Similarly writing $\tilde{\mathbf{w}}^+(z) = \tilde{\mathbf{w}}_+^+(z) - \tilde{\mathbf{w}}_-^+(z)$ with $\tilde{\mathbf{w}}_\pm^+(z) = (\mathcal{C}_\pm^{\Sigma_U}\tilde{\mathbf{w}}^+)(z)$ and applying similar reasoning, one finds that the fourth term in $\mathcal{T}_{\tilde{\mathbf{w}},\mathbf{w}}$ can be written as

$$(\mathcal{T}_{\tilde{\mathbf{w}},\mathbf{w}}^{(4)}\mathbf{m})(z) = +(\mathcal{C}_-^{\Sigma_U} \circ [\mathcal{C}_+^{\Sigma_U}, \mathcal{R}_{\tilde{\mathbf{w}}_-^+}] \circ \mathcal{R}_{\mathbf{w}^-}\mathbf{m})(z). \tag{A.47}$$

With each term of $\mathcal{T}_{\tilde{\mathbf{w}},\mathbf{w}}$ written in this way, it is not hard to see the compactness from more basic results already summarized. Consider $\mathcal{T}_{\tilde{\mathbf{w}},\mathbf{w}}^{(1)}$ written in terms of the commutator as (A.46). The multiplication operator $\mathcal{R}_{\mathbf{w}^+}$ is bounded from $\mathrm{Lip}^\mu(\Sigma_U)$ to A_+^μ. Then, from Lemma A.4.5, the commutator is a bounded operator from A_+^μ to the better space $\mathrm{Lip}^\nu(\Sigma_U)$. The result of this operation can be reinterpreted as an element of $\mathrm{Lip}^\mu(\Sigma_U)$, and from Lemma A.4.1, the inclusion map $I_{\nu\to\mu} : \mathrm{Lip}^\nu(\Sigma_U) \to \mathrm{Lip}^\mu(\Sigma_U)$ is a compact operator. The trivial inclusion map $\mathrm{Lip}^\mu(\Sigma_U) \to A_+^\mu$ is of course bounded, and finally composition with the operator $\mathcal{C}_-^{\Sigma_U}$, bounded from A_+^μ to $\mathrm{Lip}^\mu(\Sigma_U)$ by Lemma A.4.4, does not alter the compactness. The term $\mathcal{T}_{\tilde{\mathbf{w}},\mathbf{w}}^{(4)}$ expressed by (A.47) is handled similarly. The term $\mathcal{T}_{\tilde{\mathbf{w}},\mathbf{w}}^{(2)}$ given in (A.43) is shown to be compact as follows. The multiplication operator $\mathcal{R}_{\mathbf{w}^-}$ is a bounded map from $\mathrm{Lip}^\mu(\Sigma_U)$ to A_-^μ. By Lemma A.4.5, the commutator is then a bounded map from A_-^μ to the better space A_-^ν. Again, the inclusion map $I_{\nu\to\mu} : A_-^\nu \to A_-^\mu$ is compact by Lemma A.4.1. Finally, by Lemma A.4.4 the operator $\mathcal{C}_+^{\Sigma_U}$ is bounded from A_-^μ to $\mathrm{Lip}^\mu(\Sigma_U)$, and compactness is retained. The term $\mathcal{T}_{\tilde{\mathbf{w}},\mathbf{w}}^{(3)}$ written in the form (A.5) is handled similarly. All four terms of $\mathcal{T}_{\tilde{\mathbf{w}},\mathbf{w}}$ have thus been shown to be compact on $\mathrm{Lip}^\mu(\Sigma_U)$. □

We have the following result.

Lemma A.5.4 *The bounded operator* $1 - \mathcal{C}_{\mathbf{w}}$, *with the matrices* $\mathbf{w}^\pm(z) \in A_\pm^\nu$ *as defined following* (A.22), *is Fredholm on the Banach space* $\mathrm{Lip}^\mu(\Sigma_U)$ *for all* μ *with* $0 < \mu < \nu$ *and has index zero.*

Proof. By Lemma A.5.3 the operator $1 - \mathcal{C}_{\tilde{\mathbf{w}}}$ serves as both a left and a right pseudoinverse for $1 - \mathcal{C}_{\mathbf{w}}$, and therefore dim ker $(1 - \mathcal{C}_{\mathbf{w}})$ and dim coker $(1 - \mathcal{C}_{\mathbf{w}})$ are both finite. This proves that $1 - \mathcal{C}_{\mathbf{w}}$ is a Fredholm operator on $\mathrm{Lip}^\mu(\Sigma_U)$. To calculate the index, we invoke continuity of the index for Fredholm operators with respect to uniform operator norm in $\mathrm{Lip}^\mu(\Sigma_U)$. The same arguments as above prove that the family of operators $1 - \epsilon\mathcal{C}_{\mathbf{w}}$ is Fredholm for all $\epsilon \in \mathbb{C}$ and in particular for those real ϵ between 0 and 1. By the boundedness of $\mathcal{C}_{\mathbf{w}}$ on $\mathrm{Lip}^\mu(\Sigma_U)$, this family of operators is continuous in operator norm as a function of $\epsilon \in [0, 1]$. Since for $\epsilon = 0$ the $\mathrm{ind}(1 - \epsilon\mathcal{C}_{\mathbf{w}}) = 0$ and since the index is a continuous integer-valued function over the whole range of $\epsilon \in [0, 1]$, we conclude that $\mathrm{ind}(1 - \mathcal{C}_{\mathbf{w}}) = 0$. □

Corollary A.5.2 *With* $\mathbf{w}^\pm(z) \in A_\pm^\nu$ *as above,* $1 - \mathcal{C}_{\mathbf{w}}$ *has a bounded inverse defined on* $\mathrm{Lip}^\mu(\Sigma_U)$ *for all positive* $\mu < \nu$ *whenever* $\ker(1 - \mathcal{C}_{\mathbf{w}}) = \{0\}$.

Proof. Since $\mathrm{ind}(1 - \mathcal{C}_{\mathbf{w}}) = 0$, $\ker(1 - \mathcal{C}_{\mathbf{w}}) = \{0\}$ implies $\dim\mathrm{coker}(1 - \mathcal{C}_{\mathbf{w}}) = 0$, and then $1 - \mathcal{C}_{\mathbf{w}}$ being bounded implies via the closed graph theorem that $\mathrm{Ran}(1 - \mathcal{C}_{\mathbf{w}}) = \mathrm{Lip}^\mu(\Sigma_U)$. Therefore the inverse $(1 - \mathcal{C}_{\mathbf{w}})^{-1}$ exists and is defined on the whole space $\mathrm{Lip}^\mu(\Sigma_U)$. Since $1 - \mathcal{C}_{\mathbf{w}}$ is bounded and hence closed, the inverse is also closed and therefore bounded by the closed graph theorem. □

Combining Lemma A.5.2 with Corollary A.5.2 and Corollary A.5.1 gives the main result of this section.

THEOREM A.5.1 (Umbilical Fredholm alternative). *Let* Σ_{U} *be an umbilical contour, and let* $\mathbf{v}_{\mathrm{U}}(z)$ *be a jump matrix for* $z \in \Sigma_{\mathrm{U}}\backslash\{\alpha_i\}$ *satisfying the conditions of Definition* A.2.1. *Let* $\mathbf{N}_{\mathrm{U}}$ *be any constant matrix. Then the umbilical Riemann-Hilbert Problem* A.2.1 *with data* $(\Sigma_{\mathrm{U}}, \mathbf{v}_{\mathrm{U}}(z), \mathbf{N}_{\mathrm{U}})$ *has a unique solution if and only if the corresponding homogeneous problem with data* $(\Sigma_{\mathrm{U}}, \mathbf{v}_{\mathrm{U}}(z), \mathbf{0})$ *has only the trivial solution.*

A.6 APPLICATIONS TO LOCAL RIEMANN-HILBERT PROBLEMS

Here, we put together the pieces to establish the proofs of Theorems A.1.1, A.1.2, and A.1.3.

Proof of Theorem A.1.1. By Lemma A.2.1, the local Riemann-Hilbert Problem A.1.1 will have a unique solution if and only if the corresponding umbilical Riemann-Hilbert Problem A.2.1 does. By Theorem A.5.1, the latter will be the case if the only solution of the umbilical Riemann-Hilbert Problem A.2.1 that vanishes at $z = \infty$ is the trivial solution. If $\mathbf{U}_0(z)$ is such a nontrivial "vanishing" solution, then

$$\mathbf{L}_0(\zeta) := (\zeta - e^{i\theta})^{-1}\mathbf{U}_0(z(\zeta)) \tag{A.48}$$

will be a nontrivial solution of the homogeneous local Riemann-Hilbert Problem A.1.1, where $z(\zeta)$ is the transformation in Lemma A.2.1. Conversely, if $\mathbf{L}_0(\zeta)$ is a nontrivial solution of the homogeneous local Riemann-Hilbert Problem A.1.1, then with $\zeta(z)$ as in Lemma A.2.1,

$$\mathbf{U}_0(z) := (z - 1)^{-1}\mathbf{L}_0(\zeta(z)) \tag{A.49}$$

will be a nontrivial solution of the homogeneous umbilical Riemann-Hilbert Problem A.2.1. Therefore, the unique solvability of the umbilical problem and therefore of the local problem by Lemma A.2.1 is guaranteed by the condition $\{\mathbf{L}_0(\zeta)\} = \{\mathbf{0}\}$, which proves the theorem. □

Proof of Theorem A.1.2. The fact that under the additional symmetry condition (A.6) on the contour Σ_{L} and jump matrix $\mathbf{v}_{\mathrm{L}}(\zeta)$ all solutions of the homogeneous local Riemann-Hilbert Problem A.1.1 with data $(\Sigma_{\mathrm{L}}, \mathbf{v}_{\mathrm{L}}(\zeta), \mathbf{0})$ are trivial follows from an argument given by Zhou in [Z89]. This result is combined with Theorem A.1.1 to complete the proof. Note that Zhou's argument relies on Cauchy's theorem, and in the L^2 context of [Z89] a rational approximation argument is required to pull the paths of integration away from the boundary. But in the Hölder spaces, no such argument is needed, since the boundary values are automatically uniformly continuous. □

Proof of Theorem A.1.3. We begin with the representation of the solution $\mathbf{L}(\zeta)$ in terms of the fractional linear transformation $\zeta(z)$ and the solution $\mathbf{m}(z)$ of the integral equation on the corresponding umbilical contour Σ_{U} by means of the formula (A.33). Since $\zeta = \infty$ corresponds to $z = 1$, we have

$$\mathbf{L}(\zeta) - \mathbb{I} = \mathbf{U}(z(\zeta)) - \mathbf{U}(1) = \frac{i}{\pi\zeta e^{-i\theta} + \pi}(I_1^+(\zeta) + I_1^-(\zeta) + I_2^+(\zeta) + I_2^-(\zeta)), \tag{A.50}$$

where

$$I_1^{\pm}(\zeta) := \int_{\Sigma_{\mathrm{U}}} \mathbf{w}^{\pm}(s)\frac{ds}{(s - z(\zeta))(s-1)},$$
$$I_2^{\pm}(\zeta) := \int_{\Sigma_{\mathrm{U}}} (\mathbf{m}(s) - \mathbb{I})\mathbf{w}^{\pm}(s)\frac{ds}{(s - z(\zeta))(s-1)}. \tag{A.51}$$

Note that we have both

$$\|\mathbf{w}^{\pm}(z)\| = O(|z-1|) \quad \text{and} \quad \|\mathbf{m}(z) - \mathbb{I}\| = O(|z-1|^{\mu}) \quad \text{for all } \mu < 1, \tag{A.52}$$

with the latter estimate following from the Hölder theory of the integral equation (A.32).

First we show that the $O(|\zeta|^{-1})$ decay estimate holds subject to the condition (A.9) uniformly for ζ outside of all of the parallel strips S_k surrounding each ray. By (A.50), it suffices to show that the four integrals $I_1^{\pm}(\zeta)$ and $I_2^{\pm}(\zeta)$ remain bounded. Now, changing variables in the integrals by $\xi = \zeta(s)$, we find

$$I_2^{\pm}(\zeta) = -\frac{1}{2}\int_{\Sigma_{\mathrm{L}}} (\mathbf{m}(z(\xi)) - \mathbb{I})\mathbf{w}^{\pm}(z(\xi))\frac{\zeta e^{-i\theta} + 1}{\xi - \zeta}\, d\xi. \tag{A.53}$$

Now observe that

$$-\frac{1}{2}\frac{\zeta e^{-i\theta} + 1}{\xi - \zeta} = \frac{e^{-i\theta}}{2} - \frac{1}{2}\frac{1}{\xi - \zeta} - \frac{e^{-i\theta}}{2}\xi\frac{1}{\xi - \zeta}. \tag{A.54}$$

Since the estimates (A.52) imply

$$(\mathbf{m}(z(\xi)) - \mathbb{I})\mathbf{w}^{\pm}(z(\xi)) \in L^1(\Sigma_{\mathrm{L}}) \cap L^2(\Sigma_{\mathrm{L}}) \tag{A.55}$$

and

$$(\mathbf{m}(z(\xi)) - \mathbb{I})\mathbf{w}^{\pm}(z(\xi))\xi \in L^2(\Sigma_{\mathrm{L}}) \tag{A.56}$$

and since uniformly for all ζ outside the union of the strips S_k we have

$$\frac{1}{\xi - \zeta} \in L^2(\Sigma_{\mathrm{L}}) \tag{A.57}$$

we can apply the Cauchy-Schwarz inequality to find

$$\|I_2^{\pm}(\zeta)\| = O(1) \tag{A.58}$$

uniformly for ζ outside all strips S_k.

To study the remaining terms, we change variables from s to $\xi = \zeta(s)$ in $I_1^{\pm}(\zeta)$ as before and find

$$I_1^{\pm}(\zeta) = -\frac{1}{2}I_{1A}^{\pm}(\zeta) - \frac{e^{-i\theta}}{2}I_{1B}^{\pm}(\zeta), \tag{A.59}$$

where

$$I_{1A}^{\pm}(\zeta) := \int_{\Sigma_{\mathbf{L}}} \mathbf{w}^{\pm}(z(\xi)) \frac{d\xi}{\xi - \zeta} \quad \text{and} \quad I_{1B}^{\pm}(\zeta) := \int_{\Sigma_{\mathbf{L}}} \mathbf{w}^{\pm}(z(\xi)) \frac{\zeta\, d\xi}{\xi - \zeta}. \tag{A.60}$$

Because $\mathbf{w}^{\pm}(z(\cdot))$ and $1/(\xi - \zeta)$ are both in $L^2(\Sigma_{\mathbf{L}})$, the latter with a norm that is uniformly bounded with respect to ζ as long as ζ remains outside of all strips S_k, $I_{1A}^{\pm}(\zeta)$ is uniformly bounded by Cauchy-Schwarz.

To handle $I_{1B}^{\pm}(\zeta)$, we first split off the part $\Sigma_{\mathbf{L}}^{\text{in}}$ of the contour that is bounded by $|\xi| \leq R$ (i.e., the part interior to and including the circle component of $\Sigma_{\mathbf{L}}$) and recall the notation $\Sigma_{\mathbf{L}}^{(k)}$ for the rays that make up the rest of $\Sigma_{\mathbf{L}}$. Now recall the relationship $\mathbf{v}_{\mathbf{L}}(\zeta) = (\mathbb{I} - \mathbf{w}^{-}(z(\zeta)))^{-1}(\mathbb{I} - \mathbf{w}^{+}(z(\zeta)))$. This implies that for $\zeta \in \Sigma_{\mathbf{L}}^{(k)}$,

$$\Psi_k(\zeta) := \zeta \mathbf{w}^{+}(z(\zeta)) + \zeta \mathbf{w}^{-}(z(\zeta)) = \Phi_k(\zeta) - \frac{1}{\zeta} \cdot \zeta \mathbf{w}^{-}(z(\zeta)) \cdot \Phi_k(\zeta). \tag{A.61}$$

Because $\zeta \mathbf{w}^{-}(z(\zeta))$ is bounded on $\Sigma_{\mathbf{L}}^{(k)}$, it follows from our hypotheses that

$$\langle \Psi_k \rangle := \lim_{L \to \infty} \frac{1}{L} \int_{Re^{i\phi_k}}^{(R+L)e^{i\phi_k}} \Psi_k(\zeta)\, d\zeta = \langle \Phi_k \rangle, \tag{A.62}$$

where ϕ_k is the angle of the ray $\Sigma_{\mathbf{L}}^{(k)}$. Upon setting $\delta\Psi_k(\zeta) := \Psi_k(\zeta) - \langle \Psi_k \rangle$, we then find by direct integration that

$$\int_{Re^{i\phi_k}}^{\zeta} \delta\Psi_k(s)\, ds = O(\log|\zeta|), \quad \text{as} \quad \zeta \to \infty, \zeta \in \Sigma_{\mathbf{L}}^{(k)}. \tag{A.63}$$

Note that $O(\log|\zeta|)$ is the best estimate we can get generally, because even if the integral of $\delta\Phi(\zeta)$ along the ray grows more slowly than $\log|\zeta|$, the term that is explicitly inversely proportional to ζ will generally be no smaller. With these observations in hand, we may now bound the sum of $I_{1B}^{+}(\zeta) + I_{1B}^{-}(\zeta)$:

$$\begin{aligned} I_{1B}^{+}(\zeta) + I_{1B}^{-}(\zeta) &= \int_{\Sigma_{\mathbf{L}}^{\text{in}}} [\mathbf{w}^{+}(z(\xi)) + \mathbf{w}^{-}(z(\xi))] \frac{\zeta\, d\xi}{\xi - \zeta} \\ &\quad + \sum_{\text{all rays } k} \langle \Psi_k \rangle \int_{\Sigma_{\mathbf{L}}^{(k)}} \frac{\zeta\, d\xi}{\xi(\xi - \zeta)} \\ &\quad + \sum_{\text{all rays } k} \int_{\Sigma_{\mathbf{L}}^{(k)}} \delta\Psi_k(\xi) \frac{\zeta\, d\xi}{\xi(\xi - \zeta)} \\ &= A(\zeta) + B(\zeta) + C(\zeta). \end{aligned} \tag{A.64}$$

The first term on the right-hand side, $A(\zeta)$, is bounded for large ζ because the contour $\Sigma_{\mathbf{L}}^{\text{in}}$ is compact. For the integral in the sum making up the third term, $C(\zeta)$, we integrate by parts:

$$\begin{aligned} \int_{\Sigma_{\mathbf{L}}^{(k)}} \delta\Psi_k(\xi) \frac{\zeta\, d\xi}{\xi(\xi - \zeta)} &= \int_{\Sigma_{\mathbf{L}}^{(k)}} \left(\int_{Re^{i\phi_k}}^{\xi} \delta\Psi_k(t)\, dt \right) \frac{d\xi}{(\xi - \zeta)^2} \\ &\quad - \int_{\Sigma_{\mathbf{L}}^{(k)}} \left(\int_{Re^{i\phi_k}}^{\xi} \delta\Psi_k(t)\, dt \right) \frac{d\xi}{\xi^2}. \end{aligned} \tag{A.65}$$

Because the integral of $\delta\Psi_k(t)$ is bounded by $\log|\xi|$ and since $\log|\xi|/|\xi-\zeta|$ is integrable on $\Sigma_{\mathbf{L}}^{(k)}$ uniformly for all ζ outside all strips S_k, we see that $C(\zeta)$ is uniformly bounded for ζ outside all strips S_k. To study the remaining term, $B(\zeta)$, we compute the integral exactly and find that

$$\int_{\Sigma_{\mathbf{L}}^{(k)}} \frac{\zeta\, d\xi}{\xi(\xi-\zeta)} = -\log|\zeta| + O(1) \tag{A.66}$$

as $\zeta \to \infty$. The condition

$$\sum_{\text{all rays } k} \langle \Psi_k \rangle = \sum_{\text{all rays } k} \langle \Phi_k \rangle = \mathbf{0} \tag{A.67}$$

therefore guarantees that $B(\zeta)$ is bounded uniformly for ζ outside all strips S_k and therefore that $I_1^+(\zeta) + I_1^-(\zeta)$ is similarly bounded.

This proves that the estimate stated in the theorem holds for ζ bounded away from the contour rays by avoiding the parallel strips. But now we can use the analyticity of the jump matrix $\mathbf{v}_{\mathbf{L}}(\zeta)$ along with the uniformity afforded by (A.8) in a simple deformation argument to prove that the estimate in fact holds right up to the (analytic) boundary values taken on $\Sigma_{\mathbf{L}}$, which completes the proof of the theorem. □

Appendix B

Near-Identity Riemann-Hilbert Problems in L^2

In this appendix, we collect those results from the theory of Riemann-Hilbert problems and Cauchy integral operators in L^2 that we use in §4.5 to characterize and estimate the error of our approximations.

Let Σ be a compact contour in the complex λ-plane, which consists of a union of a finite number of smooth arcs and is oriented so that it forms the positively oriented boundary of a multiply connected open region Ω^+ whose complement is the disjoint union $\Sigma \cup \Omega^-$. Fix any matrix norm $\|\cdot\|$, and let $L^2(\Sigma)$ denote the space of matrix-valued functions $\mathbf{f}(\lambda)$ defined almost everywhere on Σ such that the norm defined by

$$\|\mathbf{f}\|_{L^2(\Sigma)}^2 := \int_\Sigma \|\mathbf{f}(\lambda)\|^2 \, |d\lambda| \tag{B.1}$$

is finite. Now we recall some facts [D99] about Cauchy integrals in this space. For $\mathbf{f} \in L^2(\Sigma)$, the Cauchy integral

$$(\mathcal{C}^\Sigma \mathbf{f})(\lambda) := \frac{1}{2\pi i} \int_\Sigma (s-\lambda)^{-1} \mathbf{f}(s) \, ds \tag{B.2}$$

defines a piecewise-analytic function for $\lambda \notin \Sigma$. The left and right nontangential boundary values

$$(\mathcal{C}_\pm^\Sigma \mathbf{f})(\lambda) := \lim_{z \to \lambda, z \in \Omega^\pm} (\mathcal{C}^\Sigma \mathbf{f})(z) \tag{B.3}$$

exist for almost all $\lambda \in \Sigma$ and may be identified with unique elements of $L^2(\Sigma)$. The linear operators $\mathcal{C}_\pm^\Sigma$ thus defined on $L^2(\Sigma)$ are bounded, with norms depending on the contour Σ.

Let $\mathbf{w}^\pm(\lambda)$ be defined for almost all $\lambda \in \Sigma$ and uniformly bounded. Define an operator on $L^2(\Sigma)$ by

$$(\mathcal{C}_\mathbf{w} \mathbf{m})(\lambda) := (\mathcal{C}_+^\Sigma (\mathbf{m}\mathbf{w}^-))(\lambda) + (\mathcal{C}_-^\Sigma (\mathbf{m}\mathbf{w}^+))(\lambda). \tag{B.4}$$

Then $\mathcal{C}_\mathbf{w}$ is bounded according to the simple estimate

$$\|\mathcal{C}_\mathbf{w}\|_{L^2(\Sigma)} \le \|\mathcal{C}_+^\Sigma\|_{L^2(\Sigma)} \sup_{s\in\Sigma} \|\mathbf{w}^-(s)\| + \|\mathcal{C}_-^\Sigma\|_{L^2(\Sigma)} \sup_{s\in\Sigma} \|\mathbf{w}^+(s)\|. \tag{B.5}$$

Therefore, we have the following lemma.

Lemma B.0.1 *Assume that*

$$\sup_{s\in\Sigma} \|\mathbf{w}^-(s)\| < \frac{1}{2} (\|\mathcal{C}_+^\Sigma\|_{L^2(\Sigma)})^{-1}, \qquad \sup_{s\in\Sigma} \|\mathbf{w}^+(s)\| < \frac{1}{2} (\|\mathcal{C}_-^\Sigma\|_{L^2(\Sigma)})^{-1}. \tag{B.6}$$

Then, for any $\mathbf{f} \in L^2(\Sigma)$, *the equation*

$$\mathbf{m}(\lambda) - (\mathcal{C}_\mathbf{w} \mathbf{m})(\lambda) = \mathbf{f}(\lambda) \tag{B.7}$$

has a unique solution $\mathbf{m} \in L^2(\Sigma)$, *and*

$$\|\mathbf{m}\|_{L^2(\Sigma)} \le \frac{\|\mathbf{f}\|_{L^2(\Sigma)}}{1 - (\|\mathcal{C}_+^\Sigma\|_{L^2(\Sigma)} \sup_{s\in\Sigma} \|\mathbf{w}^-(s)\| + \|\mathcal{C}_-^\Sigma\|_{L^2(\Sigma)} \sup_{s\in\Sigma} \|\mathbf{w}^+(s)\|)}. \tag{B.8}$$

Proof. This follows from standard functional analysis results concerning the invertibility of bounded perturbations of the identity operator. The solution of the equation is furnished by Neumann series. □

Lemma B.0.2 *Along with the conditions of Lemma B.0.1, assume that* $\mathbb{I} - \mathbf{w}^-(\lambda)$ *is invertible for* $\lambda \in \Sigma$ *and set* $\mathbf{v}(\lambda) := (\mathbb{I} - \mathbf{w}^-(\lambda))^{-1}(\mathbb{I} + \mathbf{w}^+(\lambda))$. *Let* $\mathbf{m}(\lambda)$ *be the unique solution of (B.7) with* $\mathbf{f}(\lambda) \equiv \mathbb{I}$. *Define, for* $\lambda \notin \Sigma$,

$$\mathbf{R}(\lambda) := \mathbb{I} + (\mathcal{C}^\Sigma(\mathbf{m}(\mathbf{w}^+ + \mathbf{w}^-)))(\lambda). \tag{B.9}$$

Then $\mathbf{R}(\lambda)$ *has boundary values* $\mathbf{R}_\pm(\lambda)$ *in* $L^2(\Sigma)$ *that satisfy almost everywhere*

$$\mathbf{R}_+(\lambda) = \mathbf{R}_-(\lambda)\mathbf{v}(\lambda). \tag{B.10}$$

Also, $\mathbf{R}(\infty) = \mathbb{I}$, *and*

$$\|\mathbf{R}(\lambda) - \mathbb{I}\| \le \frac{1}{2\pi}\|\mathbf{m}\|^2_{L^2(\Sigma)} \cdot |\Sigma|^2 \cdot \sup_{s\in\Sigma} |s-\lambda|^{-1} \cdot \sup_{s\in\Sigma} \|\mathbf{w}^+(s) + \mathbf{w}^-(s)\|, \tag{B.11}$$

where $|\Sigma|$ *is the total arc length of* Σ.

Proof. It follows directly from the integral equation satisfied by $\mathbf{m}(\lambda)$ and the Plemelj formula that everywhere the boundary values taken on Σ by $\mathbf{R}(\lambda)$ are finite, they satisfy the jump relation (B.10). The estimate (B.11) also follows directly from the representation (B.9). □

We say that $\mathbf{R}(\lambda)$ solves the Riemann-Hilbert problem with data $(\Sigma, \mathbf{v}(\lambda), \mathbb{I})$ in the $L^2(\Sigma)$ sense. In [D99], the following uniqueness result is proved.

Lemma B.0.3 *Suppose all matrices are* 2×2 *and that* $\mathbf{w}^\pm(\lambda)$ *are smooth functions on each arc of* Σ *for which* $\det \mathbf{v}(\lambda) = 1$. *Then the solution* $\mathbf{R}(\lambda)$ *of the Riemann-Hilbert problem with data* $(\Sigma, \mathbf{v}(\lambda), \mathbb{I})$, *posed in the* $L^2(\Sigma)$ *sense, is unique if it exists.*

Bibliography

[AS65] M. Abramowitz and I. A. Stegun, eds., *Handbook of Mathematical Functions*, Dover, New York, 1965.

[ASK66] S. A. Akhmanov, A. P. Sukhorukov, and R. V. Khokhlov, "Self-focusing and self-trapping of intense light beams in a nonlinear medium," *Sov. Phys. JETP*, **23** (1996): 1025–1033.

[B75] G. A. Baker Jr., *Essentials of Padé Approximants*, Academic Press, New York, 1975.

[BDT88] R. Beals, P. Deift, and C. Tomei, *Direct and Inverse Scattering on the Line*, Math. Surveys Monogr., 28, Amer. Math. Soc., Providence, R.I., 1988.

[B96] J. C. Bronski, "Semiclassical eigenvalue distribution of the Zakharov-Shabat eigenvalue problem," *Phys. D*, **97** (1996): 376–397.

[B01] ———, "Spectral instability of the semiclassical Zakharov-Shabat eigenvalue problem," *Phys. D*, **152–153** (2001): 163–170.

[BK99] J. C. Bronski and J. N. Kutz, "Numerical simulation of the semiclassical limit of the focusing nonlinear Schrödinger equation," *Phys. Lett. A*, **254** (1999): 325–336.

[CM02] S. R. Clarke and P. D. Miller, "On the semiclassical limit for the focusing nonlinear Schrödinger equation: sensitivity to analytic properties of the initial data," *Proc. Roy. Soc. Lond. Ser. A*, **458** (2002): 135–156.

[D99] P. A. Deift, *Orthogonal Polynomials and Random Matrices: A Riemann-Hilbert Approach*, Courant Lecture Notes in Mathematics, Vol. 3, Courant Institute of Mathematical Sciences, New York, 1999.

[DIZ97] P. Deift, A. Its, and X. Zhou, "A Riemann-Hilbert approach to asymptotic problems arising in the theory of random matrix models, and also in the theory of integrable statistical mechanics," *Ann. of Math.* (2), **146** (1997): 149–235.

[DKKZ96] P. Deift, S. Kamvissis, T. Kriecherbauer, and X. Zhou, "The Toda rarefaction problem," *Comm. Pure Appl. Math.*, **49** (1996): 35–83.

[DKM98] P. Deift, T. Kriecherbauer, and K. T.-R. McLaughlin, "New results on the equilibrium measure for logarithmic potentials in the presence of an external field," *J. Approx. Theory*, **95** (1998): 388–475.

[DKMVZ97] P. Deift, T. Kriecherbauer, K. T.-R. McLaughlin, S. Venakides, and X. Zhou, "Asymptotics for polynomials orthogonal with respect to varying exponential weights," *Internat. Math. Res. Notices*, **1997**, 759–782.

[DKMVZ99A] ———, "Strong asymptotics of orthogonal polynomials with respect to exponential weights," *Comm. Pure Appl. Math.*, **52** (1999): 1491–1552.

[DKMVZ99B] ———, "Uniform asymptotics for polynomials orthogonal with respect to varying exponential weights and applications to universality questions in random matrix theory," *Comm. Pure Appl. Math.*, **52** (1999): 1335–1425.

[DM98] P. Deift and K. T.-R. McLaughlin, "A continuum limit of the Toda lattice," *Mem. Amer. Math. Soc.*, **131**, 1998.

[DVZ94] P. Deift, S. Venakides, and X. Zhou, "The collisionless shock region for the long-time behavior of solutions of the KdV equation," *Comm. Pure Appl. Math.*, **47** (1994): 199–206.

[DVZ97] ———, "New results in small dispersion KdV by an extension of the steepest descent method for Riemann-Hilbert problems," *Internat. Math. Res. Notices*, **1997**, 285–299.

[DZ93] P. Deift and X. Zhou, "A steepest descent method for oscillatory Riemann-Hilbert problems: asymptotics for the mKdV equation," *Ann. of Math.* (2), **137** (1993): 295–368.

[DZ95] ———, "Asymptotics for the Painlevé II equation" *Comm. Pure Appl. Math.*, **48** (1995): 277–337.

[D81] B. A. Dubrovin, "Theta functions and non-linear equations," *Russian Math. Survey*, **36** (1981): 11–92.

[EJLM93] N. M. Ercolani, S. Jin, C. D. Levermore, and W. D. MacEvoy, Jr., "The zero-dispersion limit for the odd flows in the focusing Zakharov-Shabat hierarchy," preprint, 1993. (Preprint archive reference: `nlin.SI/0302003`)

[FT87] L. D. Faddeev and L. A. Takhtajan, *Hamiltonian Methods in the Theory of Solitons*, Springer-Verlag, New York, 1987.

[FL86] M. G. Forest and J.-E. Lee, "Geometry and modulation theory for the periodic nonlinear Schrödinger equation" in *Oscillation Theory, Computation, and Methods of Compensated Compactness*, C. Dafermos et al. eds., IMA Volumes in Mathematics and Its Applications, Vol. 2, Springer-Verlag, New York, 1986, 35–69.

[FM98] M. G. Forest and K. T.-R. McLaughlin, "Onset of oscillations in nonsoliton pulses in nonlinear dispersive fibers," *J. Nonlinear Sci.*, **8** (1998): 43–62.

[GR87] A. A. Gonchar and E. A. Rakhmanov, "Equilibrium distributions and the rate of rational approximation of analytic functions" (in Russian) *Math. Sb. (N.S.)*, **134/176** (1987): 306–352, 447; transl. in *Math. USSR-Sb.*, **62** (1989): 305–348.

[JLM99] S. Jin, C. D. Levermore, and D. W. McLaughlin, "The semiclassical limit of the defocusing NLS hierarchy," *Comm. Pure Appl. Math.*, **52** (1999): 613–654.

[K95] S. Kamvissis, "The collisionless shock phenomenon for the focusing nonlinear Schrödinger equation," *C. R. Acad. Sci. Paris Ser. I*, **321** (1995): 1525–1531.

[K96] ———, "Long time behavior for the focusing nonlinear Schrödinger equation with real spectral singularities," *Comm. Math. Phys.*, **180** (1996): 325–343.

[KS02] M. Klaus and J. K. Shaw, "Purely imaginary eigenvalues of Zakharov-Shabat systems," *Phys. Rev. E*, **65** (2002): 36607–36611.

[LL83] P. D. Lax and C. D. Levermore, "The small dispersion limit of the Korteweg-de Vries equation, I, II, III," *Comm. Pure Appl. Math.*, **36** (1983): 253–290, 571–593, 809–830.

[M01] P. D. Miller, "Some remarks on a WKB method for the nonselfadjoint Zakharov-Shabat eigenvalue problem with analytic potentials and fast phase," *Phys. D*, **152–153** (2001): 145–162.

[M02] ———, "Asymptotics of semiclassical soliton ensembles: rigorous justification of the WKB approximation," *Internat. Math. Res. Notices*, **2002**, 383–454.

[MK98] P. D. Miller and S. Kamvissis, "On the semiclassical limit of the focusing nonlinear Schrödinger equation," *Phys. Lett. A*, **247** (1998): 75–86.

[M53] N. I. Muskhelishvili, *Singular Integral Equations*, Dover, New York, 1992. Second edition published by P. Noordhoff N. V., Groningen, Holland, 1953.

[PR94] E. A. Perevozhnikova and E. A. Rakhmanov, "Variations of the equilibrium energy and S-property of compacta of minimal capacity," preprint, 1994.

[P78] S. Prössdorf, *Some Classes of Singular Equations*, North-Holland, Amsterdam, 1978.

[RS78] M. Reed and B. Simon, *Methods of Modern Mathematical Physics, IV: Analysis of Operators*, Academic Press, Boston, 1978.

[ST97] E. B. Saff and V. Totik, *Logarithmic Potentials with External Fields*, Springer-Verlag, New York, 1997.

[SY74] J. Satsuma and N. Yajima, "Initial value problems of one-dimensional self-modulation of nonlinear waves in dispersive media," *Supp. Prog. Theo. Phys.*, **55** (1974): 284–306.

[SW86] D. H. Sattinger and O. L. Weaver, *Lie Groups and Algebras with Applications to Physics, Geometry, and Mechanics*, Springer-Verlag, New York, 1986.

[S50] M. Schiffer, "Some recent developments in the theory of conformal mapping," appendix to R. Courant, *Dirichlet's Principle, Conformal Mapping, and Minimal Surfaces*, Interscience, New York, 1950.

[TV00] A. Tovbis and S. Venakides, "The eigenvalue problem for the focusing nonlinear Schrödinger equation: new solvable cases," *Phys. D*, **146** (2000): 150–164.

[W74] G. B. Whitham, *Linear and Nonlinear Waves*, Wiley Interscience, New York, 1974.

[ZS72] V. E. Zakharov and A. B. Shabat, "Exact theory of two-dimensional self-focusing and one-dimensional self-modulation of waves in nonlinear media," *Sov. Phys. JETP*, **34** (1972): 62–69.

[Z89] X. Zhou, "The Riemann-Hilbert problem and inverse scattering," *SIAM J. Math. Anal.*, **20** (1989): 966–986.

Index